D1212687

STRUCTURE AND BONDING

Volume 17

Editors: J. D. Dunitz, Zürich
P. Hemmerich, Konstanz · J. A. Ibers, Evanston
C. K. Jørgensen, Genève · J. B. Neilands, Berkeley
D. Reinen, Marburg · R. J. P. Williams, Oxford

With 77 Figures

QD
461
.S92
Vol. 17

Springer-Verlag
New York · Heidelberg · Berlin 1973

ISBN 0-387-06458-3 Springer-Verlag New York · Heidelberg · Berlin
ISBN 3-540-06458-3 Springer-Verlag Berlin · Heidelberg · New York

The use of general descriptive names, trade marks, etc. in this publication, even if the former are not especially identified, is not to be taken as a sign that such names, as understood by the Trade Marks and Merchandise Marks Act, may accordingly be used freely by anyone.

This work is subject to copyright. All rights are reserved, whether the whole or part of the material is concerned, specifically those of translation, reprinting, re-use of illustrations, broadcasting, reproduction by photocopying machine or similar means, and storage in data banks. Under § 54 of the German Copyright Law where copies are made for other than private use, a fee is payable to the publisher, the amount of the fee to be determined by agreement with the publisher. © by Springer Verlag Berlin Heidelberg 1973 · Library of Congress Catalog Card Number 67-11280. Printed in Germany. Typesetting and printing: Meister-Druck, Kassel

Contents

Structural Aspects and Biochemical Function of Erythrocuprein

Ulrich Weser

Physiologisch-Chemisches Institut der Universität Tübingen, BRD

Table of Contents

1. Introduction

Our knowledge of the structural properties and enzymic function of a large number of copper proteins has accumulated during the last few decades. The main results have been comprehensively reviewed (*1—4*) or presented at symposia (*5—54*). This survey is devoted to erythrocuprein, one of the most actively studied copper proteins. Erythrocuprein is sometimes called haemocuprein, hepatocuprein, cerebrocuprein, cytocuprein, or erythro-cupro-zinc protein. Alternatively, the name superoxide dismutase has been suggested as descriptive of its activity: the enzyme-catalyzed disproportionation of anionic monovalent superoxide radicals. However, whether or not the enzymic reaction is specific for $O_2^{-}\cdot$ still needs to be investigated[1]). Thus, the name erythrocuprein is used throughout this review.

Erythrocuprein was first isolated from bovine erythrocytes and bovine liver in 1939 (*55*). The preparations contained approximately 0.34% copper and the molecular weight was approximately 35,000. In contrast to the bluish-green copper protein isolated from erythrocytes, a colourless copper protein of the same molecular weight and Cu content was found in liver. This result was challenged by Mohamed and Greenberg (*56*) who prepared the coloured protein from horse liver. Twenty years later the isolation and characterization of erythrocuprein from normal human erythrocytes have been described (*57—61*). Cu balance studies revealed that 60% of erythrocyte copper is present in erythrocuprein (*60*). The same copper concentration was found in a soluble copper protein from normal human brain called cerebrocuprein I (*62—64*). It was already apparent that a striking similarity existed between all these soluble, copper-containing tissue proteins having a molecular weight of 32,000 ±2,000.

During the last seven years intensive studies of both human and bovine erythrocuprein have been performed (*65—83*). *Carrico* and *Deutsch* were able to demonstrate the identity of erythro-, cerebro-, and hepatocuprein using copper proteins from human tissues: the amino-acid analysis, immunochemical behaviour, and physicochemical properties were identical in all three. In 1970 they found, in addition to the 2 copper atoms, a second metallic component, namely 2 atoms of zinc (*69*).

McCord and *Fridovich* (*70*) proposed an enzymic function for erythrocuprein. They demonstrated that erythrocuprein is able to disproportionate monovalent superoxide anion radicals into hydrogen peroxide and oxygen. Detailed enzyme studies (*146, 147*) led to the pro

---

[1]) see Note added in proof

posal of another, even more important function for erythrocuprein, namely the scavenging of singlet oxygen in metabolism. When erythrocuprein was exposed to mercaptoethanol and some protein-unfolding agents, such as urea or sodium dodecylsulphate, subunits of molecular weight 16,000 were observed after electrophoretic separation (*74, 82*). Furthermore the zinc content was confirmed by two different groups (*72, 74*) using bovine erythrocuprein. The preparation, and the structural and chemical properties of erythrocuprein are discussed below in detail.

2. Preparation

Two main preparation procedures are employed. Treatment with organic solvents such as chloroform/ethanol and acetone (*55—57*) was the original method, which is still successfully used today (*70, 72, 74*). Of course, substantial modifications arose from the introduction of gel and ion-exchange chromatography. In the alternative procedure only aqueous solutions are employed and subjected to different chromatographic processes (*67, 68*). There is no doubt that the latter method represents the most gentle treatment of a protein. However, erythrocuprein has been found to be *one of the most stable proteins* (*79*) and survives a brief treatment with organic solvents without any measurable damage. An essential advantage of the first isolation procedure is the shorter time required, offering less opportunity for possible denaturation of the protein due to microbial growth or other undesired side reactions.

2.1. Isolation from Erythrocytes

Erythrocytes were harvested at 1,800 x *g* using citrated blood. The cells were washed 4 times with isotonic NaCl solution. Lysis of erythrocytes was achieved within 12 h at 4 °C by adding an equal volume of water. The haemolysate was then treated in the cold with a mixture of 0.25 volume of ethanol and 0.15 volume of chloroform (*70, 72, 74*). After dilution with 0.1% volume of water (*70, 72*) or isotonic NaCl (*72*), a pale yellow supernatant was obtained upon centrifugation at 1,800 x *g* for 30 min. This supernatant was treated with 0.05 volume of saturated lead acetate (*57—60, 72*) or with solid K_2HPO_4 (300 g per litre) (*70*). With lead acetate almost all erythrocuprein was found in the precipitate and had to be eluted with 0.33 M phosphate buffer. The use of phosphate gave two liquid phases, leaving the erythrocuprein in the upper phase. After the collection and centrifugation of the upper phase, 0.75 volume of cold acetone was added .The extracted crude erythrocuprein containing precipitates obtained after lead acetate or acetone treatment was subjected to DEAE-23 chromatography or, as performed in the author's laboratory, separated on a Sephadex G-75 column (*74*) prior to DEAE—23 cellulose chromatography. Typical elution patterns are depicted in Fig. 1 and Fig. 2. The overall enrichment was some 3000-times compared to the erythrocuprein content of the haemolysate.

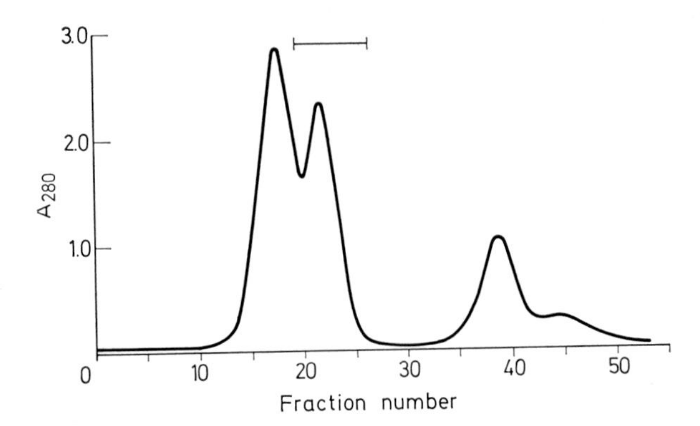

Fig. 1 (*74*). Elution pattern of crude bovine erythrocuprein separated on Sephadex-G-75. The second peak contained most of the erythrocuprein

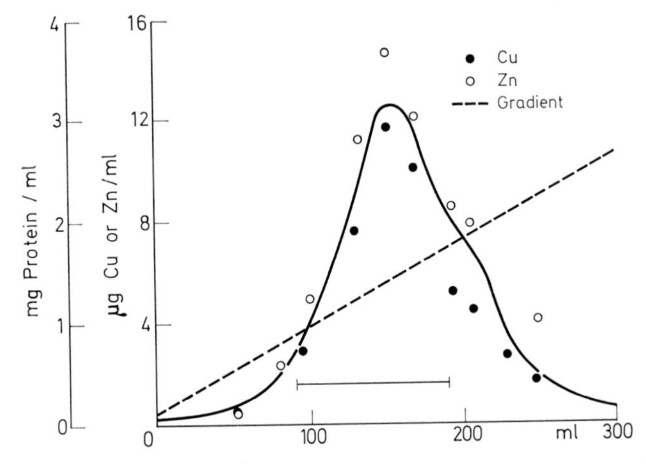

Fig. 2 (*84*). Separation of erythrocuprein on DEAE-23 cellulose. Cu and Zn are montiored using atomic absorption

In the alternative method batch adsorption of the haemolysate was carried out at 25 °C (*67*) using DEAE-23 cellulose previously equilibrated with sodium cacodylate buffer. The DEAE columns containing the adsorbed proteins were eluted with 0.15 M NaCl. The copper-containing fractions were monitored either by spot test (*85*) or atomic absorption (*86*). Some authors (*60, 85*) have used an immunochemical assay for the detection of erythrocuprein in the different fractions. The eluted copper proteins were then subjected to Sephadex-G-75 filtration and the erythrocuprein-containing fractions separated on two consecutive DEAE-23 columns having two linear gradients, one ranging from 0.03 to 0.23 $\tau/2$ and the other from 0.01 to 0.085 $\tau/2$. A final Sephadex-G-75 separation was employed to yield a highly purified product.

2.2. Isolation from Liver, Heart or Brain

Heart (*82*) and liver (*87*) homogenates (1 part tissue +2 parts buffer) were clarified by centrifugation at 13,700 × *g* for one hour and the supernatant treated with 0.25 volume ethanol and 0.05 volume chloroform. After centrifugation at 25,400 x *g* the supernatant was treated with phosphate and acetone as described above. Concentration of dilute solutions was performed either on DEAE-23 cellulose or by membrane filtration. Again, we thought it more appropriate to perform the gel exlcusion chromatography on Sephadex-G-75 directly after the acetone treatment (*74, 87*). The group of *Porter* (*64*) homogenized liver tissues in 0.25 M sucrose followed by differential acetone precipitation, treatment with chloroform-ethanol, DEAE-23 chromatography and preparative paper electrophoresis. However, the treatment with organic solvents was rather lengthy in this isolation procedure.

Carrico and *Deutsch* (*68*) used the aqueous isolation procedure. Liver and brain were homogenized and extracted three times with cacodylate buffer (1 part buffer + 1 part tissue). The extract was dialysed and clarified by centrifugation at 18,000 × *g* for one hour. Further treatment was essentially as described for the isolation from erythrocytes.

2.3. Conversion into the Apoprotein

Usually the apoprotein was prepared by employing excessive dialysis against cyanide, EDTA or 1,10-phenanthroline (*69, 70, 71, 72*). However, these apoproteins still contained considerable amounts of copper and zinc (5—20% of the original content). We have devised a new method (*76, 78*) using chelator-equilibrated gel columns. EDTA proved most convenient, as already observed (*69, 72*). Concentrated erythrocuprein samples were layered on top of a Sephadex-G.-25 column which was previously equilibrated with 10 mM EDTA, pH 3.8. The migration rate of the protein was adjusted in a way such that 8—10 hours elapsed before

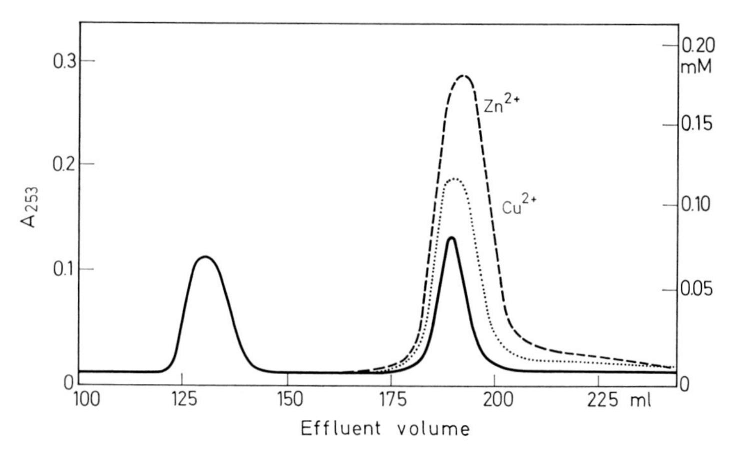

Fig. 3 (*76*). Elution pattern of bovine apoerythrocuprein by gel filtration using Sephadex-G-25

the apoprotein fractions appeared. The apoprotein was clearly separated from the metal chelates of smaller molecular weight. No Cu^{2+} or Zn^{2+} could be detected in the apoprotein (Fig. 3).

The reconstitution of the apoprotein with Cu^{2+} and Zn^{2+} was most successful under anaerobic conditions using the two-column technique (*78*). The first column contained the chelator-equilibrated gel; it was connected to a second column as soon as the apoprotein started to appear. The upper 20 mm of the second G-25-Sephadex column was previously equilibrated with lmM Cu^{2+} and Zn^{2+}. The transfer of the apoprotein to the second column was monitored at 253 nm and the two columns were disconnected when it was complete. The elution of the second column was continued using 5 mM potassium phosphate buffer, *p*H 7.2, previously saturated with N_2.

2.4. Preparation of the Subunit

Subunits of molecular weight 16,000 may be readily obtained after incubating native erythrocuprein or the apoprotein (*74, 80, 82*) together with 1% mercaptoethanol in 4 M urea or with 0.25 M boron hydride in 4 M urea. However, these subunits proved unstable when exposed to air. Recombination products having molecular weight 24,000, 32,000 and 64,000 are observed. Even the reaction of free SH moieties with iodoacetamide did not prevent these associations, which suggests a considerable contribution from hydrogen bonding and/or electrostatic forces (Fig. 4).

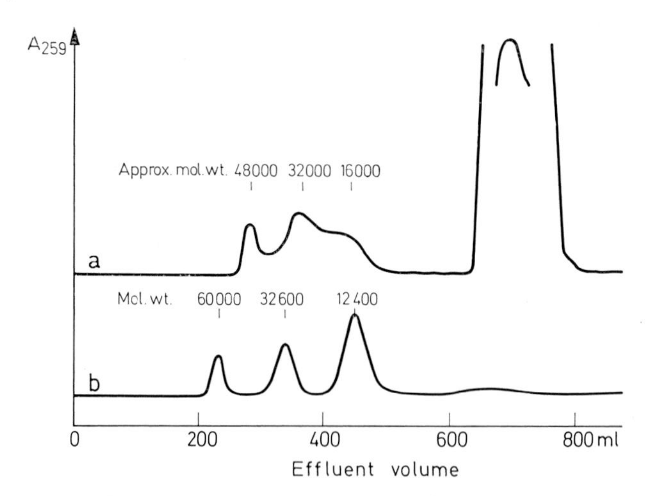

Fig. 4 (*80, 84*). Sephadex-G-75 chromatography of alkylated erythrocuprein subunit (A). Test chromatography using catalase (60,000), erythrocuprein (32,600) and cyt. c (12,000) (B)

3. Characterization

3.1. Molecular Weight, Ultracentrifugation and Electrophoretic Data

Molecular weight has been determined in several ways, the most commonly employed being determination by sedimentation equilibrium. In general, the molecular weight of human erythrocuprein is reported to be 33,600 (*60*) and the corresponding value for bovine erythrocuprein 32,600 (*70*). These values can be considered identical since they are within the experimental error. A summary of different determinations is given in Table 1.

The sedimentation pattern of erythrocuprein from either source gave one single boundary throughout, indicating a high homogeneity (*64, 70, 72*) of the protein. An example is given below (Fig. 5) using cuprein isolated from bovine liver (hepatocuprein).

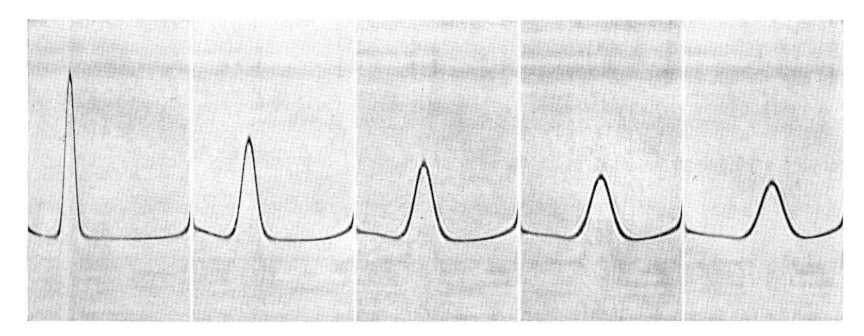

Fig. 5 (*87*). Sedimentation velocity Schlieren patterns of Cuprein isolated from bovine liver (hepatocuprein). Protein concentration was 7 mg/ml in 100 mM potassium phosphate buffer, *p*H 7.3. Photographs were taken at 16-min intervals after reaching 60000 rpm. Sedimentation from left to right

Electrophoretic studies of the isolated erythrocuprein employing starch or polyacrylamide gels have been carried out by most of the above-cited authors. The homogeneity was not as satisfactory as in the sedimentation experiments; a major component and a slightly faster component were always detectable. At the moment it is not known whether there is a genuine second component or whether the erythrocuprein decomposes during the disc-electrophoretic separation. The electrophoretic separation pattern of bovine erythrocuprein is essentially the same for erythrocuprein from human tissues (*64, 68*).

Table 1. *Molecular weight of erythrocuprein. Data taken from: a (67), b (68), c (73), d (82), e (64), f (70), g (72), h (66), i (58), k (87)*

	Human Erythrocuprein			Bovine Erythrocuprein		
	Liver	Brain	Erythrocytes	Heart	Liver	Erythrocytes
Sedimentation Constant $S^{\circ}_{20,\omega}$	2.84 [b] 3.3 [e]	2.79 [b]	2.94 [a], 3.02 [i] 2.77 [h]		3.1 [k]	3.06 [c], 3.35 [f]
Molecular weight from:						
Sedimentation equilibrium			32 500 ± 690 [a] (0.157% protein)	32 500 [d]	32 600 [k]	32 800 [c] 32 600 [f]
			32 000 ± 530 [a] (0.037% protein)			
			32 700 ± 670 [a] (0.021% protein)			
	35 600 [b] ±1800	35 700 [b] ±1100	33 200 ± 1800 [a] (0.015% protein)			
copper content			33 500 [a]			33 500 [c]
zinc content						33 500 [c]
amino acid analysis			33 760 [a]			32 585 [c]
D and [η]			28 000 ± 2 000 [i]			33 000 [c]
S and [η]						33 900 [c]
S and D						34 300 [c]
gel filtration						33 000 [c]
electrophoresis				35 000 [d]		
Partial specific volume (ml/g)						0.736 [c], 0.720 [f]
Isoelectric Point	4.76 [b]	4.74 [b]	4.75 [a], 4.6 [h], 5.2 [i]			4.95 [g]

Table 2. *Amino-acid residues per mole of erythrocuprein. Data taken from the references given in brackets*

	Human Erythrocuprein							Bovine Erythrocuprein			
	Erythrocytes				Liver		Brain	Erythrocytes			Heart
	(*58*)	(*68*)	(*67*)	(*66*)	(*68*)	(*64*)[a]	(*68*)	(*72*)	(*82*)	(*84*)	(*82*)
Lysine	20.3	23.2	22.2	22.9	22.5	17.7	22.1	20.9	22.0	22.8	22
Histidine	14.5	12.8	15.8	16.5	13.3	10.4	13.7	15.3	16	16.8	16
Arginine	9.1	7.8	6.8	8.6	8.4	7.1	8.1	7.2	10	8.7	9
Aspartic acid	41.3	36.2	36.5	40.2	36.9	27.8	36.6	35.1	35	34.1	36
Threonine	18.3	16.2	15.8	16.9	16.1	12.0	16.4	24.9	26	22.4	25
Serine	19.7	20.3	19.8	19.8	20.9	14.1	20.7	14.1	20	15.4	21
Glutamic acid	28.2	27.4	26.3	18.1	28.7	22.8	27.6	27.8	24	22.5	27
Proline	11.1	11.3	10.4	11.3	10.7	8.9	10.0	11.6	14	14.6	15
Glycine	45.6	47.1	50.1	52.2	48.0	36.6	47.5	51.6	50	50.4	54
Alanine	20.8	20.6	19.8	21.4	21.2	16.1	20.9	18.4	21	19.6	22
Valine	23.0	29.6	27.6	28.4	28.2	19.3	29.7	28.3	28	28.4	26
Methionine	—	0.1	0.0	—	0.3	0.6	0.0	2.0	0	2.1	0
Isoleucine	11.0	16.5	16.4	16.0	16.1	10.7	16.8	19.0	17	18.1	17
Leucine	17.8	19.4	17.4	19.7	19.4	15	18.5	16.8	20	17.5	18
Tyrosine	1.9	0.1	0.0	0.0	0.3	1.5	0.0	1.8	2	2.3, 2.0(*78*)	2
Phenylalanine	9.9	8.9	7.8	8.1	8.2	—	8.4	8.0	10	8.5	10
Half cyctine	11.3	6.7	6.6	5.1	7.7	4.3	6.3	6.2	—	5.9	—
Cystein	—	—	0.7	—	—	—	—	1.0[b]	—	2	—
Tryptophan	0.6	—	—	— 1.25(*92*)	—	—	—	0.4	0	0 (*78*)	0
NH_3	24.5	—	23.2	28.0	—	24.1	—	18.6	—	23.1	—

[a]) Assuming 33 600 m. wt.
[b]) After treatment with 8 M guanidine hydrochloride.

3.2. Amino-Acid Analysis

The molar ratios of the amino-acid residues were determined after erythrocuprein was subjected to acid hydrolysis at 110 °C for 20 to 96 hours. The method of *Moore* and *Stein* was used throughout (*88, 89*). The sulfhydryl content of human erythrocuprein was determined using the spectrophotometric assay of *Boyer* (*90*). Total half-cystine was analysed as S-carboxy methyl cysteine in acid hydrolysates (*91*) of reduced and alkylated erythrocuprein, or as cysteic acid in acid hydrolysates oxidized with performic acid. The data are summarized in Table 2.

The amino-acid analysis revealed some similarity between human and bovine protein. Striking differences can only be seen in the threonine content, which was approximately one third higher in the bovine protein, and in the absence of tyrosine in the human protein,

The amino-acid composition reported by *Porter, Sweeny* and *Porter* (*64*) can be considered as a preliminary result. Thus, the values given by *Hartz* and *Deutsch* (*67*) for human erythrocuprein and by *Carrico* and *Deutsch* (*8*) for the corresponding proteins from human liver and brain should be closest to the actual amino-acid composition. It is still uncertain how many tryptophan residues are present in human erythrocuprein. From measurements of the decrease of the fluorescence emission in the presence of N-bromsuccinimide it was concluded that 1.25 moles of tryptophan per mole of protein are present (*92*). With MCD studies (*78*) and chemical analysis by the method of *Spies* and *Chambers* (*93*) the absence of tryptophan residues was demonstrated using bovine erythrocuprein. Other spectrophotometric methods developed by *Edelhoch* (*94*) or *Beaven* and *Holiday* (*95*) detected 0.4—0.9 mole of tryptophan per mole of the bovine protein (*72*).

3.3. Metal Content

The presence of copper in erythrocuprein was first demonstrated by *Mann* and *Keilin* (*55*). In 1970 zinc was found in human erythrocuprein (*69*). The metal ions were measured by different spectrophotometric assay procedures using 2,2-biquinoline, bis cyclohexanone oxalyldihydrazone, and dithizone as chelating ligands (*85, 97—100*). Alternatively, atomic absorption spectroscopy (*86*), neutron activation analyses, and emission spectroscopy (*69*) were successfully employed. From the neutron activation analyses it became apparent that metals other than zinc and copper were present in amounts less than 0.1 g-atom per 33,000 g of protein. From the different analyses (Table 3) it can be concluded that erythrocuprein contains 2 g-atoms of each of copper and zinc.

Table 3. *Copper and zinc content of erythrocuprein isolated from human and bovine tissues. Data taken from: a (64), b (69), c(72), d (74), e (82), f (55), and converted values using a molecular weight of 34,000 from g (66), h (58), i (87)*

	Copper (g atoms per mole of protein)	Zinc
Human Erythrocuprein		
Liver	1.70[a], 1.83[b]	1.66[b]
Brain	1.96[b]	1.91[b]
Erythrocytes	2.0[g], 2.02[b], 1.9[h]	1.94[b]
Bovine Erythrocuprein		
Liver	1.88[f], 1.99[i]	1.88[i]
Heart	1.64[e]	1.76[e]
Erythrocytes	1.88[f], 2.0[c], 2.05[d], 1.94[e]	2.0[c], 2.01[d], 1.7[c], 1.76[e]

Carrico and *Deutsch* (*69*) reported a very poor exchange of the metals when using ^{64}Cu or ^{65}Zn. However, a slow transfer of radioactive copper into erythrocuprein can be observed if ^{64}Cu is added to whole blood. Approximately 25% of the radioactivity added was present in erythrocuprein after 12 hours (*60*).

3.4. Spectra

3.4.1. Ultraviolet and Visible Absorption Spectra

Although many studies have been performed on human erythrocuprein, no comprehensive and detailed spectrum of this metalloprotein in the visible and ultraviolet region has yet been reported. *Carrico* and *Deutsch* (*66, 68, 69*) measured the absorption spectra manually and, due to this technique, a fine structure was not observed. *Keele, McCord* and *Fridovich* (*82*) published a comparison of the UV absorption spectra of human and bovine erythrocuprein in the wavelength region 240 to 310 nm. In Fig. 6a distinct fine structure can be seen in the absorption spectrum of human erythrocuprein. Absorption bands appear at 254, 260, 266, 326 nm and shoulders at 270, 282 and 289 nm.

It is interesting to note the unusual diminished absorption in the 280 nm region ($\varepsilon_{265} = 17{,}000$). This was attributed at the low content of aromatic amino-acid residues, especially typosine and tryptophan. The absorption in the visible region was extremely low. Apart from the broad absorption band at 675 $\pm$20 nm ($\varepsilon_{675} = 275$), a shoulder can be observed at 440 nm (Fig. 7). According to the definition given by *Malkin* and *Malmström* (*3*), erythrocuprein is classified as a non-blue copper protein.

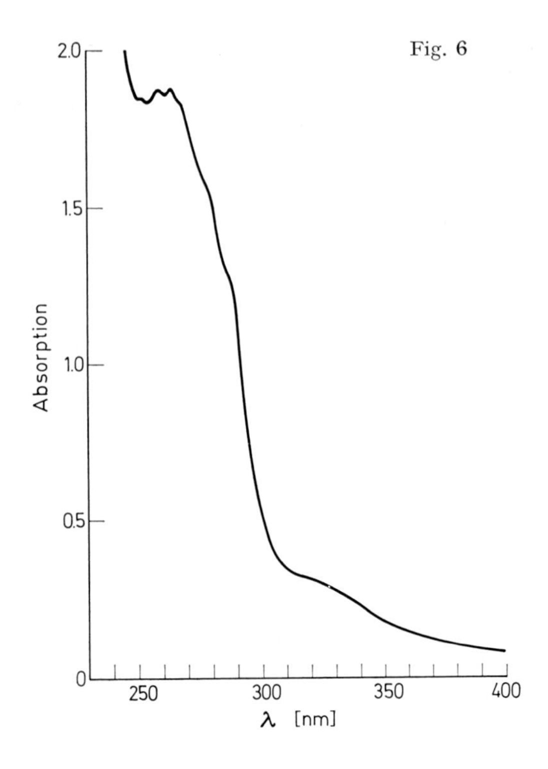

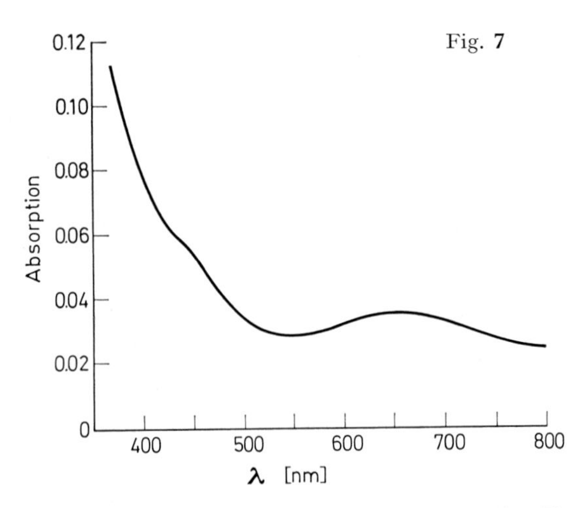

Figs. 6 und 7. Absorption spectrum of human erythrocuprein. The protein was isolated using the method given in Ref. (*74*). Recording was carried out with a Unicam SP 1800.

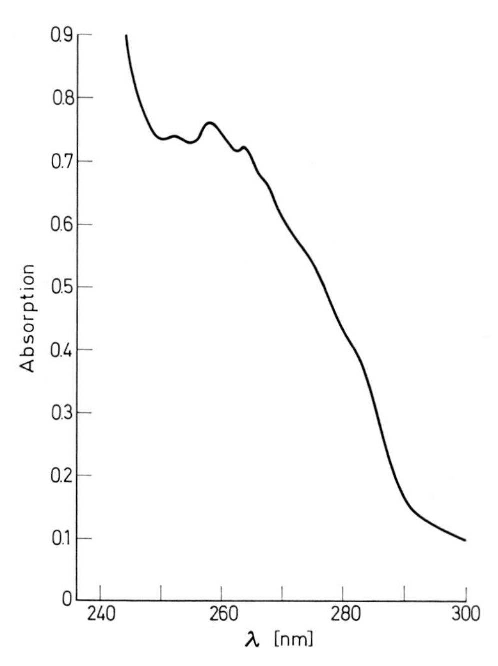

Fig. 8. Spectrum of bovine erythrocuprein in the UV region (*74*).

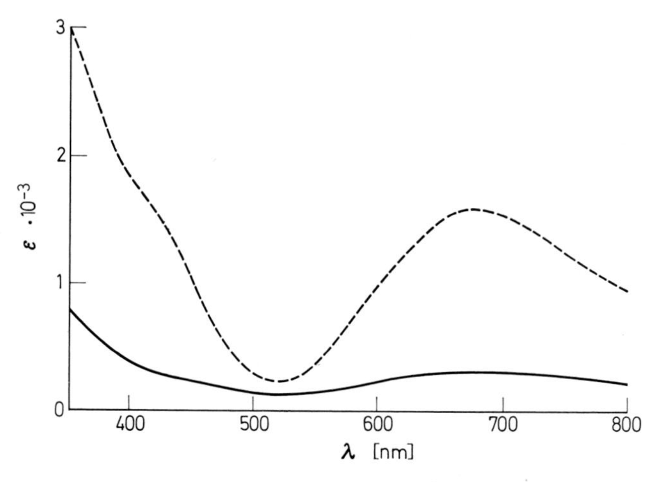

Fig. 9. Spectrum of bovine erythrocuprein in the visible region at room temperature (solid line) and at 77 °K (dashed line) (*74*).

A striking difference of both the absorption profile and the molar absorption coefficient in the UV region can be recognized using bovine erythrocuprein (Figs. 8, 9). The molar coefficients of absorption were calculated as 9,840 at 259 nm and 313 at 680 nm (*74*). The ratios of these absorbancies A_{259}/A_{680} was 31.5, which agreed with the values obtained in other laboratories (*70*). This ratio differs substantially from that of human erythrocuprein ($A_{265}/A_{675} = 65$). It was suggested that the marked decrease of absorbtion in the UV region might be accounted for the absence of tryptophan in the bovine erythrocuprein (*82*). However, this point must remain open until it is clear whether or not the human protein really contains two tryptophan residues.

Low-temperature absorption studies promised valuable information in the visible region (Fig. 9). As expected, the absorption of erythrocuprein at 680 nm was increased by a factor of approximately 5 and the shoulder at 430 nm was more distinct than at room temperature. The absorption band at 680 nm and the shoulder at 430 nm cannot be precisely assigned at present, though they are presumably due to either a pure $d \rightarrow d$ transition or a charge transfer band, or a mixture of both. Furthermore, the possibility that the 680 nm band represents more than one $d \rightarrow d$ transition was not excluded (*73*). This thought gained substantial support from CD and especially MCD data.

3.4.2. Optical Rotatory Dispersion (ORD) and Magnetic Optical Rotatory Dispersion (MORD)

The ORD and MORD spectra were measured from 220 to 600 nm (Fig. 10). The optical rotation remained negative throughout the spectrum; a trough is seen at 225 nm. A magnetic field of 8 kgauss produced no distinct changes in the ORD spectrum of erythrocuprein from 250—600 nm. However, a positive MORD was detectable at wavelengths < 240 nm.

3.4.3. Circular Dichroism (CD) and Magnetic Circular Dicroism (MCD)

Little information was obtained from ORD and MORD measurements concerning structural aspects of erythrocuprein. The application of CD and MCD spectroscopy promised to be a better tool (Figs. 11 and 12).

The positive CD Cotton effect at 260 nm and the negative Cotton effect at 274 nm may be at least partly attributed to amino-acid residues. The side-chain residues of L-cystine, L-phenylalanine. L-tyrosine or L-tryptophan cause Cotton effects at 250—310 nm (*101*). However, copper complexes of simple amino acids show CD extremes at 245—290 nm. Therefore, some contribution to the 260 nm band of erythro-

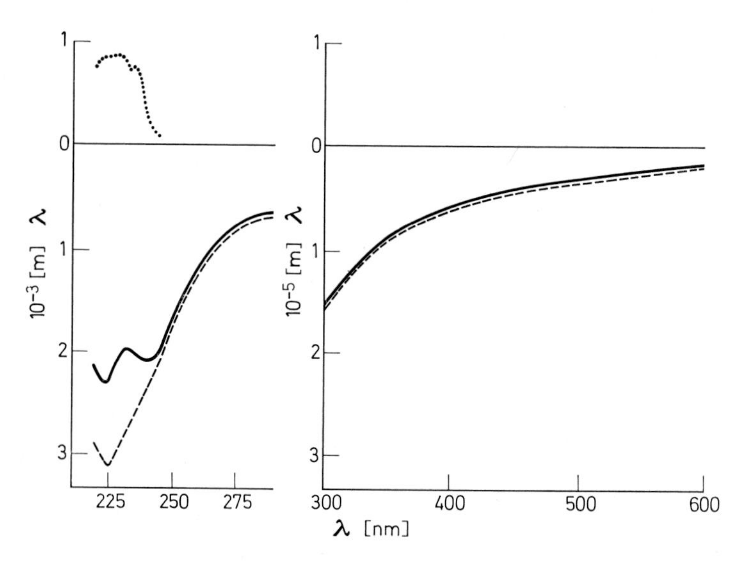

Fig. 10 (*74*). ORD (---), MORD (—) and corrected MORD (. . . .) spectra of bovine erythrocuprein. MORD was corrected by substracting the ORD portion. Molar rotations were calculated from 220—290 nm on the basis of mean residue weight, and from 300—600 nm on the basis of the molecular weight of erythrocuprein. The values are not corrected with respect to the refraction index of the solvent

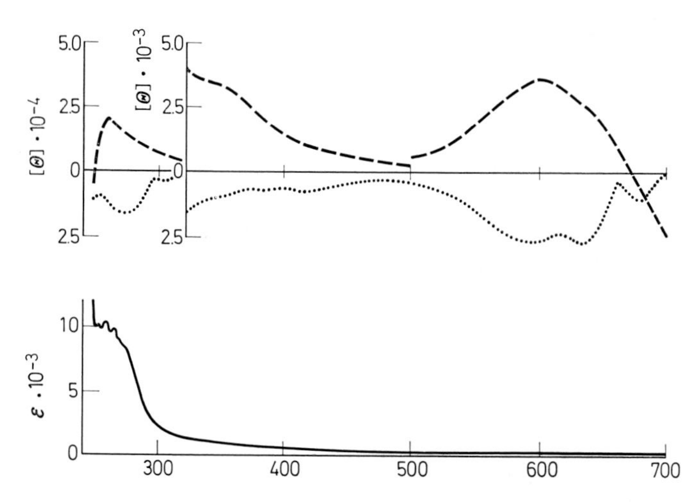

Fig. 11 (*74*). Absorption (—), CD (---) and MCD(. . .) spectra of bovine erythrocuprein. Molar magnetic ellipticity is given at 49.5 KG and was corrected for natural CD

cuprein may arise from the chelated copper (*102*). Reconstitution experiments with the apoprotein and Cu^{2+} support this view. Furthermore, the CD spectra of 1:1 cupric chelates of glycyl amino acid dipeptides show CD bands at 310—320 and 260—280 nm. The bands at 260 nm are thought to arise from ligand to Cu^{2+} charge-transfer transitions. For the Cotton effects around 320—340 the suggestion was made that the 315 nm band may be a $n \rightarrow \pi^*$ transition of an amide group.

CD studies have been found appropriate to detect the split components of the $d \rightarrow d$ transition of cupric complexes of L-amino acids. According to these investigations, the copper complexes show four CD components in the $d \rightarrow d$ transition region from 850 to 530 nm (*103*). The composite nature of the 600 nm (610 nm in Ref. (*73*)) CD band of erythrocuprein is seen from shoulders at wavelengths $>$600 nm. When a magnetic field was applied, the resolution of the bands was apparent. Two MCD components (and possibly a third) of the $d \rightarrow d$ transitions at 595 and 640 nm were detectable. The complementary utility of MCD and CD for resolving overlapping absorption bands is obvious (*104, 105*) (Fig. 12, Table 4).

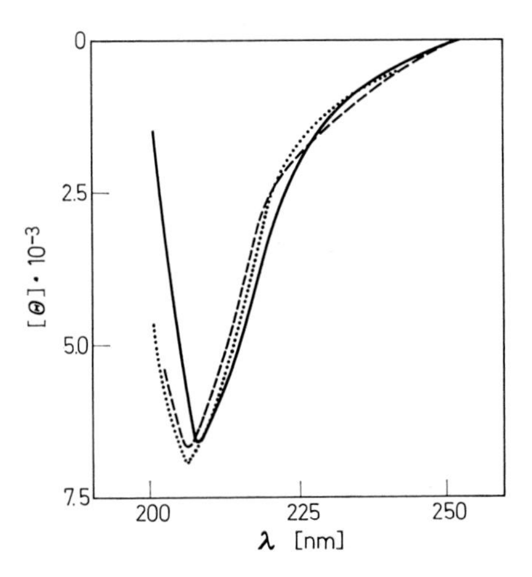

Fig. 12 (*74*). Correlation of mean residue ellipticity vs. wavelength for bovine erythrocuprein. (—) 10 mM phosphate, *p*H 7.2; (---) 9M urea and (...) 0.1M dodecyl sulphate

CD measurements were carried out in the region 185—230 nm and in the presence of urea and dodecyl sulphate to obtain some information on the protein conformation (Fig. 12, Table 4). According to Refs. (*106—112*), random-coil peptides display a characteristic negative Cotton

Table 4. *Summary of CD and MCD dat of bovine erythrocuprein (74)*

	CD ([θ] value)								Corr. MCD
Solvent:	10 mM phosphate buffer	0.1 M sodium dodecyl sulphate		9 M urea					
Reaction time:		2 h	4 days	30 min-2h	1 day	2 days	4 days	5 days	
	(208)[a]	(206)[a]	(206)[a]	(206)[a]			(210)		
	−6040	−6900	−6000	−6700			−6000		
	(261)			(261)	(261)	(261)		(261)	(275)
	19500			19 600	17 900	17 900		17 200	−16 500
				(290) (s)	(290) (s)	(290) (s)		(290) (s)	
				10 000	8 300	8 200		8 140	
	(345) (s)			(345) (s)	(345) (s)	(345) (s)		(345) (s)	
	3 500			4 300	3 520	3 460		2 920	
	(430) (s)								
	850								
	(600)								(595)
	3 150								− 2 660
									(640)
									− 2 750

CD ([θ]) values are expressed in degrees·cm^2·dmole^{-1}: values in parantheses are corresponding wavelengths. s means shoulder.
[a]) These values are calculated on the basis of mean residue weights.

effect centered at 196 nm. A negative Cotton effect is observed at 216—218 nm ($\pi \rightarrow \pi^*$ transition) and a positive CD band ($\pi \rightarrow \pi^*$ transition) at 196 nm, both being attributable to a folded-sheet protein. α-Helical peptides characteristically differ from these conformations in their CD properties: two negative CD bands are located at 222 nm and 208 nm and a positive Cotton effects is found at 191 nm. The CD spectrum of erythrocuprein exhibits only one negative Cotton effect in the wavelength region 200—250 nm of relatively low ellipticities (*73*, *74*). No double minimum such as is characteristic for α-helical content was present. The ellipticity value was only about 8% compared to peptides of helical structure such as poly-α,L-glutamate (*108*). Thus, it can be assumed that the helical portions of erythrocuprein are low. The wavelength of the negative Cotton effect at 208 nm of erytrhocuprein agrees neither with the CD spectrum of a folded-sheet peptide nor with the CD data of a random-coil peptide. Surprisingly, urea or dodecyl sulphate had

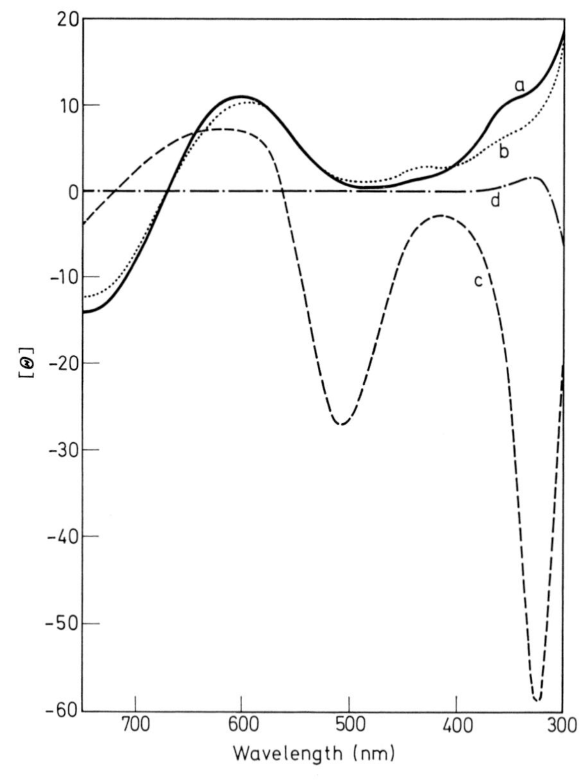

Fig. 13 (*81*). CD spectra of bovine erythrocuprein at different *p*H values. a) *p*H 6.0; b) *p*H 11.5; c) *p*H 12.5; d) readjusted with HCl to *p*H 7.5

virtually no effect on the CD spectrum. The question remains as to whether this protein has mainly random-coil or folded-sheet structure, or a protein conformation which differs from both. The *p*H dependence of the CD spectra revealed an interesting phenomenon (*81*). Up to *p*H 11.5 only very slight changes were observed in the CD bands and these were fully reversible upon readjustment to neutral *p*H. Above *p*H 12 dramatic changes (pink colour) appear in the optical properties of erythrocuprein, indicating the denaturation of the protein. These changes could not be reversed by lowering the *p*H (*81*).

A CD spectrum similar to that obtained at *p*H >12 was recorded in the presence of excessive cyanide at *p*H 8.8 except that the positive Cotton effect around 610 nm was absent. The removal of cyanide by dialysis brought back the original spectrum. Like the CD spectra, the optical spectra were different at *p*H >12 and after the addition of cyanide. The absorption band at 680 nm was shifted towards 500 nm. Again, the removal of cyanide brought back most of the original spectrum while the *p*H effect at *p*H >12 was irreversible (*81, 113*).

3.4.4. Electron Paramagnetic Resonance (EPR)

Oxidized bovine erythrocuprein revealed some EPR signals (*74, 81, 82*) very similar to those obtained with human erythrocuprein (*61, 68*). The EPR parameters were $g_{\parallel} = 2.263$, $g_{\perp} = 2.062$ and $A_{\parallel} = 0.014$ cm^{-1} (*78, 81, 114*). Four hyperfine components were centered at the $g_{\parallel}$ position. Superhyperfine splittings (approx. 14 G) were slightly detectable already at *p*H 7.5 (Fig. 14) (*74*) but were much more distinct at *p*H 11.8 where nine superhyperfine splittings could be counted. This effect was fully reversible, as seen in Fig. 15. It would be most attractive to assign these nine superhyperfine splittings to four nitrogen atoms bound to the copper. However, it has to be taken into consideration that the value of the copper hyperfine splitting constant in the perpendicular direction $|A_{\perp}|$ can be of the same order of magnitude as that of the ligand hyperfine splitting. Therefore, it could be misleading to assign the number of superhyperfine lines to the number of coordinating nitrogen atoms (*81*). On the other hand, magnetically equivalent nitrogen atoms are not necessarily structurally equivalent. Possible assignments of three to four nitrogen atoms in planar arrangement were suggested on the basis of the similarity between the visible absorption properties of the cyanide-treated protein and that of Cu(II)-triglycylglycine (*118*). Nevertheless, the x-ray analysis will be awaited with great interest.

In contrast to a number of other Cu enzymes, the erythrocuprein copper was fully detectable by EPR (*3, 81, 114*). Exposure of the protein to 6 M guanidine hydrochloride caused no measurable changes in the

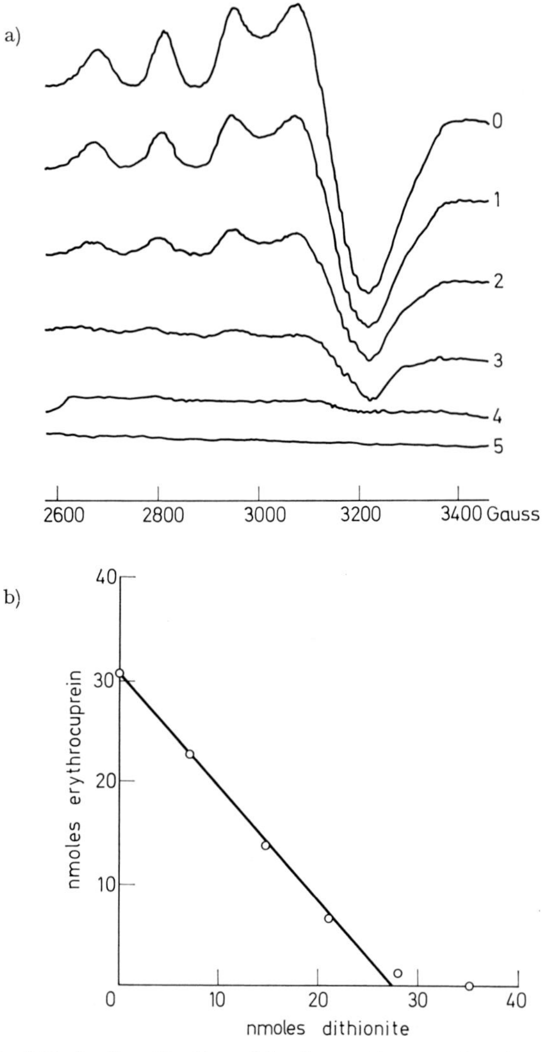

Fig. 14 (*74*). a) Reductive titration of bovine erythrocuprein using aqueous dithionite. (Erythrocuprein conc. 0.305 mM; reaction vol. 100 μl; reduction with 7.04 mM dithionite (0—5 μl) under argon for 2.5 min at 20 °C before freezing to 77 °K; microwave frequency 9.175 GHz, modulation amplitude 1 G; microwave power 5 mW.); b) Dependence of different dithionite concentrations on erythrocuprein reduction

EPR spectrum. Thus, detection of the number of transferable electrons was possible (Fig. 14). Aqueous dithionite effectively reduced erythrocuprein. Plotting the height of the $g_{\perp}$ signal versus dithionite concentra-

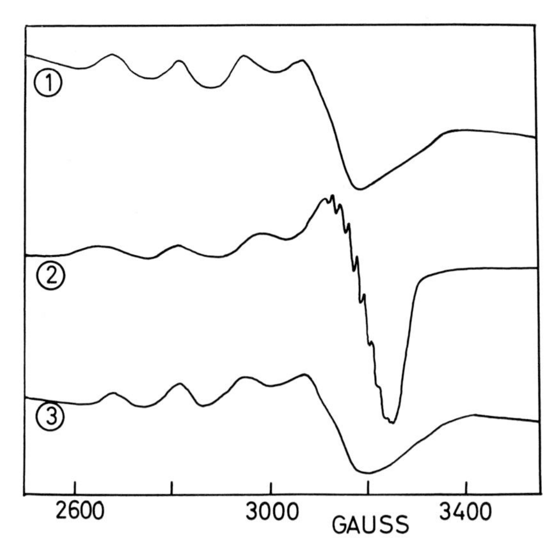

Fig. 15 (*78*). EPR spectra of bovine erythrocuprein (0.15 mM) at different *p*H values. ① *p*H 6.7; ② *p*H 11.8; ③ readjusted to *p*H 8.0. From *p*H 2—8 no measurable changes were observed. Spectra were recorded at 77 °K. Microwave frequency 9.169 GHz; modulation amplitude 2G; microwave power 5 mW

tion resulted in a straight line. One mole of erythrocuprein required one mole of dithionite, indicating the transfer of two electrons per mole of erythrocuprein. If ascorbate was employed as a reducing agent, no reduction was observed. Even reaction times up to 30 min displayed no ascorbate-induced reduction of erythrocuprein.

As in the CD measurements, dramatic changes occurred in the EPR spectrum when the *p*H was raised above 12. A copper biuret-type EPR spectrum appeared ($g_{\parallel} = 2.179$ and $A_{\parallel} = 0.019$ cm^{-1}) and a completely new copper singnal was observed after the back titration to neutral *p*H ($g_{\parallel} = 2.235$, $g_{\perp} = 2.056$ and $A_{\parallel} = 0.0175$ cm^{-1}), indicating irreversible denaturation of the native copper protein (*81*).

Short-time treatment of erythrocuprein with cyanide was found to produce considerable changes in the visible absorption spectrum and CD properties. EPR measurements of cyanide-, cyanate-, thiocyanate-, and azide-treated erythrocuprein revealed some interesting changes in the EPR parameters (*81, 113, 161, 162*). These are summarized in Table 5.

In addition, these anion-binding studies have been supported by optical titrations and by nuclear magnetic relaxation dispersion (NMRD) measurements. An example is presented of cyanide-treated erythrocuprein (Fig. 16).

Table 5. *EPR parameters of anion-erythrocuprein complexes* (*113*)

Complex	$g_{\parallel}$	$g_{\perp}$	$A_{\parallel}$
			gauss
Native erythrocuprein in water	2.27	2.09	137
Azide			
2 per mole	2.263	2.074	145
3 per mole	2.260	2.072	148
4 per mole	2.254	2.069	153
8 per mole	2.253	2.066	155
200 per mole	2.242	2.054	158
2 azide and 2 cyanide	2.22		180
Excess cyanide	2.21		188
Excess cyanate	2.268	2.068	160
Excess thiocyanate	2.27	2.076	148

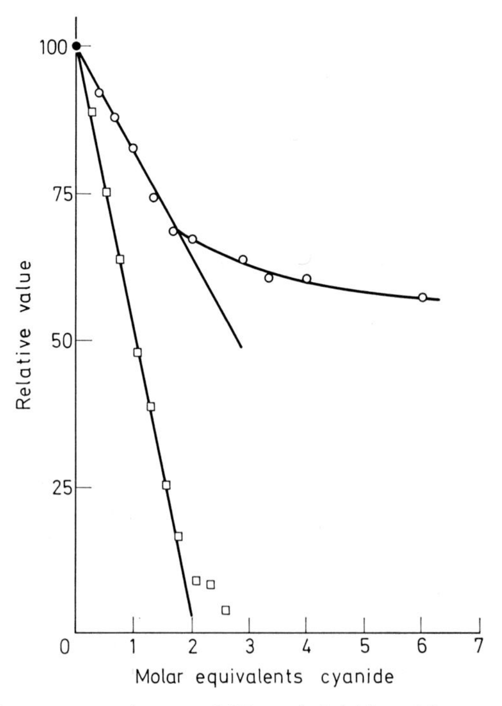

Fig. 16 (*113*). Changes occurring on addition of stoichiometric amounts of cyanide to erythrocuprein. Diminished absorption at 680 nm ○ and decay of molar relaxivity at 0.1 MHz □

From these anion-binding studies it was concluded that there are at least four anion-binding sites of different nature on the protein which can be classified into two portions. One of these portions strongly binds two moles of azide, causing a minor change in the visible absorption spectrum and an altered EPR spectrum, although the NMRD is unaffected. These two binding sites are thought to be the Zn^{2+} of the protein. The second portion of binding sites is attributed to the Cu^{2+} which binds the cyanide. In this case, substantial changes were observed in both the EPR and the absorption spectrum and the NMRD was lowered to the diamagnetic level (*113*). More experimental support is needed before it can be decided whether or not Cu^{2+} and Zn^{2+} form a ligand-bridged bimetal complex. Some evidence of the proximity of Zn^{2+} and Cu^{2+} was recently obtained by using erythrocuprein with the Zn^{2+} partially replaced by Co^{2+}. This substitution caused a marked reduction of the Cu-EPR signal indicating magnetic interaction between the two paramagnetic ions (*198*).

3.4.5. X-Ray Photoelectron Spectroscopy (XPS)

X-ray photoelectron spectroscopy proved a convenient method for studying the presence of metals in metalloproteins. In some cases it has proved possible to draw conclusions concerning the site of the metal binding (*169—171*). This method is especially suitable for those metals which cannot be studied by means of Mössbauer or EPR spectroscopy. The binding energies of Cu $2p_{3/2}$ and Zn $2p_{3/2}$ in erythrocuprein were determined and compared with the corresponding values of different Cu/Zn amino-acid complexes (*115—117*). The intensity of the signals was approximately the same whether Cu or Zn amino-acid chelates were employed. However, in the case of erythrocuprein a marked change was observed. The intensity ratio of the signals was 1:2.5 (Zn:Cu). Since two gram atoms of each metal are present in erythrocuprein, it is suggested that the Cu might be located in the outer sphere of the protein portion while Zn must be less accessible to the Mg $K\alpha$ radiation. The Zn $2p_{3/2}$ signals were 1021.5 ± 0.2 eV in the amino-acid chelates and the aquo complex, and an almost identical value of 1020.5 eV was measured in erythrocuprein. Considerable differences are observed in the Cu $2p_{3/2}$ signals for the Cu amino-acid chelates (934.2 ± 0.3 eV) and erythrocuprein (931.9 eV) (Fig. 17). Surprisingly, two different types of Cu can be characterized in the Cu[Cu(Asp)$_2$] complex. The extraneous Cu^{2+} had a signal at 932 eV while the binding energy of the inner Cu $2p_{3/2}$ was monitored at 934.1 eV. It is interesting to note the virtual identity of the binding energy of the extraneous copper in Cu[Cu(Asp)$_2$] and in erythrocuprein.

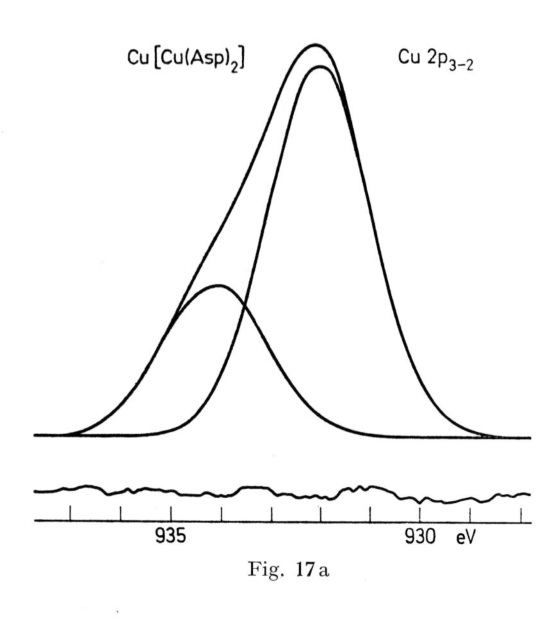

Fig. 17a

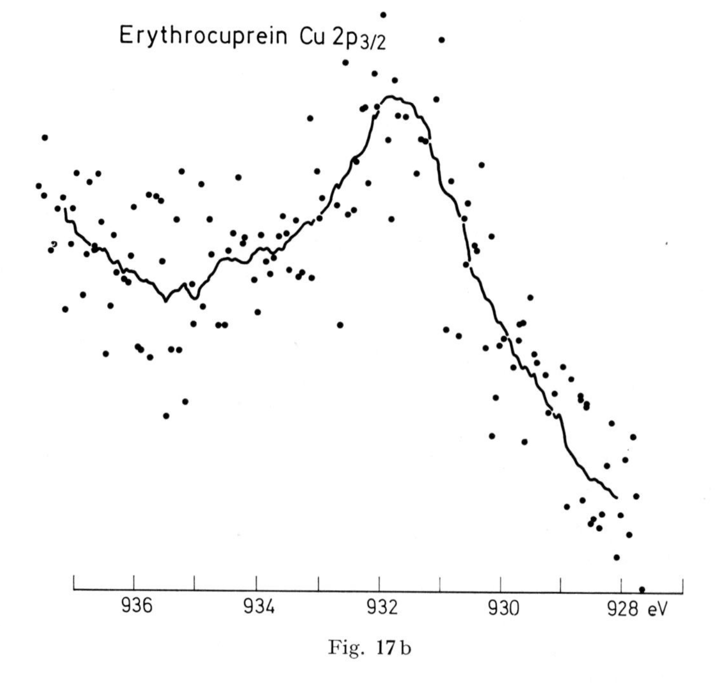

Fig. 17b

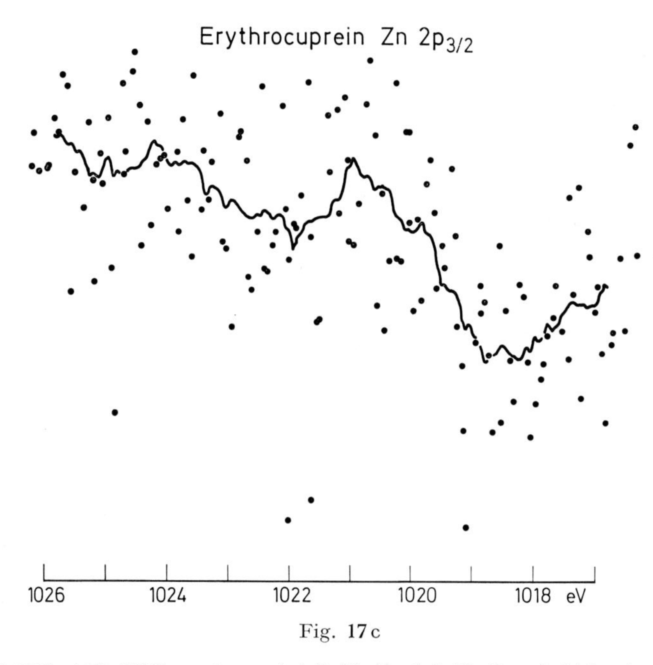

Fig. 17c

Fig. 17 (*115—117*). XPS spectrum of a) Cu[Cu(Asp)$_2$], (Cu 2p$_{3/2}$); b) bovine erythrocuprein Cu 2p$_{3/2}$; c) bovine erythrocuprein Zn 2p$_{3/2}$. The temperature was 25 °C. Pressure 10^{-5} Torr. Measurements were performed in a Varian V-IEE 15 photoelectron spectrometer. The electron binding energy was determined using the C ls signal (284 eV) obtained from Scotch tape

3.5. Immunochemical Properties

Immunochemical assay, employing either the Oudin agar diffusion technique (*119*) or the Ouchterlony agar diffusion test (*120*), was used to examine the homogeneity and identity of the protein portion of erythrocuprein from different sources. According to these techniques a high purity of erythrocuprein was reported already in 1959 (*57*) and 1961 (*60*). However, the sensitivity does not appear to be very satisfactory. Erythrocytes from pigs, rabbits, chicken, and steers were tested for the presence of erythrocuprein by the agar double-diffusion technique. No reaction was found between antihuman erythrocuprein serum and haemolysate form these species (*60*). At higher concentrations of erythrocuprein form human tissues, a precipitin line was observed in the Ouchterlony test using rabbit anti-erythrocuprein serum (*67*). The correlation of the precipitin reaction between the purified erythrocuprein from human blood, liver and brain is depicted in Fig. 18. Reactions of complete identity were obtained for the three proteins.

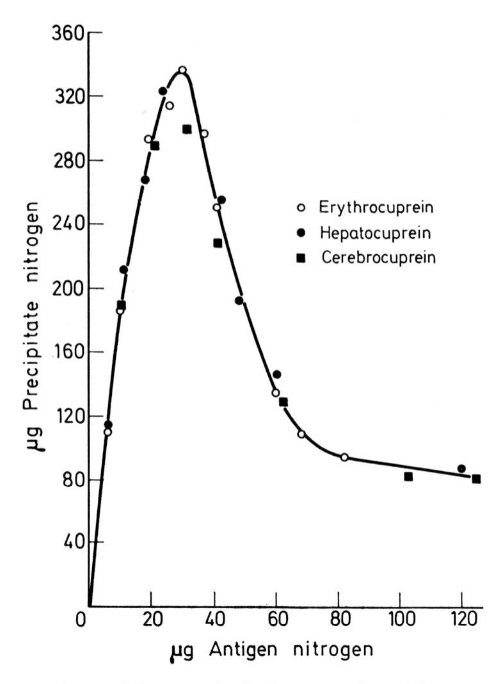

Fig. 18 (*68*). Comparison of immunological properties of human cuprein isolated from erythrocytes (○), liver (●), and brain(■). Quantitative precipitin reactions of rabbit antibody to erythrocuprein were performed

3.6. The Apoprotein

The apoproteins made from human and bovine erythrocuprein have been examined in several laboratories (*69, 70, 72, 74, 76, 78*). The purest apoerythrocuprein was obtained using the gel-filtration technique (*76*). This bovine apoerythrocuprein was metal-free and homogeneous as seen from several parameters. The numerical value of the molar absorption coefficient at 259 nm was only 37% of the corresponding value of the native protein (*72, 74*) and the fine structure of the apoprotein was well resolved. Six maxima were detectable at 250, 258, 262, 264, 268, and 275 nm. Apoproteins prepared by the dialysis procedure displayed a much higher absorption coefficient; for example, apoerythrocupreins prepared from human and bovine erythrocuprein showed 70% (*69*) and 54% (*72*), respectively, of the absorption of the native protein (Fig. 19).

The polyacrylamide gel electrophoresis showed a much lower contamination compared to other preparations (*69, 72*). The slightly faster movement of the apoprotein compared to the native erythrocuprein is obvious. Sedimentation velocity measurements revealed one symmetrical

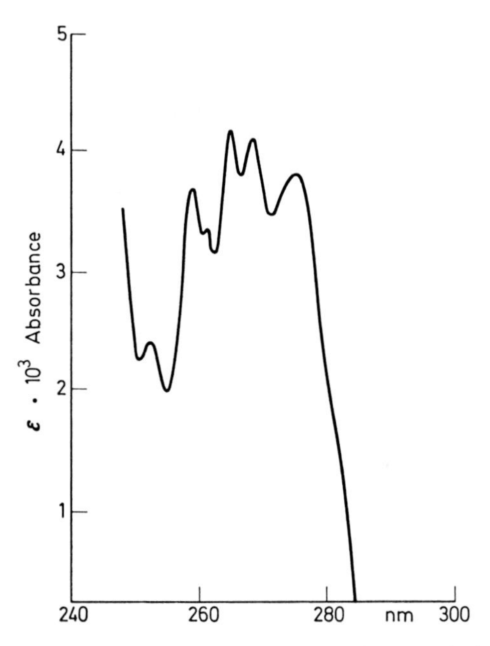

Fig. 19 (*122*). Absorption spectrum of bovine apoerythrocuprein in the UV region. The protein concentration was 5 mg; potassium phosphate buffer 10 mM, *p*H 7.2. Temperature 25°. Recording with a Unicam SP 1800 spectrophotometer

boundary of the apoprotein. The $s_{20\omega}$ value was 3.0 (*78*), which is in very close agreement with the corresponding sedimentation constant of the native protein. *Carrico* and *Deutsch* (*69*) reported a considerably lower $s_{20\omega}$ value of 2.86. They also recognized variable portions of a slower sedimentation component among different preparations which were completely absent in the apoprotein prepared by gel filtration (Fig. 20).

The *p*H stability of the native erythrocuprein was intriguing (*121*). Exposure to *p*H values from 2—12 for 5 min had no damaging effect on enzyme activity and absorption spectra (*122*). However, the apoprotein appeared to be less stable. In contrast to the native protein, the fine structure of the absorption spectrum of the apoprotein was present only between *p*H 2.5—9.5. At *p*H 7.2 no measurable amounts of Cu^{2+} or Zn^{2+} could be removed from the native or the partially reconstitued protein by means of EDTA (*69, 70, 72, 122*). A rather low *p*H value, *p*H 3.8, was necessary to remove the Cu^{2+} and Zn^{2+} completely. This low *p*H probably made the metal ions more accessible to the EDTA or may have caused a drastic reduction of the kinetic stability of the metal ions bound in the native metalloenzyme.

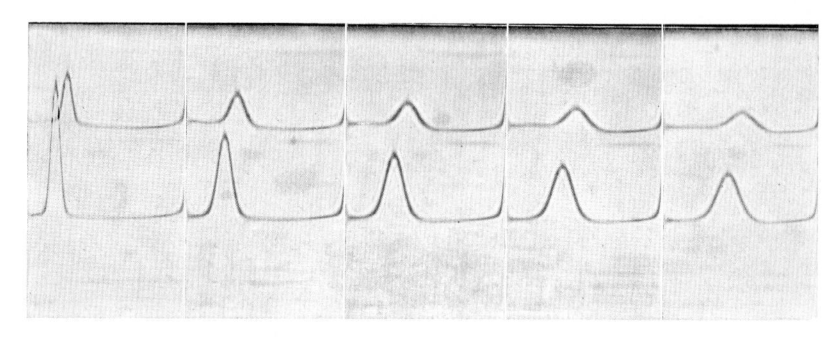

Fig. 20 (*78*). Sedimentation diagram of bovine apoerythrocuprein. The protein concentration was 2.1 mg and 4.1 mg dissolved in 5 mM phosphate buffer, *p*H 7.2. Direction of sedimentation left to right, speed 60,000 rpm. Schlieren bar angle 50°, 12 mm Al-valve synthetic boundary cell. Photographs were taken at 16-min intervals after reaching full speed

3.7. CD and MCD Data of Apoerythrocuprein

The CD and MCD data for bovine apoerythrocuprein are summarized in Table 6.

Table 6. *UV, CD, and corrected MCD data of purified apo-erythrocuprein. The employed magnetic field was 49.5 kG applied parallel to the light beam. The corrected MCD data are expressed in molar magnetic ellipticity values at 49.5 kG; the values are corrected for natural CD* (*78*)

Solvent	CD([θ]-values	Corrected MCD ([θ]-value)	UV (ε-value)
10 mM phosphate	294 (1100)		
buffer, *p*H 7.2	284 (— 890)	276 (—13 600)	275 (3830)
	269 (—4450)	269 (—14 000)	268 (4130)
		262 (—10 300)	264 (4190)
			262 (3300)
			259 (3670)
			252 (2410)
			281 (4600)
6 M guanidyl hydrochloride			275 (5400)
*p*H 5.9	269 (—5520)	277 (—16 000)	
		269 (—16 100)	269 (5400)
	262 (—5710)	260 (—12 300)	261 (4200)
			259 (4200)
			252 (3300)
6 M guanidyl hydrochloride	281 (5670)	294 (—27 920)	295 (7900)
*p*H 11.7		269 (— 9 270)	
			246 (38100)

If the CD spectrum of the native protein is compared with the corresponding spectrum of the apoprotein (*78*) a dramatic change can be observed at wavelengths >250 nm (*73, 74*). In contrast to the cupro-zinc protein the apoprotein does not show any ellipticity above 310 nm. The most characteristic feature of the CD curve of the metalloprotein in the wavelength region from 250 to 320 nm is the broad positive Cotton effect at 261 nm ($[\theta] = +19{,}500$). However, the apoprotein spectrum obtained in phosphate buffer exhibits only a small positive Cotton effect in that range and at much higher wavelength within this range a much smaller ellipticity. Two further electronic transitions can be detected from the two dichroic shoulders of relatively low intensity at 269 and 284 nm. An attempt was made to assign the 295 nm Cotton effect to the contribution of tryptophan (*73*), though this Cotton effect appears at somewhat too high wavelengths for tryptophan derivatives and peptides (*123, 124*). There is no doubt that the CD bands from 260 to 300 nm of the apoprotein are caused by aromatic amino acids or side-chain residues of L-cystine. In erythrocuprein most probably ligand to Cu^{2+} charge-transfer transitions overlap with the Cotton effects of these amino acids (*102, 125*). A further change in the CD spectrum of the apoprotein at wavelengths >250 nm occurs on dissolving the apoprotein in 6 M guanidyl hydrochloride (*p*H 5.9). The small positive Cotton effect at 294 nm disappears and two negative bands are located at 262 and 269 nm. If the *p*H of the guanidyl hydrochloride solution is adjusted to 11.7 a broad positive band at 281 nm is observed for the apoprotein.

It was demonstrated that MCD is an excellent tool for the quantitative determination of tryptophan (*126, 127*). It was found that the ^1L_b transition of tryptophan causes a positive B-term MCD band at 290 nm. At that wavelength no bands of other amino acids interfere with this MCD Cotton effect of tryptophan. Also, the intensity of this band seems to be almost independent of the conformation of a protein and, therefore, is suitable for the calculation of the tryptophan content of proteins.

The MCD spectra of erythrocuprein (*74*) and apoerythrocuprein do not show the characteristic positive MCD Cotton effect of tryptophan. According to these MCD measurements erythrocuprein and the apoprotein should not contain any tryptophan residue. This result is in contrast to the report of *Bannister, Wood* and *Dalgleish* (*72, 73*). These authors measured the tryptophan content according to *Edelhoch* (*94*) and they found 0.4 mole per mole of protein. However, this method becomes less reliable when the tyrosine/tryptophan ratio is high and when corrections for the UV absorption and the cystine content are taken into consideration.

According to preliminary results (*128*), the negative MCD band at 277 nm and the shoulder at 283 nm arise from electronic transitions of the tyrosine ring. The phenyl residue of phenylalanine causes negative MCD Cotton effects at 269 and 260 nm. On dissolving apoerythrocuprein in 6 M guanidine hydrochloride instead of phosphate buffer, the circular dichroism spectrum changes significantly; in particular, the positive Cotton effect at 294 nm disappears completely. However, the MCD spectra of apoerythrocuprein are essentially similar in phosphate buffer or guanidine hydrochloride. The two MCD Cotton effects of phenylalanine and the Cotton effect and shoulder of tyrosine are present in both spectra. As with the UV and CD spectra, a dramatic change occurs in the MCD data of apoerythrocuprein if the guanidine hydrochloride solution is adjusted to *p*H 11.7. A broad negative MCD band is observed, at 294 and the broad positive CD band at 281 nm, both being characteristic for the tyrosyl anion. The magnetic ellipticity value at 293 nm of the dipeptide amide leucine-tyrosine is 11,700, that of apoerythrocuprein 27,900. According to these MCD measurements, the protein contains two tyrosine residues; this is in good agreement with the results found by amino-acid determination using automatic amino-acid analysis of the hydrolyzed protein (*72*, *82*).

3.8. Reconstitution Studies

Bovine apoerythrocuprein was reconstituted employing different Cu^{2+} or Zn^{2+} concentrations (Fig. 21, a, b). The protein concentration was kept constant and the metal concentration varied. The titration of apoerythrocuprein with Zn^{2+} confirmed the results by *Fee* and *Gaber* (*113*) that four preferred metal binding sites are present in the protein portion. It was interesting to note that the fine structure of the UV absorption spectra of apoerythrocuprein and the Zn protein remained essentially constant (*72*, *78*, *122*). This was not the case if Cu^{2+} was employed. One Cu^{2+} alone was able to produce both a profound increase of absorption and a change in the absorption profile (*72*, *78*). In fact, the 2 Cu apoprotein displayed almost the same characteristics as the fully reconstituted protein although λ_{max} was shifted in all cases some 50 nm to lower wavelengths. It might be concluded that some hidden $d \rightarrow d$ transitions (*102*, *125*) could be responsible for the strong UV absorption of the Cu-apoprotein chelates. In all reconstitution experiments where Cu^{2+} and Zn^{2+} were employed the Zn^{2+} was able to cause a shift of the visible absorption maxima to shorter wavelengths by 50 nm and a slight increase in the UV absorption (Table 7).

The absorption of the aerobically reconstituted erythrocuprein was somewhat lower in the UV region, and λ_{max} in the visible region was shifted to a longer wavelength. Futhermore, the shoulder at 430 nm was not distinct. No spectral changes as compared to the native erythro-

Table 7. *Absorption of reconstituted erythrocuprein and the Cu^{2+}-apoprotein chelates in the visible region (122)*

Reconstitution Technique	λ max nm	(ε-value)	shoulder at 430 nm
Native erythrocuprein	680	(313)	distinct
The apoprotein	—		—
Aerobic reconstition			
2 Cu^{2+} + apoerythrocuprein	750	(240)	—
2 Cu^{2+} + 2 Zn^{2+} + apoerythrocuprein	700	(315)	—
Anaerobic reconstitution			
2 Cu^{2+} + apoerythrocuprein	730	(290)	—
2 Cu^{2+} + 2 Zn^{2+} + apoerythrocuprein	680	(360)	detectable
Anaerobically reconstituted erythrocuprein (two-column technique)	680	(300)	distinct

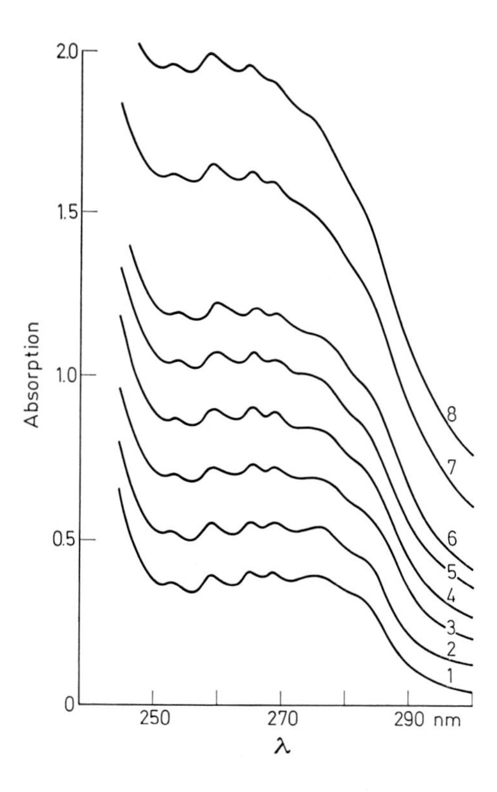

Fig. 21a

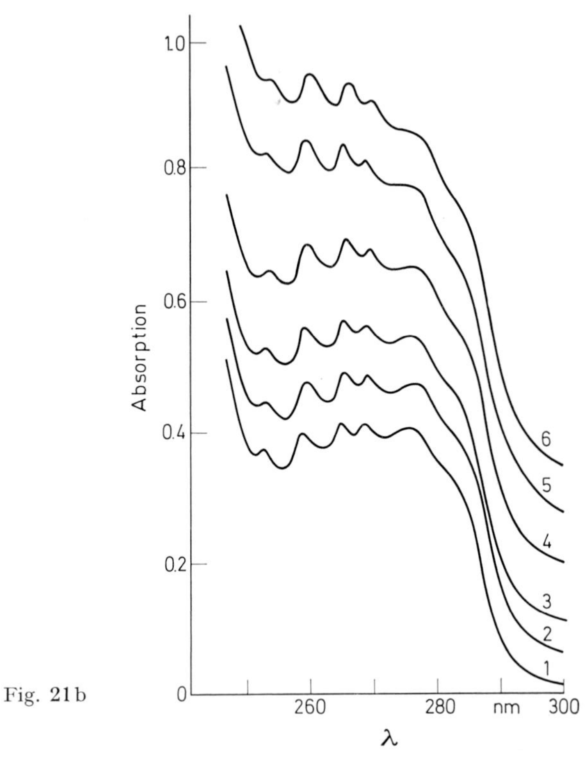

Fig. 21b

Fig. 21 (*78*). a) UV spectra of different Cu^{2+}-apoerythrocuprein chelates recorded at room temperature using a Unicam SP 1800 spectrophotometer. The apoprotein was prepared using EDTA equilibrated G-25 Sephadex columns. 3.54 mg of the protein were dissolved in 1 ml 5 mM potassium phosphate buffer, *p*H 7.2, and titrated with increasing Cu^{2+} concentrations. ① the apoprotein; for the following spectra the concentration is expressed: $Cu^{2+} \times$ (mole apoerythrocuprein)$^{-1}$, ② 0.43; ③ 0.86; ④ 1.30; ⑤ 1.73; ⑥ 2.10; ⑦ 3.10; ⑧ 4.00

b) same experimental setting with Zn^{2+} instead of Cu^{2+}. ① 3.54 mg apoprotein; for the following spectra the concentration is expressed: $Zn^{2+} \times$ (mole apoerythrocuprein)$^{-1}$; ② 1.90; ③ 3.80; ④ 14.00; ⑤ 33.00; ⑥ 51.00

cuprein were observed when the reconstitution was performed anaerobically (*78*). λ_{max} was shifted by 50 nm to 680 nm and the shoulder at 430 nm was detectable. Best results were obtained using the column technique for the reconstitution of erythrocuprein (Fig. 22, Table 7).

The purity of the reconstituted erythrocuprein was examined using analytical polyacrylamide electrophoresis (*122*). All preparations were virtually free from contamination. Again, the 2 Cu-apoerythrocuprein complex, the fully restored erythrocuprein, and the native protein all

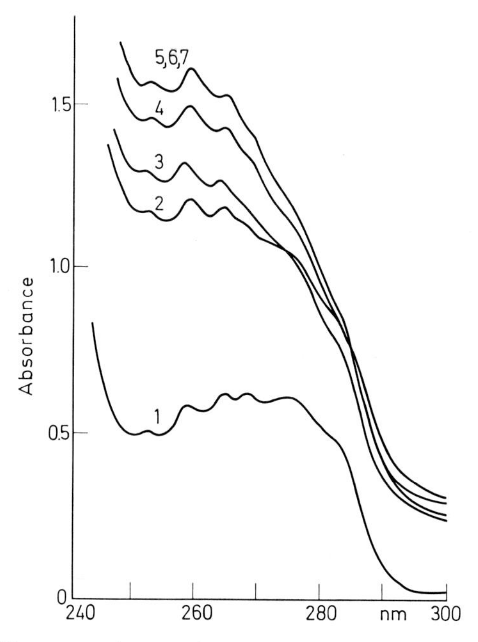

Fig. 22 (*122*). UV spectra of reconstituted bovine erythrocuprein and the 2 Cu^{2+}-apoprotein complex. ① the apoprotein; ② + 2 Cu^{2+} aerobically reconstituted; ③ + 2 Cu^{2+} anaerobically reconstituted, N_2 atmosphere; ④ aerobic reconstitution 2 Cu^{2+} + 2 Zn^{2+}; ⑤ anaerobic reconstitution 2 Cu^{2+} + 2 Zn^{2+}, N_2 atmosphere; ⑥ anaerobically reconstituted erythrocuprein using the two-colunm technique; ⑦ native erythrocuprein

displayed essentially the same electrophoretic behaviour. In contrast to the native protein, the apoprotein and the partially reconstituted Zn^{2+} protein migrated somewhat faster. From this phenomenon it might be concluded that two different binding sites for Cu^{2+} and Zn^{2+} are present in the protein moiety. This conclusion was also drawn from anion binding studies (*113*).

The EPR data supported the electrophoretic, spectroscopic, and enzymic proofs of a successfully reconstituted erythrocuprein. It was surprising that the aerobically reconstituted erythrocuprein and even the 4 Cu^{2+} apoprotein showed most of the EPR characteristics of the 2 Cu—2 Zn enzyme. Only the $A_{\parallel}$ values of the 4 Cu protein and the aerobically reconstituted erythrocuprein were different. It is suggested that in the last two metalloproteins the Cu^{2+} ligand distances are somewhat distorted or the ligands are slightly displaced due to conformational changes in the protein portion (Talbe 8).

Table 8. *EPR parameters of reconstituted bovine erythrocuprein* (*78*)

	$g_{\parallel}$	$g_{\perp}$	$A_{\parallel}$ (Gauss)	g_m
Native erythrocuprein	2.258	2.037	132 ± 2	2.080
Anaerobically reconstituted erythrocuprein	2.258	2.037	137 ± 2	2.080
Aerobically reconstituted erythrocuprein	2.259	2.037	143 ± 2	2.080
4-Cu-apoprotein	2.265	2.037	145 ± 2	2.080
2-Cu + native erythrocuprein	2.260	2.032	137 ± 2	2.080

Fig. 23 (*74*). Acrylamide-gel electrophoresis in the presence of sodium dodecyl sulphate and mercaptoethanol. a) Cytochrome *c* (mol. wt. 12,000); b) haemoglobin subunit (mol. wt. 15,500); c) ovalbumin (mol. wt. 46,000); d) catalase (mol. wt. 60,000) and e) erythrocuprein. All reference compounds (conc. 2 mg/ml) and erythrocuprein were incubated in the presence of 0.1 M sodium phosphate buffer (*p*H 7.2), 1% 2-mercaptoethanol, 4 M urea and 1% sodium dodecyl sulphate for 1 h at 45°. Aliquots of 10 μl (approx. 20 μg protein) were subjected to gel electrophoresis for 2 h at 100 V and 200 mA. The buffer of the upper electrode chamber contained 0.1% sodium dodecyl sulphate in 0.1 M sodium phosphate buffer (*p*H 7.2). The lower electrode chamber contained 0.1 M sodium phosphate (*p*H 7.2).

3.9. The Subunits

Some four years ago studies on erythrocuprein which was oxidized with mild performic acid or reduced and alkylated revealed that the protein contains a subunit of molecular weight about 12,000 (*129*). Subunits of molecular weight 16,000 were detected after the native erythrocuprein had been subjected to sodium dodecyl sulphate treatment in the presence of mercaptoethanol (*74, 82*). In addition to the 16,000 molecular-weight subunits, it was noted that a short-time treatment revealed protein portions of molecular weight of approximately 64,000 which may be attributed to tetrameric species (Fig. 23).

It was obvious that the disulphide bridges were reduced to the free SH moieties. Alternatively to mercaptoethanol, boron hydride was successfully employed as a reducing agent (*80*). As stated already under 2.4, even the stabilization of the free SH moieties using iodo-acetamide could not prevent this uncontrolled recombination. It was most attractive to assign the recombination of these alkylated monomers to electrostatic forces and/or hydrogen bonding. This thought was substantiated by heating the aggregated dimers in 4 M urea for 3 min to 95°, followed by rapid cooling in an ice bath. Surprisingly, monomers were obtained which remained stable for several days (*197*). In contrast to the native blue erythrocuprein, the monomeric Cu protein was colourless although it was EPR-active.

3.10. Microbial and Plant-Type "Cuprein"

During recent months Cu-Zn proteins of striking similarity with erythrocuprein and displaying superoxide dismutase activity were successfully isolated from plants and microorganisms. The absorption spectra of both the garden pea enzyme (*199*) (mol wt. 31,000) and the Cu–Zn protein from *Neurospora crassa* (*200*) (mol wt. 31,000) displayed almost the same features as bovine erythrocuprein. The ultraviolet absorption of the respective Cu–Zn protein isolated from *Saccharomyces cerevisiae* (*201*) was higher in the 280 nm region, which is most likely due to the higher content of aromatic amino-acid residues. Furthermore, the microbial enzymes contained only three to four half-cytine residues (Table 9). An interesting phenomenon was the splitting of the native microbial proteins into 16,000 mol. wt. subunits in the presence of protein-unfolding agents, and in the absence of disulphide reducing agents. In contrast to the bovine erythrocuprein, these microbial enzymes probably have no disulphide bridges at all.

Table 9. *A comparison of amino acid content between mammalian, plant-type and microbial "cuprein"*

	Bovine Erythrocu-prein (*84*)	Garden Pea (*199*)	*Neurospora crassa* (*200*)	*Saccharomyces cerevisiae* [a])
Lysine	22.8	10	12	21.3
Histidine	16.8	18	11	10.7
Arginine	8.7	6	9	8.1
Aspartic acid	34.1	45	36	33.7
Threonine	22.4	30	26	18.8
Serine	15.4	14	14	21.3
Glutamic acid	22.5	19	20	27.3
Proline	14.6	14	14	15.6
Glycine	50.4	56	39	40.1
Alanine	19.6	21	20	27.5
Valine	28.4	21	22	27.2
Methionine	2.1	0	0	0
Isoleucine	18.1	20	13	11.4
Leucine	17.5	21	11	14.4
Tyrosine	2.3	0	2	2.4
Phenylalanine	8.5	9	6	12.8
Tryptophan	2	—	—	—
Half-cystine	5.9	6	3	3.4
Cu	2.05	1.97	1.93	1.80
Zn	2.01	2.02	1.80	2.02

[a]) recalculated from experimental data of 201 and 214.

4. Enzymic Function

For a long time erythrocuprein was thought to act exclusively as a copper-transporting protein. This was a very attractive conclusion since over 50% of the erythrocyte copper content is present in erythrocuprein (*60*). However, in the absence of any known function of a metalloprotein, it is always tempting to assign to it the role of storage or transport of the respective metal ions. For example, caeruloplasmin was considered to be the main copper-transporting protein in blood plasma. It subsequently turned out that this copper protein is a key enzyme in iron metabolism, responsible for the oxidation of Fe^{2+} to the Fe^{3+} bound in transferrin (*130—132*).

The enzymic-catalyzed reduction of cytochrome c_{ox} using the xanthine - xanthine oxidase reaction led to the assumption that $O_2^{-}\cdot$ was the active reducing agent (*133*). Myoglobin and carboanhydrase were able to inhibit this reduction. From kinetic data it was concluded that the $O_2^-\cdot$ must be rapidly disproportionated by myoglobin or carboanhydrase. The main activity was detected in a minor protein component present in the carboanhydrase preparations (*133*). This contaminant protein was successfully enriched by employing washed bovine erythrocytes (*70*). The purified protein turned out to be identical with the long-known erythrocuprein. Further details of the enzymic function of erythruprein will be discussed in the following paragraphs.

4.1. Model Reactions Using Non-Enzymically Prepared $O_2^-\cdot$

$O_2^-\cdot$ From Tetrabutyl-Ammonium Superoxide. Tetrabutyl-ammonium superoxide was found convenient for the superoxide dismutase assay since it dissolved readily without decomposition in N,N-dimethyl formamide (*70*). Infusions of this solution into a cuvette containing aqueous oxidized cytochrome c reduced the available cytochrome c. In the presence of different erythrocuprein concentrations the reduction rate was progressively diminished. According to *McCord* and *Fridovich* (*70*) an enzyme unit was defined as 50% inhibition of the rate of reduction of cytochrome c. Alternatively tetranitromethane was found appropriate for monitoring $O_2^-\cdot$ (*70, 136—148*) where the stable nitroform anion $C(NO_2)_3^-$ is being formed (Equ. (a)):

$$O_2^-\cdot + C(NO_2)_4 = O_2 + C(NO_2)_3^- + NO_2 \quad \text{(a)}$$

$$H\dot{O}_2 + O_2^-\cdot + H^+ = O_2 + H_2O_2 \quad \text{(b)}$$

Using this reaction, *McCord* and *Fridovich* evaluated the rate constant for the reaction of superoxide dismutase with $O_2^-\cdot$ as approximately 5×10^{11} M^{-1} sec^{-1}. The corresponding value for the spontaneous disproportionation of $O_2^-\cdot$ near neutral *p*H (Eq. (b)) is assumed to be 8.5×10^7 M^{-1} sec^{-1} (second-order rate constant) (*139, 140*). *Fridovich et al.* concluded from these data that erythrocuprein accelerates the $O_2^-\cdot$ disproportionation by 3—4 orders of magnitude.

$O_2^-\cdot$ Prepared by Pulse Radiolysis. Pulse radiolysis proved most convenient to prepare $O_2^-\cdot$ in relatively high concentrations without the use of dimethylsulphoxide or N,N-dimethyl formamide as possible interfering solvents (*139, 182, 193, 194*). Moreover, the time resolution was as low as 1 μsec which allowed the determination of the kinetics and mechanism of the enzyme. Oxygen-saturated aqueous solutions containing trace amounts of ethanol (85 mM) as a scavenger for OH radicals were irradiated by 4 MeV electrons. The $O_2^-\cdot$ concentration was monitor-

ed from its absorption at 250 nm (*139, 140*). In contrast to the *p*H-dependent spontaneous disproportionation (*193*), no such dependency was observed in the presence of erythrocuprein. A second-order rate constant of $1.6 \pm 0.3 \times 10^9$ M^{-1} s^{-1} was determined for the reaction between $O_2^-\cdot$ and erythrocuprein (*182*). This value is near the upper limit expected for a diffusion-controlled enzyme-substrate reaction (*195*) and thus in marked contrast to the corresponding value obtained by *McCord* and *Fridovich*. Unfortunately, there was no indication of substantial copper reduction during the reaction with $O_2^-\cdot$.

$O_2^-\cdot$ Generated by Flavin-dependent O_2Activation. It was shown above that artificially produced $O_2^-\cdot$ is a convenient tool for the study, of the enzymic reaction of erythrocuprein in a simple model system avoiding undesired interference from other sources. However, $O_2^-\cdot$ formation, including all known intermediate states, should also be studied in the light of its possible biological relevance. Thus, attention will be focused upon the formation of $O_2^-\cdot$ using a biochemically most important compound, namely the flavin moiety (for comprehensive studies see Refs. *183—185*).

Experimental evidence for the participation of the flavin moiety in $O_2^-\cdot$ formation in nonenzymic reactions was presented by *Ballou et al.* (*151*). During the oxidation of different reduced flavins by molecular oxygen substantial yields of $O_2^-\cdot$ were obtained. Tetra-acetyl riboflavin proved most appropriate as a model compound (*153*). The respective EPR parameters for $O_2^-\cdot$ were in close agreement with those obtained in the KIO_4—H_2O_2 system, in the xanthine oxidase reaction (*149*), and in the oxidation of the reduced tetra-acetyl riboflavin. They also agree well with the EPR data obtained earlier (*154—156*). The time course of the appearance and decay of $O_2^-\cdot$ produced by the reaction of tetra-acetyl riboflavin is shown in Fig. 24.

Measurements made during the first 50 msec after passing the maximum and in the presence of either erythrocuprein or cytochrome c show an accelerated decay of the $O_2^-\cdot$ From Fig. 24 it can further be concluded that the superoxide anion is present in much higher concentrations at *p*H 10.6 which is probably due to the higher stability of $O_2^-\cdot$ at elevated *p*H values (*193*). On the other hand $O_2^-\cdot$ is involved in autocatalytic reoxidation of reduced flavins, as demonstrated in stopped-flow studies. For example, in the presence of 0.1 μM erythrocuprein the reoxidation of tetra-acetyl riboflavin by O_2 was diminished by a factor of 4 (*151*).

It was suggested that the reduced flavins yield a flavin-oxygen complex called HFlOOH which may exist in two isomeric forms (Fig. 25). Either of these intermediate compounds can follow the different decay modes depicted in Fig. 25. Only the intermediate type V yields the protonated superoxide radical $H\dot{O}_2$. However, it seems of great interest to examine the question, to what extent does $O_2^-\cdot$ or $H\dot{O}_2$ represent the sole substrate for erythrocuprein. Why should not erythrocuprein be able to react with some of the other intermediate flavin residues? This question will be discussed in more detail in Sections 4.5 and 4.6.

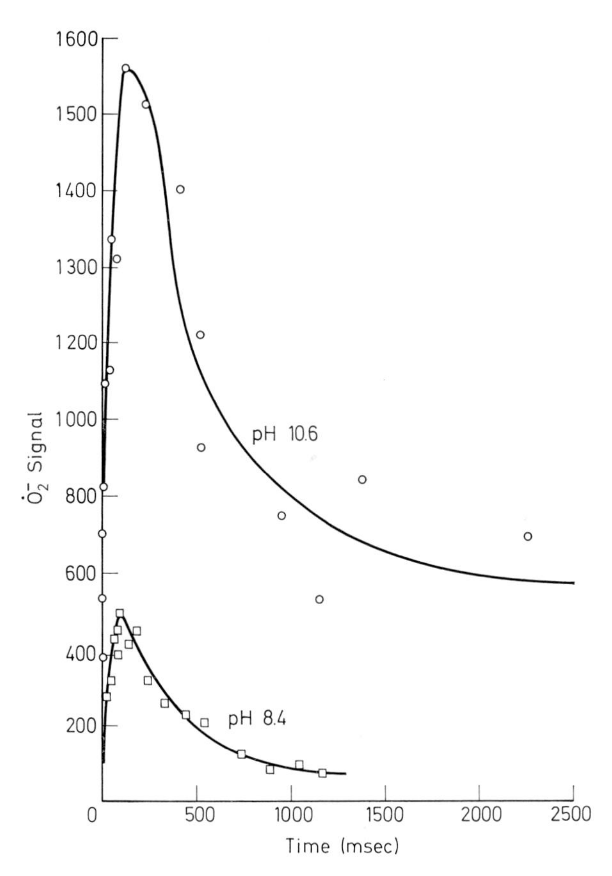

Fig. 24 (*151*). Time course of the appearance and decay of $O_2^-\cdot$. EPR signal produced by reaction of 1.25×10^{-4}M reduced tetra-acetyl riboflavin and 6.5×10^{-4}M O_2 at 20 °C. Top curve, in 0.1 M glycine *p*H 10.6. Bottom curve, in 0.1 M glycylglycine *p*H 8.4. All times include the 5 msec quenching time. Maximum intensity at *p*H 10.6 corresponds to 1.19×10^{-4} M oxygen radical. Modulation amplitude 10 gauss at 100 kHz. Microwave power 12 mW. Microwave frequency 9.2 GHz Temperature 84 °K

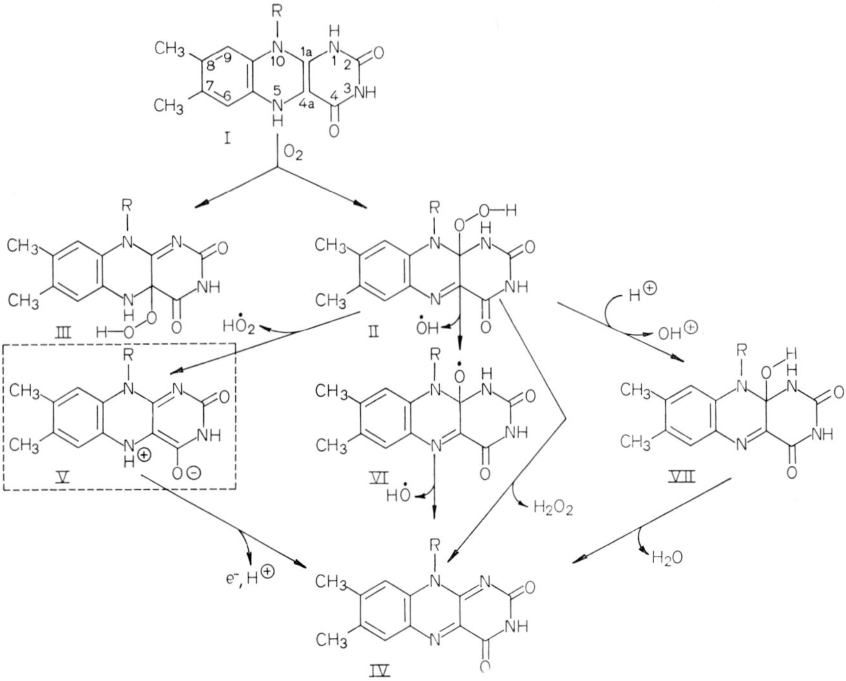

Fig. 25 (*184*). Decay of the reduced flavin-oxygen complex HFlOOH

4.2. Enzymically Produced $O_2^{-}\cdot$

The most established source for $O_2^{-}\cdot$ in enzymic systems proved to be the flavoproteins (*183—185*). Already in 1949 *Horecker* and *Heppel* (*141*) described some oxygen dependency of the cytochrome-c reduction using the xanthine-xanthine oxidase reaction. It was concluded (*141, 142*) that reduced oxygen species might be the reducing agents. Catalase was unable to inhibit this reaction, therefore univalently charged oxygen was presumed to be the active agent. $O_2^{-}\cdot$ was indirectly detected using chemiluminescence during the aerobic action of xanthine oxidase in the presence of xanthine and luminol or lucigenin (*143—147*). The actual presence of $O_2^{-}\cdot$ was demonstrated in EPR studies using the enzymic reactions of flavoproteins (*148—151*). *Knowles et al.* (*149*) employed the rapid-freezing technique and were able to show that $O_2^{-}\cdot$ is indeed a product of the oxidation of the reduced xanthine oxidase (Fig. 26).

As in the case of the reduced flavin-oxygen complex HFlOOH, where any of a number of possible intermediates are being formed

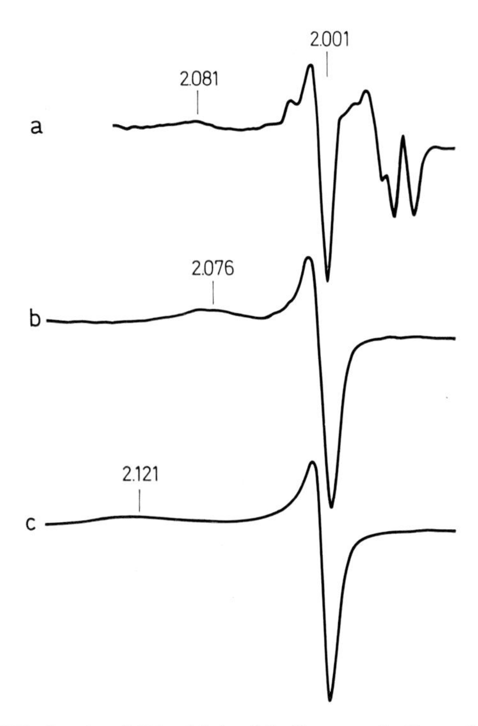

Fig. 26 (*149*). EPR signals of $O_2^-\cdot$ obtained both enzymically and non-enzymically. Recording at -170 °C, rapid freezing technique (*152*); a) xanthine—xanthine oxidase reaction. Note the molybdenum signals (overmodulated). The sloping base line is attributed to an iron signal; b) non-enzymically induced $O_2^-\cdot$ from H_2O_2 and $NaIO_4$, *p*H 9.9; c) same as (b) but at *p*H 13.2

($O_2^-\cdot$, $\dot{O}H$ or OH^+), a similar mode of action can be taken into consideration for the native flavoproteins. The specific action of different flavin oxidases, dehydrogenases and hydroxylases can most likely be attributed to specific proteins which are bound with the flavin moiety and are in a position to dictate the particular breakdown of HFlOOH. For example, the oxidases were thought to react specifically, yielding oxidized flavin and H_2O_2, while the dehydrogenases were supposed to yield both flavin and superoxide radicals ($HFl\cdot + H\dot{O}_2$). The hydroxylases were assumed to account for the reaction of one of the activated oxygen intermediates with the respective substrate.

Only those flavoproteins where $O_2^-\cdot$ is being formed are able to display cytochrome-c reductase activity which can be inhibited by erythrocuprein (Table 10). However, it cannot be fully excluded that erythrocuprein may react with the above-mentioned intermediate

Table 10. *Effect of erythrocuprein on cytochrome-c reductase activity of flavoproteins. The cytochrome-c reductase activity was measured in air-equilibrated solutions containing 0.1 M pyrophosphate, pH 8.5, in the presence of 3.33×10^{-5} M cytochrome c and 10 μg bovine catalase. The concentration of erythrocuprein in this assay mixture was 0.62 μM. The temperature was 25° (150)*

Enzyme	Reductant	Molecular Activity[a])		
		In absence of erythrocuprein	In presence of erythrocuprein	Under anaerobic conditions
Glucose oxidase	0.1 *M* glucose	0.038	0.034	
D-amino oxidase	$1.86 \times 10^{-2}M$ D-alanine	0.054	0.060	
L-amino acid oxidase	$1.25 \times 10^{-3}M$ L-leucine	0.127	0.126	
Glycollate oxidase	$6.7 \times 10^{-4}M$ glycollate	0.68	0.50	
Lactate oxidase	$3.33 \times 10^{-4}M$ L-lactate	<0.010	<0.010	
P-hydroxy benzoate hydroxylase[b])	$1.5 \times 10^{-4}M$ TPNH + $3.3 \times 10^{-4}M$ p-hydroxybenzoate	213	213	
Melilotate hydroxylase[b])	$2 \times 10^{-4}M$ DPNH + $1.5 \times 10^{-4}M$ melilotate	72	60	
Old yellow enzyme	$2 \times 10^{-4}M$ TPNH	5.3	0.37	1.27
Ferredoxin TPN reductase	$2 \times 10^{-4}M$ TPNH	4.6	1.25	2.3
Flavodoxin[c])	$2 \times 10^{-4}M$ TPNH + $3.2 \times 10^{-8}M$ ferredoxin TPN reductase	0.102	0.047	
Lipoyl dehydrogenase	$2 \times 10^{-4}M$ DPNH	2.7	0.26	0.21
Glutathione reductase	$2 \times 10^{-4}M$ TPNH	0.9	0.072	0.10

[a]) Molecules of cytochrome *c* reduced per min. per molecule of enzyme-bound flavin.
[b]) The quoted values are due in part (and possibly completely) to the nonenzymic reduction of cytochrome *c* by the hydroxylated products of these enzymes.
[c]) A coupled system containing 5×10^{-5} *M* flavodoxin. Results are corrected for the blank rate due to the ferredoxin TPN reductase alone.

flavin compounds. In this case the cytochrome-c reductase assay is useless and a completely different assay ought to be developed for the evaluation of the enzymic activity of erythrocuprein.

Massey and coworkers were able to demonstrate that superoxide is formed at the reduced flavin site of the flavoproteins (*183, 184*). When they used "deflavo-xanthine oxidase", no $O_2^{-}\cdot$ was detectable. In other laboratories (*148, 156, 158*) the non-heme iron portion of xanthine oxidase was made more or less responsible for the $O_2^{-}\cdot$ formation. In fact, ferredoxins are able to produce superoxide radicals (*148, 156—158, 196*) although much more slowly (about one fourth) compared to $O_2^{-}\cdot$ formation using the reduced flavin-complex. A recent study by *Misra* and *Fridovich* (*158*) suggests that the non-heme iron portion of xanthine oxidase supports the $O_2^{-}\cdot$ formation together with the reduced flavin site. However, since deflavo-xanthine oxidase displays no superoxide formation, the production of superoxide at the flavin moiety of xanthine oxidase should be given priority. Nevertheless, the ubiquity of non-heme iron proteins of ferredoxin structure, especially in mitochondrial metabolism, is opening up a new field of interest for both $O_2^{-}\cdot$ and erythrocuprein.

4.3. Enzymic Properties of the Native Protein, the Partially Reconstituted Protein and the Fully Reconstituted Erythrocuprein

McCord and *Fridovich* (*70*) defined a unit for the enzymic activity of erythrocuprein as the protein concentration necessary to decrease the velocity of the cytochrome-c reduction by a factor of 0.5, provided their standard mixture is employed. The reduction velocity

$$v = d\,\text{Cyt}\ c_{red} \times dt^{-1} = k[\text{Cyt}\ c_{ox}] \times [O_2^{-}\cdot]$$

is diminished when the concentration of either component decreases. However, the suppression of both the enzymically catalysed production of $O_2^{-}\cdot$ and the velocity of the cytochrome-c reduction may also be observed using enzyme inhibitors; for example, p-mercuribenzoic acid. This inhibitor was shown to have a pseudo-erythrocuprein activity due to the diminished reduction velocity of cytochrome *c* (Fig. 27).

In awareness of this problem, both the reduction velocity of cytochrome c and the actual concentration of cytochrome c were studied using the xanthine - xanthine oxidase reaction. To achieve the complete reduction of cytochrome c, sufficient $O_2^{-}\cdot$ had to be available; on the other hand, added erythrocuprein was still able to diminish this reduction. According to the results given in refs. (*77*) and (*79*), a cytochrome c concentration which was 10—20 times lower than the concentration of xanthine and xanthine oxidase met these requirements. Under these conditions both the initial reduction velocity and the plateau height were proportional to the erythrocuprein concentration. The advantage of using the plateau height was, firstly, convenience of measurement, since the plateau remained constant over 30 min, and secondly, false interpretations due to pseudoerythrocuprein activity were minimized.

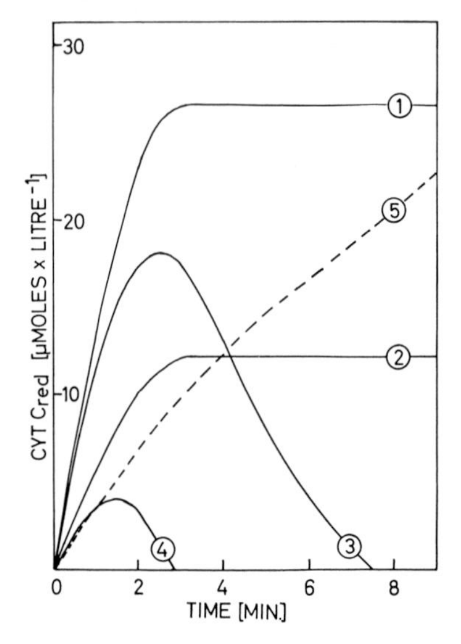

Fig. 27 (*79*). Reduction of oxidized cytochrome c by $O_2^{-}\cdot$ produced during the enzymic-catalysed oxidation of xanthine. ① in the presence of catalase (16 nM), erythrocuprein omitted; ② after addition of erythrocuprein (25 nM) and catalase (16 nM); ③ without either erythrocuprein or catalase; ④ in the presence of erythrocuprein (25 nM), catalase omitted; ⑤ in the presence of both catalase (16 nM) and p-chloro-mercuri-benzoic acid (400 μM). The assay was performed at 25 °C in a volume of 0.76 ml. The concentration of the different components dissolved in 50 mM phosphate buffer, *p*H 7.8 was: Xanthine 0.33 mM; cytochrome c_{ox}, 27 μM; EDTA, 0.1 mM; xanthine oxidase, 0.21 μM. The reduction of cytochrome c was recorded in a Unicam SP 1800 spectrophotometer at 550 nm

The diminished initial reduction velocity as a measure for the presence of erythrocuprein may be conveniently used in model reactions where no enzyme is involved in the $O_2^{-}\cdot$ production (*70, 151*). However, in a biological system where the formation of $O_2^{-}\cdot$ is enzymically catalysed, measurement of the plateau hight appears to be more appropriate. At least the formation of urate has to be controlled if the initial velocity of the cytochrome-c reduction is measured using the xanthine - xanthine oxidase system.

Erythrocuprein was found to be extraordinarily stable with regard to temperature, *p*H, and storage conditions. An interesting phenomenon was that enzyme activity was stimulated by 300% after erythrocuprein had been exposed for 5 min to *p*H 10.8 ± 1.0 and was accompanied by profound changes in the EPR spectra (see also 3.4.4.) at *p*H$\sim$11. From the EPR data it may be concluded that the ligand conformation around the Cu^{2+} is quickly adjusted to the prevailing *p*H conditions. However, it may be assumed that conformational changes in the protein portion are not as quickly reversible, indicating some metastable situation. Profound conformational changes in the tertiary structure of a Zn enzyme have already been described (*165*) and should be considered here in this context.

Table 11. *Enzymic activity of different Cu^{2+} and/or Zn^{2+} apo-erythrocuprein chelates. The reciprocal concentrations of these different metal apoprotein complexes were compared under equilibrium conditions ($[Cyt\ c_{red}] \times [Cyt\ c_{ox}]^{-1} = 1$). For each metal protein 4 different assays were performed. Incubations were carried out at 25°. The assay mixture was composed of: xanthine, $3.3 \times 10^{-4} M$; beef-heart cytochrome c_{ox}, $2.7 \times 10^{-5} M$; catalase, $1.6 \times 10^{-8} M$; xanthine oxidase, $2.1 \times 10^{-7} M$; HEPES buffer, $5 \times 10^{-2} M$, pH 7.8* (*78, 122*)

Metal apoerythrocuprein chelates (aerobically reconstituted) [Moles × (mole protein)$^{-1}$]	Required metal apoprotein concentration to yield $[Cyt\ c_{red}] \times [Cyt\ c_{ox}]^{-1} = 1$ [μmoles^{-1} × litre]
Apoerythrocuprein	0
+1 Zn^{2+}	0
+2 Zn^{2+}	0
+3 Zn^{2+}	0
+4 Zn^{2+}	0
+1 Cu^{2+}	1.0
+2 Cu^{2+}	3.1
+3 Cu^{2+}	9.1
+4 Cu^{2+}	20.0
+1 Cu^{2+} +1 Zn^{2+}	2.2
+1 Cu^{2+} +2 Zn^{2+}	3.5
+1 Cu^{2+} +3 Zn^{2+}	6.7
+2 Cu^{2+} +1 Zn^{2+}	14.3
+2 Cu^{2+} +2 Zn^{2+}	17.0
+3 Cu^{2+} +1 Zn^{2+}	20.0
Native erythrocuprein stored at 4 °C	33.4
Native erythrocuprein stored at 22 °C	27.2
Anaerobic reconstitution N_2,	
2 Cu^{2+} +2 Zn^{2+}	26.5
two column technique	33.5
Aerobic reconstitution	17.2

4.3.1. Differently Reconstituted Metal-Apoprotein Chelates

Enzymic studies using different Cu^{2+} and/or Zn^{2+} apoprotein chelates shed some light on the problem of the nature and number of metal ions involved in the enzymic disproportionation of superoxide anions (*78*). The highest specific enzyme activity was obtained when two Cu^{2+} were present and in addition two Zn^{2+} or two other Cu^{2+} ions (Table 11). The need for two distinct Cu^{2+} ions can be seen by comparing the activity of the 2 Cu^{2+}—1 Zn^{2+} protein with the activity of the 3 Zn^{2+}, and the 1 Cu^{2+}—2 Zn^{2+} proteins, respectively. Only the 2 Cu^{2+}—1 Zn^{2+} protein was active in a similar way to the 2 Cu^{2+}—2 Zn^{2+} protein. The enzymic properties of the 4 Cu^{2+}, the 2 Cu^{2+}—2 Zn^{2+}, and the 3 Cu^{2+}—1 Zn^{+} proteins were essentially identical. Considering the fact that the 4 Zn^{2+} protein is enzymically inactive while the 2 Cu^{2+}—1 Zn^{2+} protein is almost fully active, the conclusion can be drawn that the 2 Cu^{2+} are responsible for the electron transfer while the two additional Cu^{2+} ions (substituted for Zn^{2+}) act as essential cofactors. A two-electron transfer has already been shown in Ref. (*74*) where dithionite was used as a reducing agent.

4.4. Evidence of Accelerated H_2O_2 Formation in the Presence of Erythrocuprein

In comparing the spontaneous and the enzymically catalyzed disproportionation of $O_2^{-}\cdot$ into H_2O_2 and O_2, the initial velocity of the H_2O_2 formation would be expected to increase in the presence of erythrocuprein. *McCord* and *Fridovich* (*70*) demonstrated that H_2O_2 is a product of the erythrocuprein-catalyzed reaction, but they did not find that the initial velocity of the H_2O_2 formation is higher. Accelerated H_2O_2 formation can be seen, provided the assay for H_2O_2 is sufficiently fast (*79, 114, 166*). In the presence of 10 μM erythrocuprein (Curve ④) the rate of H_2O_2 formation after 1 min was as high as 240% compared to the control (Curve ①) (Fig. 28).

It is possible that even more H_2O_2 is detectable. The erythrocuprein-catalyzed disproportionation of $O_2^{-}\cdot$ is probably fast enough to minimize other reactions of $O_2^{-}\cdot$ in the assay system, while in the slower spontaneous disproportionation some of the $O_2^{-}\cdot$ may be lost, for example, via the cycle of *Haber* and *Weiss* (*167*).

$$O_2^{-}\cdot + H_2O_2 + H^+ = O_2 + H_2O + \cdot OH$$

$$\cdot OH + H_2O_2 = H_2O + O_2^{-}\cdot + H^+$$

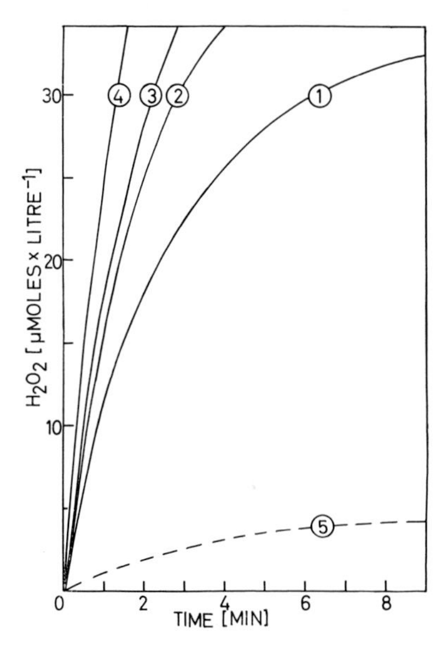

Fig. 28 (*79*). Time course of hydrogen peroxide formation in the presence of erythrocuprein and p-mercuri-benzoic acid. ① control, ② 100 nM, ③ 500 nM, ④ 10,000 nM erythrocuprein, ⑤ 130 μM p-mercuri-benzoic acid. 2.25 ml of the assay mixture was composed of: homovanillic acid. 0.9 mM; horseradish peroxidase, 0.5 μM; EDTA, 0.1 mM; xanthine, 0.33 mM; 50 mM potassium phosphate buffer (*p*H 7.8) served as the solvent. Fluorescence readings were taken at 425 nm using 1 cm light-path quartz cells

4.5. Specificity of the Substrate

In the preceding chapters — with the exception of the discussion of different flavoproteins — no such discussion was started regarding the specificity of the superoxide anion as the sole substrate for erythrocuprein. In this context let us recall some basic principles of the chemical and physical properties of oxygen (for extensive reviews see Refs. (*172–178a*)).

The O_2 molecule can be considered as a stable biradical. In contrast to the N_2 molecule, two π^*-antibonding molecular orbitals are occupied according to Hund's rule, each by one electron. Both electrons have the same spin direction. This binding situation represents the ground state of the electronic triplet expressed as:

$$^3\Sigma_g^- \, [\pi_{x,y}\,(4)\; \overset{*}{\pi}_x\,(\uparrow)\; \overset{*}{\pi}_y\,(\uparrow)]$$

Upon excitation of this triplet ground state, two metastable singlet states

$$^1\Sigma_g^+ [\pi_{xy}(4)\, \overset{*}{\pi}_x(\uparrow)\, \pi^*(\downarrow)] \quad \text{and} \quad ^1\Delta_g[\pi_{xy}(4)\, \overset{*}{\pi}_x(\downarrow\uparrow)\, \overset{*}{\pi}_y(\mathrm{o})]$$

and two excited triplet states

$$^3\Sigma_u^- [\pi_{xy}(3)\, \overset{*}{\pi}_{xy}(3)] \quad \text{and} \quad ^3\Sigma_u^+ [\pi_{xy}(3)\, \overset{*}{\pi}_{xy}(3)]$$

are possible. Because of the high excitation energy required (36,000 and 49,000 cm^{-1}), the excited triplet states seem rather unlikely to occur in biochemical systems. However, evidence for singlet-type oxygen in biochemical processes has been reported (*146, 147, 160, 179—181*). The excitation energy of $^1\Delta_g O_2$ is 22 kcal (8,000 cm^{-1}), and 37 kcal (13,000 cm^{-1}) is required to produce $^1\Sigma_g^+ O_2$. These two singlet states differ in their electronic occupation of the π^*-antibonding molecular orbitals. The $^1\Delta_g O_2$ has both antibonding electrons located with antiparallel spin in one π^*-antibonding molecular orbital, leaving the second unoccupied. In the $^1\Sigma_g^+$ state, the two electrons of the triplet ground state remain in the two π^*-antibonding orbitals, but with antiparallel spin. In principle, the conversion of $^1\Sigma_g^+$ into $^1\Delta_g$-oxygen is possible. According to *Khan* (*159*) O_2^{2-} is an established source of $^1\Delta_g$ oxygen while $^1\Sigma_g^+$ oxygen, apparently, is preferentially formed from $O_2^-\cdot$ (*146*).

A simple method for the production of singlet oxygen using a saturated solution of potassium superoxide in dimethylsulphoxide has been described by *Khan* (*159*). The resulting $^1\Sigma_g^+$ oxygen was monitored by the fluorescence of added anthracene. However, the exact mode of formation of singlet oxygen is still unknown. A pronounced quenching of the anthracene fluorescence was observed after the addition of water. Nevertheless, the generation of singlet oxygen in aqueous media, for example, during the xanthine-oxidase catalyzed oxidation of xanthine, was taken into consideration. *Arneson* (*146*) concluded from his measurements employing xanthine oxidase and aldehyde oxidase that singlet oxygen is formed from $O_2^-\cdot$, which was measured by chemiluminescence. This chemiluminescence was quenched in the presence of erythrocuprein, sulphite, cysteine and cytochrome *c* (*147*). Singlet oxygen may even be formed in the absence of $O_2^-\cdot$ (*147, 160*). During the reaction of lipoxidase on its substrate linoleic acid, a "singlet oxygen-like" intermediate was observed which was quenched rather specifically by erythrocuprein. In this system no cytochrome-c reductase activity was measured which could be attributed to $O_2^-\cdot$ (*147*). Thus, *Rotilio et al.* (*147*) propose that the main activity of erythrocuprein is to quench singlet oxygen rather than to accelerate the $O_2^-\cdot$ disproportionation.

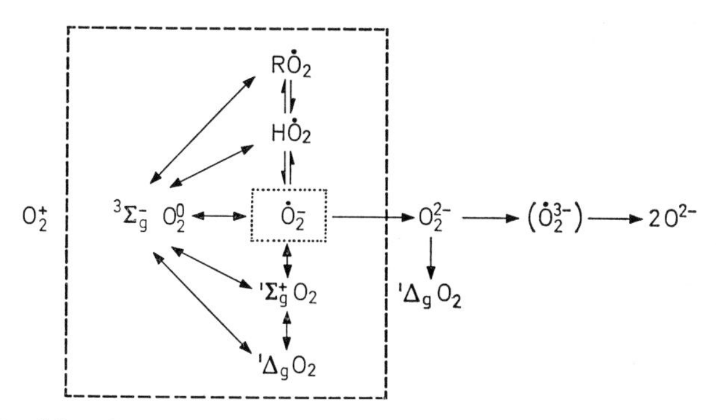

Fig. 29. Possible substrate complex for erythrocuprein

Thanks to its electron affinity, oxygen is able to accept up to four electrons (Fig. 29).

The electrons are filling the empty σ^* molecular orbital and the half-filled π^* molecular orbitals. As the number of electrons increases, the O—O distances are progressively increased, finally causing dissociation. In Fig. 29 the substrate specificity of $O_2^-\cdot$ for erythrocuprein should be critically examined. Considering the fact that erythrocuprein reacts extremely fast ($\sim 2 \times 10^9$ M^{-1} s^{-1} ((*82*) with $O_2^-\cdot$ and/or $H\dot{O}_2$ prepared by pulse radiolysis, it seems rather likely that this Cu—Zn protein could well react with some intermediates of $O_2^-\cdot$. For example, the flavoproteins form the intermediary flavin-oxygen complex called HFlOOH (*183*—*185*). It could be possible that erythrocuprein reacts already with this complex. Futhermore, no conlcusion can be drawn as to whether the metals, molybdenum and iron, involved in the electron transport are oxidized, leaving the oxygen molecule in the reduced state as $O_2^-\cdot$, or whether the metals are reduced and molecular oxygen or some singlet-type oxygen intermediates are formed. This redox mesomerism (*186*) and the various possible intermediate states of oxygen given in the dashed square of Fig. 29 makes it hard to assign a substrate specificity for erythrocuprein.

4.6. The Biochemical Reactivity of Copper in Erythrocuprein

Unfortunately, our knowledge of the structure of nearly all copper proteins including erythrocuprein is rather limited. Thus for the time being great caution is necessary in the interpretation of physical and chemical properties. Extensive studies using Cu model complexes of low molecular weight were carried out as an approach to understanding these structural and functional properties (*4*, *8*, *53*, *187*, *188*).

The biochemical properties of Cu in the different Cu proteins could be formally described as (Fig. 30):

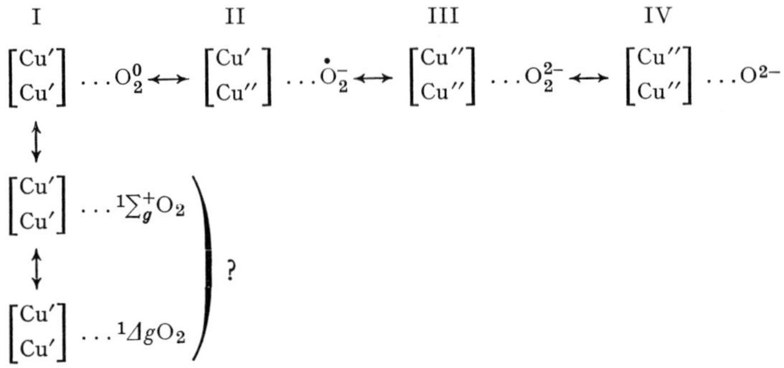

$$\overset{\text{I}}{\begin{bmatrix}Cu'\\Cu'\end{bmatrix}\ldots O_2^0} \leftrightarrow \overset{\text{II}}{\begin{bmatrix}Cu'\\Cu''\end{bmatrix}\ldots \dot{O}_2^-} \leftrightarrow \overset{\text{III}}{\begin{bmatrix}Cu''\\Cu''\end{bmatrix}\ldots O_2^{2-}} \leftrightarrow \overset{\text{IV}}{\begin{bmatrix}Cu''\\Cu''\end{bmatrix}\ldots O^{2-}}$$

$$\updownarrow$$

$$\left(\begin{matrix}\begin{bmatrix}Cu'\\Cu'\end{bmatrix}\ldots {}^1\Sigma_g^+O_2\\ \updownarrow \\ \begin{bmatrix}Cu'\\Cu'\end{bmatrix}\ldots {}^1\Delta gO_2\end{matrix}\right)?$$

Fig. 30. Redox mesomerism of Cu in different Cu proteins

In Cu compounds of high covalence no precise decision can be reached regarding the oxidation state of the copper (*53*), while in ionic systems reference to the occupancy of the 3*d* subshell is sufficient. Thus, the oxidation states given in I—IV are of merely formal nature. The interesting phenomenon in this scheme is the redox mesomerism of Cu which implies that the chelated metal ion could have different biochemical actions. Type I would represent the reversible oxygenation as found in haemocyanin, type II would be the superoxide dismutation, provided $\dot{O}_2^-\cdot$ really is the substrate, and types III and IV are represented by the catalatic and oxidative action displayed by a considerable number of copper proteins (polyphenol oxidases, amine oxidases etc.). The biochemical specificity of each chelated copper is more of less given by the macromolecular ligands.

An extensive study of the catalatic and peroxidative activity of different Cu complexes was performed by *Sigel* (*188*). It was demonstrated that one or two coordination sites of the copper had to be vacant to allow the binding of the substrate (Fig. 31).

The type III complex remained essentially inactive. Proof of an intermediate Cu^{2+}-peroxy complex was shown by an absorption shoulder at 360 nm (*189, 190*). This biochemical reactivity was induced not only by ethylene diamine and its derivatives, but also by monomeric or polymeric amino acids or nucleotides.

In this context we wanted to find out to what extent the copper chelates of, either the free amino acids or some low molecular weight peptides, would be in a position to display superoxide dismutase activity.

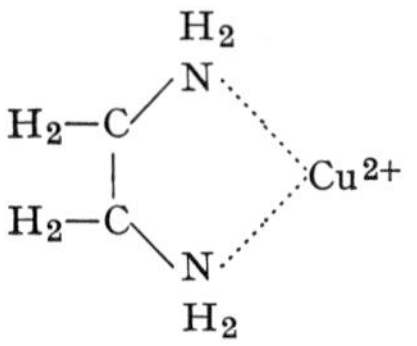

(I) 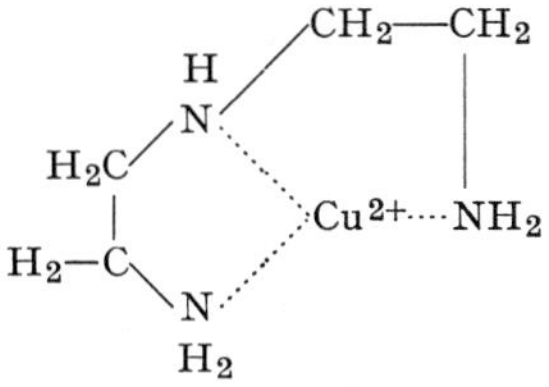 (II)

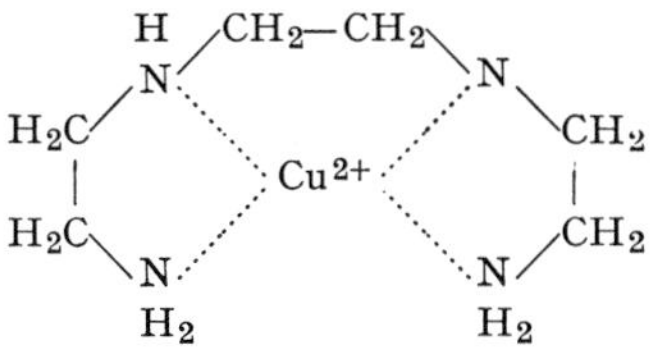

(III)

Fig. 31 (*190*). Blocking of copper coordination sites

We tested all the amino acids present in bovine erythrocuprein, but only the copper complexes of histidine, tyrosine and lysine showed enzyme activity similar to that reported for erythrocuprein, i.e. inhibition of cytochrome-c reductase activity and quenching of singlet-type oxygen (Fig. 32, Table 12). In the presence of the metal-free ligands and Cu^{2+} alone, no such biochemical activity was detected (*179*). It has to be emphasized that the Cu-chelate concentrations used were in the micromolar region, i.e. their concentration was almost 3 orders of magnitude lower than the Cu^{2+} concentrations employed by *Sigel* (*190*). As the copper amino acid concentrations required for superoxide dismutase activity are extremely low, it is hard to give a precise assignment of the exact number of coordinated amino acids with Cu^{2+} in solution.

The Cu chelates using EDTA or serum albumin display no superoxide dismutase activity. It is very attractive to conclude that — as in the case of the catalatic action of Cu^{2+} — one or more Cu-coordination sites are fully accessible to water or the substrate.

Studies concerning the residence time of water bound to Cu^{2+} have recently been performed with native erythrocuprein (*191*). A residence time of $4 \times 10^{-6} - 10^{-8}$ s was determined on each of the two Cu^{2+} per protein molecule. This constant is not so different from the second-order rate constant for the erythrocuprein—$O_2^{-}\cdot$ reaction (*182*). In other words, the coordinated water can be most rapidly exchanged by $O_2^{-}\cdot$ and/or

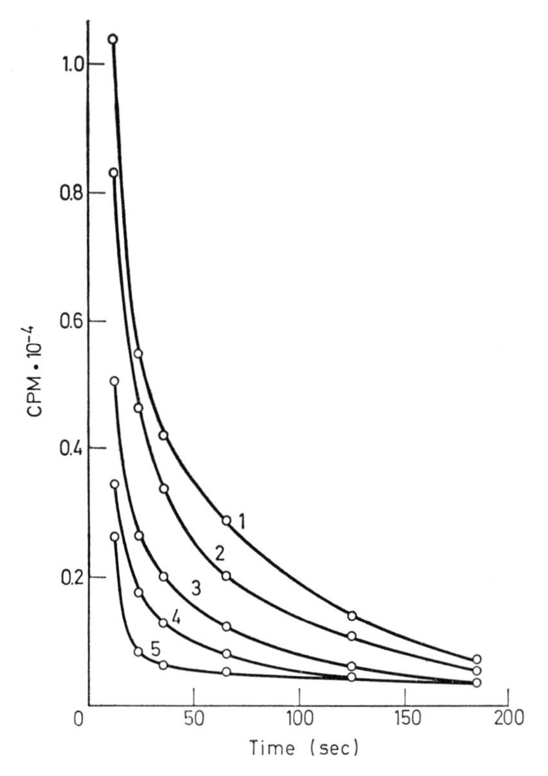

Fig. 32 (*179*). Chemiluminescence assay of bovine erythrocuprein and different Cu^{2+}-amino acid complexes at *p*H 7.8. ① None; ② $Cu(Lys)_2$; 50 nM; ③ $Cu(His)_2$, 100 nM; ④ Cu-Tyr, 145 nM; ⑤ bovine erythrocuprein, 8 nM. The assay components were pipetted in a disposable scintillation vial at room temperature. The total volume was 2.22 ml. The assay mixture was composed of; HEPES buffer, 50 mM; xanthine, 0.33 mM; catalase, 800 i.U.; luminol, 1 mM; The reaction was started with 0.08 units xanthine oxidase (definition as given by *J. Cooper, P. A. Srere, M. Tabachnick* and *E. Racker,* Arch. Biochem. Biophys. *74* (1958) 306). The first reading was taken after 10 sec. During the counting the coincidence of the Packard scintillation counter was turned on. The background was 4 ± 1 cpm

singlet-type oxygen. However, this fast exchange is only possible when the copper is readily accessible to the substrate. From XPS measurements (see Chapter 3.4.5) it was concluded that Cu^{2+} is probably located in the outer sphere of the protein molecule.

At the moment no decision can be made as to whether or not erythrocuprein is an essential enzyme in metabolism. In order to assign a specific role to it, two criteria must be fulfilled: firstly appropriate ligands bound to the Cu^{2+} must be able to induce a "superoxide dismutase"

Table 12. *Superoxide dismutase activity of different Cu^{2+}-amino acid chelates. All the amino acids and peptides used were in the L-form. The cytochrome-c reductase assay was performed as given in Table 11* (*179*)

Cu^{2+}-Chelate	Required equivalent of chelated Cu^{2+} to yield $[\text{Cyt } c_{red}] \times [\text{Cyt } c_{ox}]^{-1} = 1$ $[\mu \text{ moles}^{-1} \times \text{litre}]$
Native bovine erythrocuprein	16.7
1-Cu-apoerythrocuprein	1.0
$Cu(Lys)_2$	0.9
CuTyr	0.6
$Cu(His)_2$	0.5
$Cu(His\text{-}methylester)_2$	0.2
Cu—Leu—Tyr	0.5
Cu—Lys—Ala	0.2
Cu—His—Leu—Gly	0.6
Cu—Gly—His—Leu	0
Cu—His—Leu—Leu	0.3
Cu—Leu—His—Leu	0.04
Cu—Leu—Leu—His	0
$Cu^{2+} \cdot aq$	0
Cu—EDTA	0
Cu—bovine serum albumin	0

activity; and secondly, the protein molecule has to be in a position to protect the free coordination site of the copper ion from unspecific and undesired binding with a large number of naturally occurring ligands. Last but not least, it has to be realized that the question of substrate specificity is still open. Whether or not erythrocuprein reacts with radicals other than oxygen is another most important question.

4.7. Possible Role in Metabolism

Although erythrocuprein has been known since 1939, it was 30 years before a biochemical function was assigned to this copper protein. There is substantial evidence that erythrocuprein is involved in the disproportionation of $O_2^{-}\cdot$ and/or the quenching of singlet-type oxygen. In metabolism $O_2^{-}\cdot$ is enzymically produced. However, the only source so far established for the formation of $O_2^{-}\cdot$ is represented by the flavoprotein dehydrogenases, although evidence of $O_2^{-}\cdot$ was recently reported during the oxidation of epinephrine (*192*) and ferredoxin (*148, 156–158, 196*). On the other hand, radiation or extraneous reagents, for example, the 1,1'-ethylene-2,2'-dipyridylium ion known as the herbicide "Diquat" (*168*), give rise to the formation of $O_2^{-}\cdot$. The decay products of $O_2^{-}\cdot$ are highly active in the organism and should be rapidly destroyed. The reaction of $O_2^{-}\cdot$ can be either oxidative, as shown by the oxidation of dihydroxy phenols (*192*), or reductive, as comprehensively studied in the cytochrome-c reductase essay. The extent to which the scavenging of $O_2^{-}\cdot$, some intermediate superoxide complexes or singlet-type oxygen will have priority is a most interesting question and a challenging task for further studies. All these excited oxygen species would be extremely reactive and "burn" anything in the living cell. Singlet oxygen-like intermediates are reported to be generated directly and not via $O_2^{-}\cdot$ (*160*). *De Luca* and *Dempsey* (*180*) have suggested that singlet oxygen may be produced in the mechanism of oxidation in firefly luminescence. Nevertheless, the ubiquity of erythrocuprein in a great number — if not all — aerobic cells and its reactivity in the biochemistry of oxygen suggest that this metalloprotein may possess a most important and possibly essential function.

Acknowledgements. Some experimental portions of this review were supported by subsidiary grants from the Deutsche Forschungsgemeinschaft (DFG WE 401/4 and 401/6), and a personal EMBO long-term fellowship awarded to U.W. I wish to express my gratitude for all the cooperativity and helpful discussions of my friends and colleagues: Dr. *G. Barth, W. Bohnenkamp*, Dr. *E. Bunnenberg*, Dr. *R. Cammack*, Prof. *C. Djerassi*, Dr. *L. Flohé*, Dr. *A. M. Fretzdorff, H. J. Hartmann, E. Joester*, Dr. *G. Jung, M. Ottnad, W. Paschen*, Miss *R. Prinz*, Mrs. *A. Schallies*, Dr. *W. Schlegel, G. Thomas, W. H. Thomas, G. Voelcker*, Dr. *W. Voelter, W. Voetsch*, Prof. *G. Weitzel* and *R. Zimmermann*.

Thanks go further to Prfs. *E. Bayer, P. Hemmerich* and *V. Massey* for their generous cooperativity in providing galley proofs of recent studies. I am also in debted to Drs. *J. Bannister, H. F. Deutsch, J. Fee, I. Fridovich* and *G. Rotilio* for many stimulating discussions and for submitting some reprints.

Note added in proof. Shortly after submitting this manuscript to the publishers an overwhelming number of studies on the cupreins have appeared in the literature. The two papers anounced in 3.4.4. as being in press (*161, 162*) have become available and describe different experimental conditions of selective metal removal from erythrocuprein leaving the remaining protein portion undamaged. Physical and chemical characterization was performed using EPR, CD, optical spectra, electrophoretic behaviour and enzymic function. Proteolytic attack by carboxypeptidase B on the 2 Cu protein resulted in the loss of only one lysine residue. Both Cu of the native enzyme could be reversibly reduced using ferrocyanide. Stoichiometric addition of H_2O_2 under anaerobic conditions caused the disappearance of the Cu EPR signal. This phenomenon was somewhat later confirmed in a quantitative study by *Symonan* and *Nalbandyan* (*202*). From EPR measurements (*162*) it was further concluded that the copper sites were equivalent and rhombic. Ligand exchange studies led to the proposal that the copper binding site is composed of three nitrogen atoms as strong ligands while the fourth ligand is considerably weaker and can be easily exchanged with solvent anions (*162, 191*). An orthorhombic crystal form of bovine erythrocuprein was grown at 4 °C. From *x*-ray crystallographic investigation the dimeric molecule (4 subunits) had space group P 2_1 2_1 2_1 and $a = 51$ Å, $b = 61$ Å, and $c = 147$ Å (*203*). A monoclinic crystal form was obtained in 58% 2-methyl-2,4-pentanediol. This second crystal form had space group C_2: $a = 93.4$ Å, $b = 90.4$ Å, $c = 71.6$ Å, $\beta = 95.1°$ and four subunits (2 molecules of erythrocuprein) were also detectable per asymmetric unit.

From fluorescence emission and other physicochemical studies strong evidence was obtained that unlike in the case of bovine erythrocuprein tryptophan was found in human erythrocuprein (*204*). In a comprehensive pulse radiolysis study *Klug, Rabani* and *Fridovich* (*205*) determined the rate constant of superoxide dismutase to be 2.3×10^9 $M^{-1}s^{-1}$. They confirmed the results of *Rotilio, Bray* and *Fielden* (*182*) who measured somewhat earlier a constant of $1.6 \pm 0.3 \times 10^9$ $M^{-1}s^{-1}$. Both groups were able to demonstrate the constancy of this rate constant in the approximate *p*H region from *p*H 5 to 10.

Proof of the generation of $O_2^-\cdot$ during electrolysis of aqueous solutions was obtained in the laboratory of *Fridovich* (*206*). The $O_2^-\cdot$ catalysed oxidation of adrenaline served as a monitor which could be inhibited by superoxide dismutase. Ultrasonication of buffered aerated solutions gave rise to the formation of $O_2^-\cdot$ which could be detected using the cytochrome-c reductase assay (*207*). This sonication induced cytochrome-c reduction was also inhibited by native erythrocuprein. Another sensitive superoxide dismutase assay using the reduction of nitro blue tetrazolium by $O_2^-\cdot$ was developed by *Beauchamp* and *Fridovich* (*208*). This assay allowed the detection of erythrocuprein in the ng/ml region. During the metalloenzyme conference in Oxford, 1972, *Fridovich* summarized the basic facts on superoxide dismutase (erythrocuprein) (*209*).

The significance of superoxide during the hydroxylation of aromatic compounds was shown using the NADH-phenazine methosulphate-O_2 model system (*210*). From their observations that rat liver microsomes are in a position to catalyse an NADPH-dependent oxidation of adrenalin to adrenochrom *Aust, Roerig* and *Pederson* concluded the participation of $O_2^-\cdot$ (*210*). Another physiological function of erythrocuprein — the protection of biomembranes against oxidative breakage — was shown be *Fee* and *Teitelbaum* (*211*) who studied the peroxidation of red blood cells. *Zimmermann et al.* (*212*) supported these results in the case of the lipid peroxidation of the inner membranes of rat liver mitochondria. There was additional indication that during the oxidative breakdown of the mitochondrial membrane $O_2^{-}\cdot$ cannot be considered to be the exclusive substrate for erythrocuprein. In the author's laboratory a further study on the enzymic catalysed oxidation of xanthine using the chemiluminescence test (*147*) was carried out (*213*). At higher *p*H the initial light emission was followed by a distinct maximum which was shifted towards a longer time. In the presence of erythrocuprein this maximum appeared much earlier. Cupreins isolated from bovine liver and *Saccharomyces cerevisiae* were compared in a detailed study (*214*) and three short contributions deal with *x*-ray photoelectron measurements of the native and the Co-substituted erythrocuprein (*215—217*). The Co substituted erythrocuprein was already discussed by *Rotilio* on the metalloenzyme conference in Oxford last year (*218* and *198*).

Singlet Oxygen Decontaminase (S.O.D.). It has been demonstrated that $^1\Sigma_g^+O_2$ can be generated from potassium superoxide (*159*) or during photochemical reactions (*219*). Unfortunately, the unequivocal answer whether or not singlet oxygen can be formed directly during a chemical reaction in the dark avoiding the superoxide mediated formation of singlet oxygen was not possible (*147, 160, 179, 212, 213, 220*). This was due to the complexities of the employed assay mixtures. With the report by *Peters et al.* (*221*) final evidence was presented that $^1\Delta_gO_2$ was exclusively formed in the dark using an aqueous solution of CrO_8^{3-} at neutral *p*H.

$$4\,CrO_8^{3-} + 2\,H_2O = 4\,CrO_4^{2-} + 7\,O_2 + 4OH^-$$

There was also unequivocal proof that no superoxide at all was generated during this chemical reaction. Thus, this result supports earlier observations of a direct singlet oxygen generation in biological systems avoiding the pathway via superoxide.

The evidence that during the reaction of CrO_8^{3-} with water $^1\Delta_gO_2$ is exclusively generated (*219*) prompted us (*222, 224*) to use this chromium peroxocomplex as a substrate for erythrocuprein. *Stomberg* and *Brosset* (*223*) showed in 1960 that the chromium ion is in the oxidation state +5 and surrounded by four O_2^{2-} groups. No evidence from X-ray crystallographic data so far was obtained for the presence of superoxide. The possibility that O_2^- may be formed during the aqueous decomposition of

CrO_8^{3-} was examined using the cytochrome *c* reductase assay and the nitro blue tetrazolium staining method. Both assays for superoxide proofed absolutely negative (*224*). CrO_4^{2-} and CrO_8^{3-} did not disturb these assays since additional potassium superoxide caused a strong positive reaction for the presence of O_2^-.

Using CrO_8^{3-} as a substrate we were able to demonstrate the specificity and efficiency of the cupreins to scavenge highly energetic $^1\Delta_gO_2$ (Table 13). Even in concentrations up to 2×10^{-10}M erythrocuprein was able to quench the $^1\Delta_gO_2$ mediated chemiluminescence.

Table 13. *Singlet oxygen decontaminase activity of Cu-chelates and erythrocuprein. The erythrocuprein concentration required to produce a 50% inhibition of the chemiluminescence was taken an enzyme unit for quantitating the singlet oxygen decontaminase activity. The assay mixture contained: HEPES buffer, 0.1M, pH 7.8; luminol, 0.3 mM; crystalline K_3CrO_8, 0.3 mM; the total volume was 5.1 ml. Counting was performed as described in the legend to Fig. 32*

Cu^{2+}-Chelate	Required equivalent of chelated copper to yield 50% inhibition of $^1\Delta_gO_2$ induced chemiluminescence Moles $\times$ [10^6]	Specific activity . Units $\times$ mole^{-1} $\times$ [10^{-5}]
Native erythrocuprein	0.0036	2800
Boiled erythrocuprein	260	0.038
Apoerythrocuprein	0.42	24
Cu-EDTA	300	0.033
$Cu(His)_2$	10	1
$Cu(Lys)_2$	23	0.43

The reactivity of native erythrocuprein was higher compared to the superoxide dismutase activity shown in Table 12. Of utmost importance was the observation that the model chelates and $CuSO_4$ were virtually inactive compared to the native enzyme. The difference was 4 orders of magnitude which implies a much higher specificity for this enzymic reaction of the cupreins. The powerful reactivity of erythrocuprein is further demonstrated by the fact that the apoprotein displayed a detectable enzymic activity due to traces of copper which were undetectable by atomic absorption measurements or EPR spectroscopy. No such difference between apoprotein and the boiled native enzyme was observed using the superoxide dismutase assay.

The long known phenomenon that singlet oxygen species are being evolved during biochemical reactions led to the assumption that these oxygen species were the actual physiological substrates for erythrocuprein (*147, 179, 213, 219*). However, the question remained open concerning the superoxide mediated formation of these oxygen species. No final decision could be made due to the rather complex nature of the employed assay systems. With the present study unequivocal evidence was obtained that superoxide was not involved in the singlet oxygen production.

The high specificity of erythrocuprein with regard to scavenge singlet oxygen supports the statement of the entatic state of metal ions in metalloenzymes (*225, 226*). The protein portion of the native erythrocuprein is providing the proper ligand for the metal ions producing this entatic situation. In contrast, the low molecular weight Cu-chelates are virtually inactive because the copper would not be in this entatic state required to induce this singlet oxygen decontaminase activity.

Considering all the presented data the conclusion must be made that the main physiological function of the cupreins is the scavenging of singlet oxygen species rather than catalyzing the superoxide disproportionation. The occurrence of superoxide would be just one specific pathway for singlet oxygen formation. The protective action of a singlet oxygen decontaminase upon membrane lipids, all sorts of unsaturated compounds including nucleic acids and a great number of reducing agents can be considered essential in the metabolism of the aerobic cell.

5. References

1. *Dawson, C. R., Mallette, M. F.*: Advan. Protein Chem. *2*, 179 (1945).
2. *Hamilton, G. A.*: Advan. Protein Chem. *32*, 55 (1969).
3. *Malkin, R., Malmström, B. G.*: Advan. Enzymol. *33*, 177 (1970).
4. *Freeman, H. C.*: Advan. Protein Chem. *22*, 257 (1967).
5. *Jørgensen, C. K.*: In: The biochemistry of copper, p. 1 (*J. Peisach, P. Aisen* and *W. E. Blumberg*, eds.). New York: Academic Press 1966.
6. *Hemmerich, P.*: In: The biochemistry of copper, p. 15 (*J. Peisach, P. Aisen* and *W. E. Blumberg*, eds.). New York: Academic Press 1966.
7. *Gould, D. C., Mason, H. S.*: In: The biochemistry of copper, p. 35 (*J. Peisach, P. Aisen* and *W. E. Blumberg*, eds.). New York: Academic Press 1966.
8. *Blumberg, W. E.*: In: The biochemistry of copper, p. 49 (*J. Peisach, P. Aisen* and *W. E. Blumberg*, eds.). New York: Academic Press 1966.
9. *Brill, A. S., Venable, J. H. Jr.*: In: The biochemistry of copper, p. 67 (*J. Peisach, P. Aisen* and *W. E. Blumberg*, eds.). New York: Academic Press 1966.
10. *Freeman, H. C.*: In: The biochemistry of copper, p. 77 (*J. Peisach, P. Aisen* and *W. E. Blumberg*, eds.). New York: Academic Press 1966.

11. *Gurd, F. R. N., Bryce, G. F.:* In: The biochemistry of copper, p. 115 (*J. Peisach, P. Aisen* and *W. E. Blumberg*, eds.), New York: Academic Press 1966.
12. *Williams, R. J. P.:* In: The biochemistry of copper, p. 131 (*J. Peisach, P. Aisen* and *W. E. Blumberg*, eds.). New York: Academic Press 1966.
13. *Breslow, E.:* In: The biochemistry of copper, p. 149 (*J. Peisach, P. Aisen* and *W. E. Blumberg*, eds.). New York: Academic Press 1966.
14. *Porter, H.:* In: The biochemistry of copper, p. 159 (*J. Peisach, P. Aisen* and *W. E. Blumberg*, eds.). New York: Academic Press 1966.
15. *Peters, S. P., Sir:* In: The biochemistry of copper, p. 175 (*J. Peisach, P. Aisen* and *W. E. Blumberg*, eds.). New York: Academic Press 1966.
16. *Sarkar, B., Kruck, T. P. A.:* In: The biochemistry of copper, p. 183 (*J. Peisach, P. Aisen* and *W. E. Blumberg*, eds.). New York: Academic Press 1966.
17. *Petering, H. G., Van Giessen, G. J.:* In: The biochemistry of copper, p. 211 (*J. Paisach, P. Aisen* and *W. E. Blumberg*, eds.). New York: Academic Press 1966.
18. *Peisach, J.:* In: The biochemistry of copper, p. 149 (*J. Peisach, P. Aisen* and *W. E. Blumberg*, eds.). New York: Academic Press 1966.
19. *Beinert, H.:* In: The biochemistry of copper, p. 213 (*J. Peisach, P. Aisen* and *W. E. Blumberg*, eds.). New York: Academic Press 1966.
20. *Wharton, D. C., Gibson, Q. H.:* In: The biochemistry of copper, p. 235 (*J. Peisach, P. Aisen* and *W. E. Blumberg*, eds.). New York: Academic Press 1966.
21. *Gelder, B. F., Slater, E. C.:* In: The biochemistry of copper, p. 245 (*J. Peisach, P. Aisen* and *W. E. Blumberg*, eds.). New York: Academic Press 1966.
22. *Tzagoloff, A., MacLennan, D. H.:* In: The biochemistry of copper, p. 253 (*J. Peisach, P. Aisen* and *W. E. Blumberg*, eds.). New York: Academic Press 1966.
23. *Yamanaka, T.:* In: The biochemistry of copper, p. 275 (*J. Peisach, P. Aisen* and *W. E. Blumberg*, eds.). New York: Academic Press 1966.
24. *Chance, B.:* In: The biochemistry of copper, p. 293 (*J. Peisach, P. Aisen* and *W. E. Blumberg*, eds.). New York: Academic Press 1966.
25. *Dawson, C. R.:* In: The biochemistry of copper, p. 305 (*J. Peisach, P. Aisen* and *W. E. Blumberg*, eds.). New York: Academic Press 1966.
26. *Brooks, D. W., Dawson, C. R.:* In: The biochemistry of copper, p. 343 (*J. Peisach, P. Aisen* and *W. E. Blumberg*, eds.). New York: Academic Press 1966.
27. *Kertesz, D.:* In: The biochemistry of copper, p. 359 (*J. Peisach, P. Aisen* and *W. E. Blumberg*, eds.). New York: Academic Press 1966.
28. *Levine, W. G.:* In: The biochemistry of copper, p. 371 (*J. Peisach, P. Aisen* and *W. E. Blumberg*, eds.). New York: Academic Press 1966.
29. *Nakamura, T., Ogura, Y.:* In: The biochemistry of copper, p. 389 (*J. Peisach, P. Aisen* and *W. E. Blumberg*, eds.). New York: Academic Press 1966.
30. *Katoh, S., San Pietro, A.:* In: The biochemistry of copper, p. 407 (*J. Peisach, P. Aisen* and *W. E. Blumberg*, eds.). New York: Academic Press 1966.
31. *Nara, S., Yasunobu, K. T.:* In: The biochemistry of copper, p. 423 (*J. Peisach, P. Aisen* and *W. E. Blumberg*, eds.). New York: Academic Press 1966.
32. *Goldstein, M.:* In: The biochemistry of copper, p. 443 (*J. Peisach, P. Aisen* and *W. E. Blumberg*, eds.). New York: Academic Press 1966.
33. *Lontie, R., Witters, R.:* In: The biochemistry of copper, p. 455 (*J. Peisach, P. Aisen* and *W. E. Blumberg*, eds.). New York: Academic Press 1966.
34. *Vanneste, W., Mason, H. S.:* In: The biochemistry of copper, p. 465 (*J. Peisach, P. Aisen* and *W. E. Blumberg*, eds.). New York: Academic Press 1966.
35. *Walshe, J. M.:* In: The biochemistry of copper, p. 475 (*J. Peisach, P. Aisen* and *W. E. Blumberg*, eds.). New York: Academic Press 1966.

36. *Aspin, N., Sass-Kortsak, A.:* In: The biochemistry of copper, p. 503 (*J. Peisach, P. Aisen* and *W. E. Blumberg*, eds.). New York: Academic Press 1966.
37. *Scheinberg, H.:* In: The biochemistry of copper, p. 513 (*J. Peisach, P. Aisen* and *W. E. Blumberg*, eds.). New York: Academic Press 1966.
38. *Poillon, W. N., Bearn, A. G.:* In: The biochemistry of copper, p. 525 (*J. Peisach, P. Aisen* and *W. E. Blumberg*, eds.). New York: Academic Press 1966.
39. *Walaas, E., Walaas, O., Løvstad, R.:* In: The biochemistry of copper, p. 537 (*J. Peisach, P. Aisen* and *W. E. Blumberg*, eds.). New York: Academic Press 1966.
40. *Curzon, G., Cumings, J. N.:* In: The biochemistry of copper, p. 545 (*J. Peisach, P. Aisen* and *W. E. Blumberg*, eds.). New York: Academic Press 1966.
41. *Osaki, S., McDermott, J. A., Johnson, D. A., Frieden, E.:* In: The biochemistry of copper, p. 559 (*J. Peisach, P. Aisen* and *W. E. Blumberg*, eds.). New York: Academic Press 1966.
42. *Redfield, A. C.:* The Haemocyanins. Biol. Rev. *9*, 176 (1934).
43. — Haemocyanin in "Copper Metabolism: A Symposium on Animal, Plant and Soil Relationships", p. 174. Baltimore, Maryland: Johns Hopkins Press 1950.
44. *Frieden, E., McDermott, J. A., Osaki, S.:* Oxidases and related redox systems, p. 240. New York: John Wiley 1965.
45. *Redmond, J. R.:* In: Physiology and biochemistry of haemocyanins, p. 5 (*F. Ghiretti*, ed.). New York: Academic Press 1968.
46. *Osaki, S., McDermott, J. A., Frieden, E.:* In: Physiology and biochemistry of haemocyanins, p. 25 (*F. Ghiretti*, ed.). New York: Academic Press 1968.
47. *van Bruggen, E. F. J.:* In: Physiology and biochemistry of haemocyanins, p. 37 (*F. Ghiretti*, ed.). New York: Academic Press 1968.
48. *Gruber, M.:* In: Physiology and biochemistry of haemocyanins, p. 49 (*F. Ghiretti*, ed.). New York: Academic Press 1968.
49. *Witters, R., Lontie, R.:* In: Physiology and biochemistry of haemocyanins, p. 61 (*F. Ghiretti*, ed.). New York: Academic Press 1968.
50. *Rombauts, W. A.:* In: Physiology and biochemistry of haemocyanins, p. 75 (*F. Ghiretti*, ed.). New York: Academic Press 1968.
51. *Elliot, F. G., Hoebeke, J.:* In: Physiology and biochemistry of haemocyanins, p. 81 (*F. Ghiretti*, ed.). New York: Academic Press 1968.
52. *Gould, D. C., Ehrenberg, A.:* In: Physiology and biochemistry of haemocyanins, p. 95 (*F. Ghiretti*, ed.). New York: Academic Press 1968.
53. *Williams, R. J. P.:* In: Physiology and biochemistry of haemocyanins, p. 113 (*F. Ghiretti*, ed.). New York: Academic Press 1968.
54. *Mahler, H. R.:* In: Trace elements, p. 311 (*C. A. Lamb, O. G. Bently* and *J. M. Beattle*, eds.). New York: Academic Press 1958.
55. *Mann, T., Keilin, D.:* Proc. Roy. Soc. (London), Ser. B *126*, 303 (1939).
56. *Mohamed, M. S., Greenberg, D. M.:* J. Gen. Physiol. *37*, 433 (1954).
57. *Markowitz, H., Cartwright, G. E., Wintrobe, M. M.:* J. Biol. Chem. *234*, 40 (1959).
58. *Kimmel, J. R., Markowitz, H., Brown, D. M.:* J. Biol. Chem. *234*, 46 (1959).
59. *Nyman, P. O.:* Biochim. Biophys. Acta *45*, 387 (1960).
60. *Shields, G. S., Markowitz, H., Klassen, W. H., Cartwright, G. E., Wintrobe, M. M.:* J. Clin. Invest. *40*, 2007 (1961).
61. *Malmström, B. G., Vänngard, T.:* J. Mol. Biol. *2*, 118 (1960).
62. *Porter, H., Folch, J.:* J. Neurochem. *1*, 260 (1957).
63. — *Ainsworth, S.:* J. Neurochem. *5*, 91 (1959).
64. — *Sweeny, M., Porter, E.:* Arch. Biochem. Biophys. *105*, 319 (1964).

65. *Stansell, M. J., Deutsch, H. F.:* J. Biol. Chem. *240,* 4299 (1965).
66. — — J. Biol. Chem. *240,* 4306 (1965).
67. *Hartz, J. W., Deutsch, H. F.:* J. Biol. Chem. *244,* 4565 (1969).
68. *Carrico, R. J., Deutsch, H. F.:* J. Biol. Chem. *244,* 6087 (1969).
69. — — J. Biol. Chem. *245,* 723 (1970).
70. *McCord, J. M., Fridovich, I.:* J. Biol. Chem. *244,* 6049 (1969).
71. — — J. Biol. Chem. *244,* 6056 (1969).
72. *Bannister, J., Bannister, W., Wood, E.:* Europ. J. Biochem. *18,* 178 (1971).
73. *Wood, E., Dalgleish, D., Bannister, W.:* Europ. J. Biochem. *18,* 187 (1971).
74. *Weser, U., Bunnenberg, E., Cammack, R., Djerassi, C., Flohé, L., Thomas, G., Voelter, W.:* Biochim. Biophys. Acta *243,* 203 (1971).
75. — Angew. Chem. *83,* 939 (1971).
76. — *Hartmann, H. J.:* FEBS Letters *17,* 78 (1971).
77. — *Voelcker, G.:* FEBS Letters *22,* 15 (1972).
78. — *Barth, G., Djerassi, C., Hartmann, H. J., Krauss, P., Voelcker, G., Voelter, W., Voetsch, W.:* Biochim. Biophys. Acta, *243,* 203 (1972).
79. — *Bohnenkamp, W., Cammack, R., Hartmann, H. J., Voelcker, G.:* Z. Physiol. Chem. *353,* 1059 (1972).
80. *Bohnenkamp, W., Weser, U.:* Z. Physiol. Chem. *353,* 695 (1972).
81. *Rotilio, G., Finazzi-Agro', A., Calabrese, L., Bossa, F., Guerrieri, P., Mondovi, B.:* Biochemistry *10,* 616 (1971).
82. *Keele, B. B., McCord, J. M., Fridovich, I.:* J. Biol. Chem. *246,* 2875 (1971).
83. *Fee, J. A., Gaber, B. B.:* The Sec. Internat. Sympos. on Oxidases and related Oxidation-Reduction Systems, Memphis, Tennessee, 1971. Organized by *T. E. King, H. S. Mason* and *M. Morrison.*
84. *Bohnenkamp, W.:* Ph. D. Thesis, Tübingen (1973).
85. *Petersen, R., Bollier, M.:* Anal. Chem. *27,* 1195 (1955).
86. *Elwell, W. T., Gidley, J. A. F.:* Atomic-absorption spectrophotometry, p. 89 and 129. Oxford: Pergamon Press 1966.
87. *Schallies, A.:* Biochemical Master Thesis, Tübingen 1972.
88. *Moore, S., Stein, W. H.:* In: Methods in enzymology, Vol. 6, p. 819 (*S. P. Colowick* and *N. O. Kaplan,* eds.). New York: Academic Press 1963.
89. *Moore, S.:* J. Biol. Chem. *238,* 235 (1963).
90. *Boyer, P. D.:* J. Am. Chem. Soc. *76,* 4331 (1954).
91. *Hirs, C. H. W.:* J. Biol. Chem. *219,* 611 (1956).
92. *Bannister, W. H., Salisbury, C. M., Wood, E. J.:* Biochim. Biophys. Acta *168,* 392 (1968).
93. *Spies, J., Chambers, D.:* Anal. Chem. *21,* 1249 (1949).
94. *Edelhoch, H.:* Biochemistry *6,* 1948 (1967).
95. *Beaven, G. H., Holiday, E. R.:* Advan. Protein Chem. *7,* 319 (1952).
96. *Lahey, M. E., Gubler, C. J., Cartwright, G. E., Wintrobe, M. M.:* J. Clin. Invest. *32,* 322 (1953).
97. *Peisach, J., Levine, W. G., Blumberg, W. E.:* J. Biol. Chem. *242,* 2847 (1967).
98. *Van De Bogart, M., Beinert, H.:* Anal. Biochem. *20,* 325 (1967).
99. *Malmström, B. G.:* Methods Biochem. Analy. *3,* 327 (1956).
100. *Felsenfeld, G.:* Arch. Biochem. Biophys. *87,* 247 (1960).
101. *Bayer, E., Bacher, A., Krauss, P., Voelter, W., Barth, G., Bunnenberg, E., Djerassi, C.:* Europ. J. Biochem. *22,* 580 (1971).
102. *Tsangaris, J. W., Chang, J. W., Martin, R. B.:* J. Am. Chem. Soc. *91,* 726 (1969).
103. *Yasui, T., Hidaka, J., Shimura, Y.:* J. Am. Chem. Soc. *87,* 2762 (1965).
104. *Voelter, W., Bunnenberg, E., Djerassi, C.:* J. Am. Chem. Soc. *90,* 6163 (1968).

105. — *Barth, G., Records, R., Bunnenberg, E., Djerassi, C.:* J. Am. Chem. Soc. *91*, 6165 (1969).
106. *Holzwarth, G., Doty, P.:* J. Am. Chem. Soc. *87*, 218 (1965).
107. *Velluz, L., Legrand, M.:* Angew. Chem. *77*, 842 (1965).
108. *Beychok, S.:* In: Poly-α-amino acids, p. 293 (*G. D. Fasman*, ed.). New York: Marcel Dekker 1967.
109. *Gratzer, W. B., Cowburn, D. A.:* Nature *222*, 426 (1969).
110. *Jirgensons, B.:* Optical rotatory dispersion of proteins and other macromolecules, p. 57. Berlin-Heidelberg-New York: Springer 1969.
111. *Li, L., Spector, A.:* J. Am. Chem. Soc. *91*, 220 (1969).
112. *Voelter, W.:* Chem. unserer Zeit in the press.
113. *Fee, J. A., Gaber, B. P.:* J. Biol. Chem. *247*, 60 (1972).
114. *Weser, U.:* Z. Physiol. Chem. *353*, 769 (1972).
115. *Jung, G., Weser, U., Voelter, W.:* Z. Physiol. Chem. *353*, 720 (1972).
116. — *Ottnad, M., Bremser, W., Bohnenkamp, W., Weser, U.:* Biochim. Biophys. Acta *295*, 77 (1973).
117. *Ottnad, M.:* Chemical Master Thesis, Tübingen 1972.
118. *Falk, K. E., Freeman, H. C., Jansson, T., Malmström, B. G., Vänngård, T.:* J. Am. Chem. Soc. *89*, 6071 (1967).
119. *Oudin, J.:* Ann. Inst. Pasteur *85*, 336 (1953).
120. *Ouchterlony, O.:* Arkiv Kemi *B26*, 1 (1949).
121. *Voelcker, G.:* Biochemical Master Thesis, Tübingen 1971.
122. *Hartmann, H. J.:* Biochemical Master Thesis, Tübingen 1971.
123. *Flohé, L., Schaich, E., Voelter, W., Wendel, A.:* Z. Physiol. Chem. *352*, 170 (1971).
124. *Strickland, E. H., Horwitz, J., Billups, C.:* Biochemistry *8*, 3205 (1969).
125. *Hartzell, C. R., Gurd, F. R. N.:* J. Biol. Chem. *244*, 147 (1969).
126. *Barth, G., Voelter, W., Bunnenberg, E., Djerassi, C.:* J. Am. Chem. Soc. *94*, 1293 (1972).
127. — *Records, R., Bunnenberg, E., Djerassi, C., Voelter, W.:* J. Am. Chem. Soc. *93*, 2545 (1971).
128. — *Djerassi, C., Bunnenberg, E.:* to be published.
129. *Hartz, J. W.:* Ph. D. Thesis, University of Wisconsin (1968).
130. *Huber, C. T., Frieden, E.:* J. Biol. Chem. *245*, 3973 (1970).
131. — J. Biol. Chem. *245*, 3979 (1970).
132. *Osaki, S., Johnson, D. A., Frieden, E.:* J. Biol. Chem. *246*, 3018 (1971).
133. *McCord, J. M., Fridovich, I.:* J. Biol. Chem. *243*, 5753 (1968).
134. *Cotton, F. A., Wilkinson, G.:* Advanced inorganic chemistry, p. 393. New York: London: John Wiley and Sons 1966.
135. *Maricle, D. L., Hodgson, W. G.:* Anal. Chem. *37*, 1562 (1965).
136. *Czapski, G. H., Bielski, H. J.:* J. Phys. Chem. *67*, 2180 (1963).
137. *Rabani, J., Mulac, W. A., Matheson, M. S.:* J. Phys. Chem. *69*, 53 (1965).
138. *Bielski, B. H. J., Allen, A. O.:* J. Phys. Chem. *71*, 4544 (1967).
139. *Czapski, G., Dorfman, L. M.:* J. Phys. Chem. *68*, 1169 (1964).
140. *Behar, D., Czapski, G., Rabani, J., Dorfman, L. M., Schwarz, H. A.:* J. Phys. Chem. *74*, 3209 (1970).
141. *Horecker, B. L., Heppel, L. A.:* J. Biol. Chem. *178*, 683 (1949).
142. *Fridovich, I., Handler, P.:* J. Biol. Chem. *237*, 916 (1962).
143. *Totter, J. R., Medina, V. J., Scoseria, J. L.:* J. Biol. Chem. *235*, 238 (1960).
144. — *de Dugros, E. C., Riveiro, C.:* J. Biol. Chem. *235*, 1839 (1960).
145. *Greenlee, L. L., Fridovich, I., Handler, P.:* Biochemistry *1*, 779 (1962).
146. *Arneson, R. M.:* Arch. Biochem. Biophys. *136*, 352 (1970).

147. *Finazzi-Agro', A., Giovagnoli, C., Del Sole, P., Calabrese, L., Rotilio, G., Mondovi, B.:* FEBS Letters *21*, 183 (1972).
148. *Handler, P., Rajagopalan, K. V., Aleman, V.:* Federation Proc. *23*, 30 (1964).
149. *Knowles, P. F., Gibson, J. F., Pick, F. M., Bray, R. C.:* Biochem. J. *111*, 53 (1969).
150. *Massey, V., Strickland, S., Mayhew, S. G., Howell, L. G., Engel, P. C., Matthews, R. G., Schuman, M., Sullivan, P. A.:* Biochem. Biophys. Res. Commun. *36*, 891 (1969).
151. *Ballou, D., Palmer, G., Massey, V.:* Biochem. Biophys. Res. Commun. *36*, 898 (1969).
152. *Bray, R. C.:* Biochem. J. *81*, 189 (1961).
153. *Hemmerich, P.:* Helv. Chim. Acta *47*, 664 (1966).
154. *Ichikawa, T., Iwasaki, M., Kuwata, K.:* J. Chem. Phys. *44*, 2979 (1966).
155. *Bennet, J. E., Mile, B., Thomas, A.:* Trans. Faraday Soc. *64*, 3200 (1968).
156. *Bray, R. C., Palmer, G., Beinert, H.:* J. Biol. Chem. *239*, 2667 (1964).
157. *Misra, H. P., Fridovich, I.:* J. Biol. Chem. *246*, 6886 (1971).
158. — — J. Biol. Chem. *247*, 188 (1972).
159. *Khan, A. U.:* Science *168*, 476 (1970).
160. *Chan, H. W. S.:* J. Am. Chem. Soc. *93*, 2357 (1971).
161. *Rotilio, G., Calabrese, L., Bossa, F., Barra, D., Finazzi-Agro', A., Mondovi, B.:* Biochemistry *11*, 2182 (1972).
162. *Rotilio, G., Morpurgo, L., Giovagnoli, C., Calabrese, L., Mondovi, B.:* Biochemistry *11*, 2187 (1972).
163. *Massey, V.:* Biochim. Biophys. Acta *34*, 255 (1959).
164. *Van Gelder, B. F., Slater, E. C.:* Biochim. Biophys. Acta *58*, 593 (1962).
165. *Coleman, P. L., Iweibo, I., Weiner, H.:* Biochemistry *11*, 1010 (1972).
166. *Tipton, K. F.:* Anal. Biochem. *28*, 318 (1969).
167. *Haber, F., Weiss, J.:* Proc. Roy. Soc. (London) *A 147*, 332 (1934).
168. *Stancliffe, T. C., Pirie, A.:* FEBS Letters *17*, 297 (1971).
169. *Siegbahn, K.:* ESCA, Atomic molecular and solid state structure by means of electron spectroscopy. Uppsala: Almquist & Wiksells 1967.
170. *Jørgensen, C. K.:* Modern aspects of ligand field theory. Amsterdam: North-Holland Press 1971.
171. *Kramer, E. N., Klein, M. P.:* In: Electron Spectroscopy (*D. A. Shirley*, ed.). Amsterdam: North Holland Publ. Co. 1972.
172. *George, P.:* In: Oxidases and related redox systems, p. 3 (*T. E. King, H. S. Mason* and *M. Morrison*, eds.). New York: J. Wiley, Inc., 1965.
173. *Ardon, M.:* Oxygen. New York: Benjamin Press 1965.
174. *Fallab, S.:* Angew. Chem. *79*, 500 (1967).
175. *Samuel, D.:* In: Biochemie des Sauerstoffs, p. 6 (*B. Hess* and *H. J. Staudinger*, eds.). Berlin-Heidelberg-New York: Springer 1968.
176. *Griffith, J. S.:* Proc. Roy. Soc. (London), Ser. *A*. *235*, 23 (1956).
177. — In: Oxygen in the animal organism, p. 141. Oxford: Pergamon Press 1964.
178. *Bayer, E., Schretzmann, P.:* Struct. Bonding. Berlin-Heidelberg-New York: Springer 1967.
178a. — *Krauss, P., Röder, A., Schretzmann, P.:* In: Sec. Intern. Symposium on Oxidases and related Oxidation-Reduction Systems, Memphis, Tennessee (1971), org. by *T. E. King, H. S. Mason* and *M. Morrison*, in the press.
179. *Joester, K. E., Jung, G., Weber, U., Weser, U.:* FEBS Letters *25*, 25 (1972).
180. *De Luca, M., Dempsey, M. E.:* Biochem. Biophys. Res. Commun. *40*, 117 (1970).
181. *McDonagh, A. F.:* Biochem. Biophys. Res. Commun. *44*, 1306 (1971).

182. *Rotilio, G., Bray, R. C., Fielden, M. E.:* Biochim. Biophys. Acta *268*, 605 (1972).
183. *Massey, V., Palmer, G., Ballou, D.:* In: Sec. Intern. Symposium on Oxidases and related Oxidation-Reduction Systems, org. by *T. E. King, H. S. Mason* and *M. Morrison*, Memphis, Tennessee (1971), in the press.
184. — — — In: Flavins and Flavoproteins, p. 349—361. (*H. Kamin*, ed.). Univ. Park Press, Butterworths 1971.
185. *Hemmerich, P., Bhaduri, A. P., Blankenhorn, G., Bürstelein, M., Haas, W., Knappe, W. R.:* In: Sec. Intern. Symposium on Oxidases and related Oxidation-Reduction Systems, org. by *T. E. King, H. S. Mason* and *M. Morrison*, Memphis, Tennessee (1971), in the press.
186. *Hemmerich, P.:* unpublished.
187. — *Beinert, H., Vänngard, T.:* Angew. Chem. Intern. Ed. Engl. *5*, 422 (1966).
188. *Sigel, H.:* Angew. Chem. *81*, 161 (1969).
189. *Brintzinger, H., Erlenmeyer, H.:* Helv. Chim. Acta *48*, 826 (1965).
190. *Sigel, H., Müller, U.:* Helv. Chim. Acta *49*, 671 (1966).
191. *Gaber, B. P., Brown, R. D., Koenig, S. H., Fee, J. A.:* Biochim. Biophys. Acta *271*, 1 (1972).
192. *Misra, H. P., Fridovich, I.:* J. Biol. Chem. *247*, 3170 (1972).
193. *Rabani, J., Nielsen, S. O.:* J. Phys. Chem. *73*, 3736 (1969).
194. *Nilsson, R., Pick, F. M., Bray, R. C., Fielden, M.:* Acta Chem. Scand. *23*, 2554 (1969).
195. *Eigen, M., Hammes, G. G.:* Advan. Enzymol. *25*, 1 (1963).
196. *Orme-Johnson, W. H., Beinert, H.:* Biochem. Biophys. Res. Commun. *36*, 905 (1969).
197. *Bohnenkamp, W., Weser, U.:* Intern. Congress of Biophysics, Moscow (1972), Abstracts *2*, 101 EV1b 3/9.
198. *Calabrese, L., Rotilio, G., Mondovi, B.:* Biochim. Biophys. Acta *263*, 827—829 (1972).
199. *Sawada, Y., Ohyama, I., Yamazaki, I.:* Biochim. Biophys. Acta *268*, 305 (1972).
200. *Misra, H. P., Fridovich, I.:* J. Biol. Chem. *247*, 3410 (1972).
201. *Weser, U., Fretzdorff, A. M., Prinz, R.:* FEBS Letters *27*, 267 (1972).
202. *Symonyan, M. A., Nalbandyan, R. M.:* FEBS Letters *28*, 22 (1972).
203. *Richardson, D. C., Bier, C. J., Richardson, J. S.:* J. Biol. Chem. *247*, 6368 (1972).
204. *Bannister, W. H., Dalgleish, D. G., Bannister, J. V., Wood, E. J.:* Intern. J. Biochem. *3*, 560 (1972).
205. *Klug, D., Rabani, J., Fridovich, I.:* J. Biol. Chem. *247*, 4839 (1972).
206. *Forman, H. J., Fridovich, I.:* Science *175*, 339 (1972).
207. *Lippitt, B., McCord, J. M., Fridovich, I.:* J. Biol. Chem. *247*, 4688 (1972).
208. *Beauchamp, C., Fridovich, I.:* Anal. Biochem. *44*, 276 (1971).
209. *Fridovich, I.:* Metalloenzyme Conference, Oxford 1972, Abstr. Commun., p. 22.
210. *Aust, S. D., Roerig, D. L., Pederson, T. C.:* Biochem. Biophys. Res. Commun. *47*, 1133 (1972).
211. *Fee, J. A., Teitelbaum, D.:* Biochem. Biophys. Res. Commun. *49*, 150 (1972).
212. *Zimmermann, R., Flohé, L., Weser, U., Hartmann, H. J.:* FEBS Letters *29*, 117 (1973).
213. *Weser, U., Paschen, W.:* FEBS Letters *27*, 248 (1972).
214. — *Prinz, R., Schallies, A., Fretzdorff, A., Krauss, P., Voelter, W., Voetsch, W.:* Z. Physiol. Chem. *353*, 1821 (1972).
215. *Jung, G., Ottnad, M., Hartmann, H. J., Rupp, H., Weser, U.:* Z. Anal. Chem. *263*, 282 (1973).

216. — — *Bremser, W., Hartmann, H. J., Weser, U.*: Z. Physiol. Chem. *354*, 341 (1973).
217. — — *Bohnenkamp, W., Weser, U.*: FEBS Letters *25*, 346 (1972).
218. *Rotilio, G.*: Metalloenzyme Conference, Oxford 1972, Abstr. Commun., p. 24.
219. *Politzer, I. R., Griffin, G. W., Laseter, J. L.*: Chem. Biol. Interactions *3*, 73 (1971).
220. *Goda, K., Chu, I. W., Kimura, T., Schaap, A. P.*: Biochem. Biophys. Res. Commun. *52*, 1300 (1973).
221. *Peters, J. W., Pitts, J. N.*, Jr., *Rosenthal, I., Fuhr, H.*: J. Am. Chem. Soc. *94*, 4348 (1972).
222. *Paschen, W., Weser, U.*: Biochim. Biophys. Acta. submitted.
223. *Stomberg, R., Brosset, C.*: Acta Chem. Scand. *14*, 441 (1960).
224. *Paschen, W.*: Ph. D. Thesis, Tübingen.
225. *Vallee, B. L., Williams, R. J. P.*: Proc. Natl. Acad. Sci. USA *59*, 498 (1968).
226. *Behnke, D., Vallee, B. L.*: Proc. Natl. Acad. Sci. USA *69*, 2442 (1972).

Received August 14, 1972

Ferritin*

Robert R. Crichton

Max-Planck-Institut für Molekulare Genetik, Berlin-Dahlem, Germany

Table of Contents

---

*) Some of the research reported in this review was supported by the Deutsche Forschungsgemeinschaft (grants Cr 45/1 and Cr. 45/2).

I. Introduction

The importance of iron in mammalian organisms is such that one requires to make no case for the study of this mineral and its metabolism. The major role of haemoglobin and myoglobin in the transport and storage of oxygen together with their characteristic red colour has made them objects of study ever since the advent of biochemistry. Indeed many of the most outstanding contributions to the development of physiological chemistry and its steady transition into biochemistry and molecular biology can be traced to investigations on these two proteins. The other essential role of iron as a component of a series of haem enzymes involved in the terminal pathway of electron transport, leading to the ultimate electron acceptor molecular oxygen and associated with the concomitant production of energy, has been studied for many years.

There are a number of other iron containing proteins in mammals as reference to Table 1 will show. Flavoprotein dehydrogenases involved in a number of oxidation reactions, catalase and peroxidase accounting for metabolism of hydrogen peroxide, and a varied number of oxidases and oxygenases catalysing the introduction of molecular oxygen into a series of substrates ranging in diversity from tryptophan to steroids represent the better known of those proteins which have well established catalytic functions. Transferrin, the β_1-globulin found in plasma, transports iron from one part of the body to another and has been extensively dealt with in a previous review in this series (*1*).

There remain two iron containing proteins which together account for one quarter of the total iron in man and it is with the best characterized of these that we will be concerned here, namely ferritin. Haemosiderin, the other of these two, will not be discussed.

Table 1. *Distribution of iron containing proteins in a normal adult man (adapted from (5))*

Protein	Molecular weight of protein	Amount of protein (g)	Amount of iron (g)	% of total body iron	Nature of iron haem (H) or non-haem (N)	Number of iron atoms bound molecule	Valence state	Function
haemoglobin	68,000	750	2.60	65	H	4	Fe^{2+}	oxygen transport in plasma
myoglobin	17,000	40	0.13	6	H	1	Fe^{2+}	oxygen storage in muscle
transferrin	76,000	20	0.007	0.2	N	2	Fe^{3+}	iron storage in plasma
ferritin	444,000	2.4	0.52	13	N	0—4,700	Fe^{3+}	iron storage in cells
haemosiderin	not known	1.6	0.48	12	N	5000	Fe^{3+}	iron storage in cells
catalase	280,000	5.0	0.004	0.1	H		Fe^{2+}	metabolism of H_2O_2
cytochrome c	12,500	0.8	0.004	0.1	H	1	Fe^{2+}/Fe^{3+}	terminal oxidation
peroxidase	44,100		} 0.02	} <0.5	H			metabolism of H_2O_2
cytochromes and oxidase					H		Fe^{2+}/Fe^{3+}	terminal oxidation
flavoprotein dehydrogenases, oxidases, and oxygenases					N		Fe^{2+}	oxidation reactions, incorporation of molecular oxygen

In 1894 *Naunyn-Schmiedeberg* (*2*), the well known German pharmacologist described a protein to which he gave the name Ferratin. This substance, isolated from pig liver, contained 6% iron and a variable amount of phosphate and remained the object of some considerable criticism until 1934. In that year the Czech scientist *Laufberger* (*3*) described the isolation of a pure protein rich in iron, by heat treatment of horse spleen followed by repeated ammonium sulphate and alcohol precipitation. The name ferritin was proposed by him, as he put it "in memory of Schmiedeberg's ferratin", and in a detailed publication (*4*) he described its purification, its high iron content (over 20% by weight), and prophetically remarked that in his opinion ferritin was a substance that served as an iron depot in the organism.

In order to simplify the treatment of the present state of our knowledge about this protein a few brief definitions seem in order. *Ferritin* refers, unless otherwise specified to the protein from horse spleen, about which most is known at present. *Apoferritin* is the term used for the protein part of ferritin, and unless specified also refers to the horse spleen protein. *Native apoferritin* is the term used to describe the nearly iron free protein which can be isolated from ferritin by centrifugal techniques (Fig. 1). *Full ferritin* is defined in Fig. 1 as is *iron-poor ferritin*: they are molecules which contain the maximum amount of iron and much less than the average amount of iron respectively.

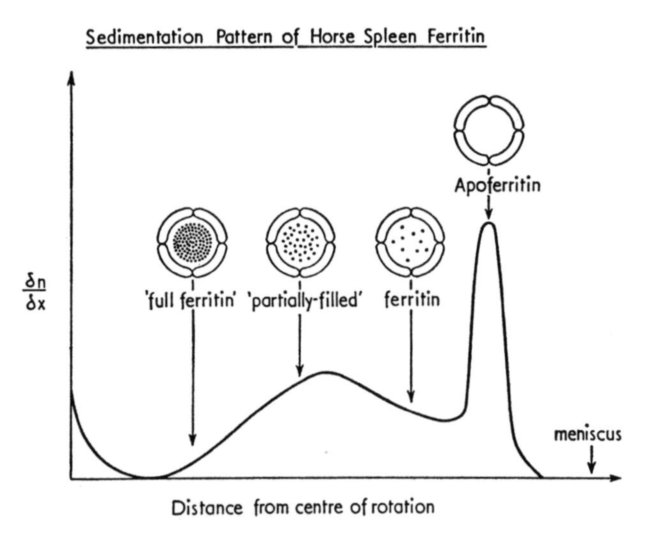

Fig. 1. Fractionation of native ferritin by ultracentrifugation. The change in concentration dC/dr is plotted against the distance from the centre of rotation (*r*)

II. Physiological Function of Ferritin

As is described in Section III below, ferritin is widely distributed throughout the various organs of mammals, with particularly high concentrations being found in liver, spleen and bone marrow. Its function can be summarized in physiological terms quite simply: it represents a depot in which surplus iron can be stored within the cell in a non-toxic form and from which it can be mobilized as and when required either within that particular cell itself, or in other cells of the organism. Although the precise function of ferritin in other species is less well characterized, this is presumably its major role there. However, in mammals we can distinguish a number of distinctive features of the function of ferritin in different organs, and these are summarized in Fig. 2 (*5*).

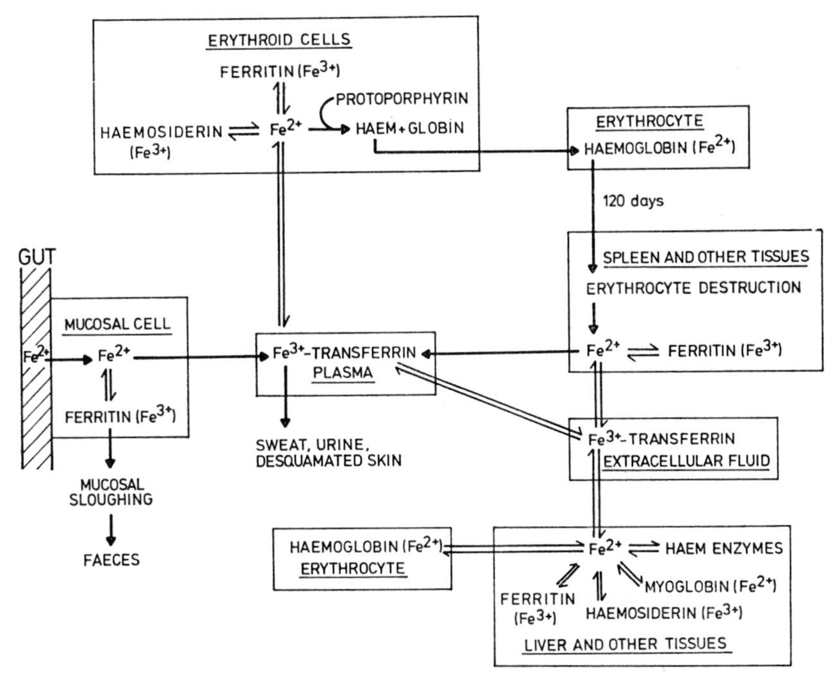

Fig. 2. Physiological role of ferritin in iron metabolism. For detailed explanation, see text

In cells of the erythroid series (that is, cells which are precursors of erythrocytes and which synthesise haemoglobin) transferrin molecules carrying Fe^{3+} bind to specific receptor sites on the cell membrane and transfer iron to the cells (*6*). This iron, probably after reduction to Fe^{2+}, is then incorporated into ferritin: the process involves oxidation to the ferric state (see Section VI). Ferritin iron then serves as a source of iron for haem synthesis, and its mobilization requires reduction to Fe^{2+} (Section VII).

At the end of their life span, erythrocytes are phagocytised by cells of the spleen, bone marrow and other tissues. Although the globin chains are degraded, and the porphyrin excreted after conversion to bilirubin in the bile, most of the iron is conserved; it is first stored in ferritin, and can then be subsequently mobilized from these tissues by transferrin.

In liver, muscle and other tissues iron is taken up by the cells when transferrin saturation levels are high and deposited first in ferritin and subsequently transferred to haemosiderin. This pool of ferritin iron is most likely used within these cells to meet requirements for synthesis of haem enzymes, myoglobin and other non-haem iron proteins. The ferritin can also store iron released from the breakdown of such iron containing proteins in the course of their turnover. Mobilisation of iron from these tissues once again probably involves reduction of iron to Fe^{2+} and its transfer across the cell membrane to transferrin. In such tissues the level of transferrin saturation seems likely to play a major role in determining the balance between deposition of iron in ferritin and its mobilization from the storage form.

Ferritin also occurs in the cells of the intestinal mucosa where it has been thought to play some role in regulating the amount of dietary iron absorbed from the gut. From evidence of increased absorption of radioactive iron in humans with iron deficiency compared with normal subjects, who absorb very little, the idea of a "mucosal block" was

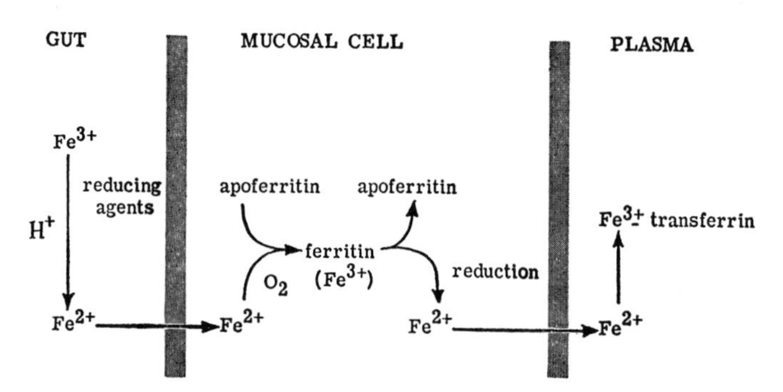

Fig. 3. The mucosal block theory of iron absorption. Iron taken up by the mucosal cell is converted to ferritin: when the ferritin becomes physiologically saturated with iron, no more is taken up by the cell until iron is released from ferritin and transferred to the plasma

advanced (*7*). Thus it was proposed that there was a mucosal acceptor for iron which, when physiologically saturated prevented further entry of iron through the mucosa into the body (Fig. 3). The suggestion (*7*) that this acceptor for iron might be apoferritin seemed to be confirmed by an experiment in which ferritin was crystallised from duodenal mucosa following administration of a large dose of iron to guinea pigs (*8*). The mucosal block theory, outlined in Fig. 3, postulates a major role for ferritin in the absorption of dietary iron. Dietary iron, after reduction to Fe^{2+} in the upper part of the intestine, is absorbed by the mucosal cells and transferred to ferritin. The requirements of plasma iron are then met from this pool of mucosal ferritin, the transfer of iron requiring reduction and oxidation. Once the mucosal ferritin is saturated no further iron is taken up.

This hypothesis has been criticised in some respects (*9*) on a number of grounds that need not concern us here. Whether in fact ferritin does serve to regulate the uptake of dietary iron, whether all the iron absorbed from the diet passes through ferritin, or whether in the initial rapid phase of iron absorption from the gut iron is transferred through the mucosal cell without entering ferritin, remains to be fully clarified. It does seem clear from experiments on dogs and man that ingestion of an adequate amount of iron prevents the absorption of any further iron fed for some time after (*10, 11*). And there can be no doubt that apoferritin is produced by mucosal cells following iron administration and takes up large amounts of iron, forming ferritin as *Granick* first showed (*8*). Some of this ferritin iron is probably slowly reduced to ferrous iron and transferred to the plasma over the course of the life span of the mucosal cell; what remains in the cell at the end of its 7—10 days of life is certainly lost when the cell is sloughed (*12*). At the very least then, it seems that ferritin is a repository for quite a large amount of the iron entering the mucosal cell: whether it is the means of getting that iron from the lumen of the intestine to the plasma remains an open question. Ferritin has been implicated in two further biological activities. On intravenous injection it inhibits the constrictor response to topical adrenaline of the precapillary sphincter and mesoarterioles in the mesenteric capillary bed of the rat (*13*) and it inhibits the flow of urine in hydrated dogs and rabbits by stimulation of the neurohypophysis to secretion of antidiuretic hormone (*14*). Thus, it is both a vasodepressor (and as such is released into the blood from liver under conditions of experimental shock) and an antidiuretic.

III. Distribution and Isolation

A. Distribution of Ferritin

The presence of ferritin in the spleen of dog, cat and jackal was established by *Kuhn et al.* (*15*); they were unable to detect it in guinea pig, rabbit or whale. *Granick* (*16*) then presented a comprehensive survey in which he showed that ferritin was present in liver, spleen, bone marrow and testes of horse, in human liver, spleen, bone marrow and in liver and spleen of dog, guinea pig, mouse, pig and rat. Dog and cat kidney as well as rabbit spleen, and testes of mouse, pig and rabbit were also shown to contain ferritin. No ferritin was found at that time in cattle, sheep, deer, chicken and bullfrogs, although as we will see below many of these species have since been shown to contain ferritin. The presence of ferritin in human cells (*17*) and in cells of the human lines Hela, KB and HEP-2 in culture (*17, 18*) raised the interesting possibility that ferritin might be rather widespread in its distribution within different organs and tissues. Indeed a recent paper reports on the demonstration by an immunological procedure of ferritin in liver, spleen, bone marrow, heart, lung, kidney, pancreas, thyroid, adrenal, brain, intestine and placenta of human (*19*). The presence of ferritin in serum had been established by its identification as the active component of vasopressor material (*13*). The development of the rat mesoappendix assay (*20*) enabled detection of ferritin in plasma of patients with diseases that were characterised by shock (*21*), hypertension (*21*) or oedema (*22, 23*). Using a quantitative immunoprecipitin method ferritinemia was demonstrated in patients with hepatocellular necrosis (*24—26*) and severe haemolytic anaemia. Indeed in patients with high serum levels of ferritin, the protein could be detected in unconcentrated urine specimens (*26*), but could not be detected in normal healthy subjects (*26*). More recently a much more sensitive radioimmunoassay has been developed for ferritin (*27, 28*) and this has enabled the unequivocal demonstration of low levels of ferritin in the plasma of normal human subjects. Moreover, using this method the levels of ferritin in the plasma of patients suffering from a variety of disorders of iron metabolism has been shown to closely reflect the state of body iron stores (*27, 28*). Thus levels of 35.6 ng/ml were found in normal subjects, 5.4 ng/ml in iron deficiency anaemias and 1700 ng/ml in iron overload (*27*). It seems likely that the presence of ferritin in serum reflects leakage from tissue stores. However, the possibility that serum ferritin has some function (other than the vasopressor activity referred to earlier) has not been explored.

As pointed out earlier (Section II) ferritin can be isolated from the mucosa of guinea pigs (*8*). It is also found in rat mucosa where a number of groups have concerned themselves with its possible role in iron metabolism (see Section II and X). Although it has not been isolated, reports have appeared concerning ferritin particles in rat macrophages and associated mast cells (*29*) as well as in guinea pig erythroblasts (*30*).

It has been observed that the content of liver ferritin in adult female rats is more than twice that in male rats (*31*). The level of ferritin in both sexes is low until about 7 weeks after birth: from then on the levels in both increase, but the level in females rises about twice as rapidly as that in males (*32*). An analysis of the distribution of ferritin in liver cells has shown that it is almost exclusively confined to hepatocytes whereas the Kupfer cells seem to be the major site of haemosiderin deposition (*33*).

Ferritin is not however confined exclusively to vertebrates nor even to mammalia. It has been isolated from dolphin and tuna fish (*34, 35*), though not to the author's knowledge as yet from whales, and its occurrence has been confirmed in both eggs and early embryos of the frog (Rana pipiens) (*36*), in snail hepatopancreas and ovotestes (*37*), in octopus vulgaris (*38*) and in invertebrates — worms (*39*) and molluscs (*40*). Following its original observations in plants by *Hyde et al.* (*41*) its presence in plants, where it seems to be extensively present in chloroplasts, has been widely reported — pea embryos, willow, Xanthium, Soyghum, to name but a few (*41—45*). Phytoferritin is similar to mammalian ferritin in its electron microscopic appearance, iron content and molecular weight and is principally found in the plastids, cotyledons and shoots of young plants and in the chloroplasts of angiosperms. As the chloroplasts mature they lose their ferritin: it is presumed that this storage iron serves as a source of iron for various haem and non-haem iron-containing enzymes required in the further development of the plant tissue. Phytoferritin is usually found within the plastids packed in a close packed semi-crystalline array, which has on occasions been confused with virus particles (*46*). Such assemblies of ferritin are seen only in tissues of mammals with a considerable degree of iron overload. Not only are there a selection of phytoferritins reported in the literature; since its discovery in a fungus (*47*) and its isolation from the fungus *Phycomyces* (*48*), a recent report of a mycoferritin in the soil fungus *Mortiarrella alpina* (*49*) together with its isolation extends the known distribution of ferritin to the borders of the microbial world.

We may conclude that the distribution of ferritin throughout Nature is quite widespread. Some preliminary comparisons between these different ferritins will be presented in Section V.

B. Isolation

The original isolation procedure for horse spleen ferritin developed by *Laufberger* (*4*) took advantage of the stability of the protein to heating at 80 °C, its precipitation by half saturation with ammonium sulphate and its ready crystallisation from 5% cadmium sulphate. Further small modifications in the method were introduced by a number of other workers (*13, 15, 50*), but the method itself remained unchanged in its basic essentials. Omission of the heat treatment step (*51*) made no difference in the purity of ferritin isolated from horse spleen. Although this method is ideally suited for horse spleen it has a number of disadvantages when used for ferritin isolation from other mammalian tissues. Despite the heat treatment and ammonium sulphate steps other proteins can be carried through the isolation procedure, especially if they become adsorbed to ferritin. Thus, for example haemoglobin when present in large amounts, is a major contaminant of ferritin. It also seems to be the case that the lower the level of ferritin in a tissue, the more likely it is that other proteins will be isolated along with the ferritin. Yet another difficulty associated with the original procedure is that not all other ferritins crystallise so well, or in such good yield, as that from horse spleen. The ferritins of human liver and spleen, as was first shown by *Granick* (*16*) form semiamorphous structures with rounded edges, which cannot really be called crystals.

A number of modifications to the original isolation procedure have been suggested. DEAE cellulose was used in the purification of rat liver ferritin (*52*), but the product was only 30% pure. *Drysdale* and *Ramsay* (*53*) used CM cellulose to separate ferritin from other iron-containing proteins, but without achieving satisfactory purification. On account of the large size of ferritin gel filtration on columns of Sephadex G-200 (*54*) or more recently of Sepharose 4B (*55*) has been used. A convenient method for small scale isolation of ferritin is to utilize its high density and to isolate the protein by repeated high speed centrifugation (*56*). This technique can also be incorporated into large scale isolation methods. A more recent method utilising polyacrylamide gel electrophoresis has been described for small scale isolation (*32*). Methods for isolation of ferritins from a number of mammalian tissues which we have found useful (*57*) are summarized in scheme 1: the scheme also incorporates a number of other methods which have been referred to above.

For the isolation of ferritin from other organisms a number of quite different problems are encountered. Thus in the preparation of ferritin from rabbit liver, dolphin and tuna spleen (*34, 35, 58*) acetone precipitation was used to purify the ferritin after ammonium sulphate treatment. In the isolation of phytoferritin and Phycomyces ferritin the problems of breaking open the cells must be satisfactory resolved. However, in Phycomyces, once that has been done extraction of the cell contents with butanol/water results in precipitation of most of the protein leaving almost pure ferritin in the aqueous phase (*48*). Whether this method

Scheme 1. Isolation procedures for ferritin (reproduced from (*57*))

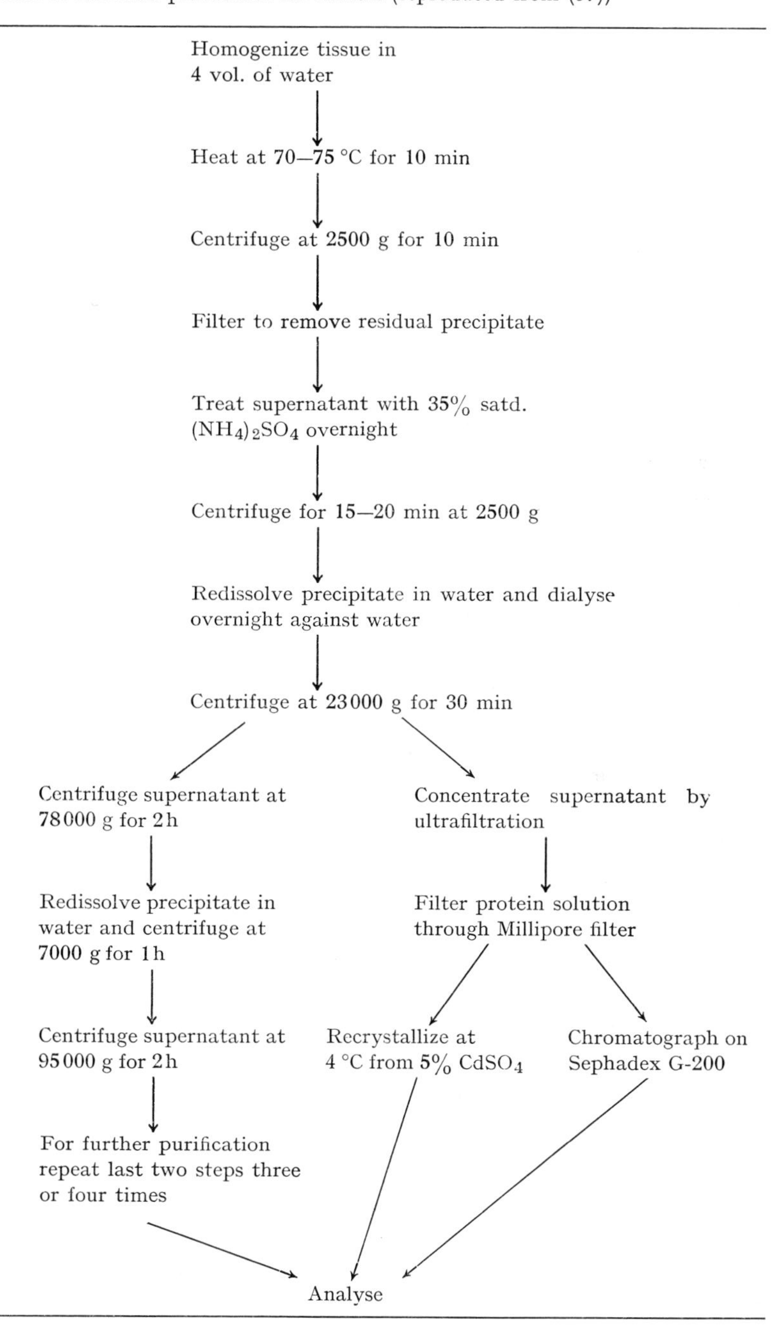

could be utilized for the isolation of other ferritins remains to be established. It is clear that horse spleen ferritin can be partitioned between butanol and water with no loss of protein at the interface (*R. R. Crichton*, unpublished observation).

C. Preparation of Apoferritin

Apoferritin can be prepared from ferritin either by chemical reduction of the ferric iron to ferrous, or alternatively by centrifugal methods. The method which was originally described (*59*) namely reduction in sodium dithionite containing α,α'-bipyridyl at *p*H 4.6 followed by removal of Fe^{2+} by dialysis against water has one great disadvantage, namely that autoxidation of dithionite can occur giving colloidal sulphur which binds to the apoferritin and renders it useless for any subsequent investigations (*60*, *61*). This problem can be avoided by carrying out the preparation in the cold under nitrogen using a *p*H of 5.0—5.2, adding dithionite (7%) to the ferritin in a closed vessel and then precipitating the protein with ammonium sulphate (*60*). This latter procedure must be repeated several times to obtain a truly iron-free preparation of apoferritin. These workers were doubtful of whether the presence of bipyridyl had any value. *Suran* and *Tarver* (*51*), like *Behrens* and *Taubert* (*60*) omited bipyridyl, and removed the dithionite by gel filtration on a column of Sephadex G-25. They also used thioglycollate at *p*H 8.6 and noted that some apoferritin was produced. Thioglycollate (1% v/v) in acetate buffer at *p*H 5.0—5.5 has been used in our laboratory for some years for apoferritin preparation and has the great advantage that not only does the thioglycollate chelate the ferrous iron produced, but there is no problem of autoxidation and the time required to remove all of the iron is typically less than 6 hr. Instead of dialysis, which requires several days to remove all of the thioglycollate, we routinely apply the apoferritin solution directly to a column of Sephadex G-25 in the buffer in which the protein is subsequently required, and can isolate the protein free of iron, thioglycollate and acetate buffer inside an hour or two. This effectively reduces the time (and labour) required for apoferritin preparations from 3—4 days to about 7—8 hr (*K. Wetz* and *R. R. Crichton*, unpublished work). Apoferritin can also be prepared using a number of other reducing agents such as ascorbate (see Section VII).

Native apoferritin can be prepared by a number of centrifugal methods. Density gradient centrifugation (*62*) in isopycnic gradients of CsCl or in sucrose have been used to fractionate native ferritin according to its iron content. The native apoferritin sediments with a buoyant density of 1.23, whilst full ferritin has a density in excess of 1.8. Typically using an SW 27 rotor 100 hr are required at 25.000 rpm with a CsCl solution of density 1.30—1.40. Shorter times can be obtained with preformed sucrose gradients in the SW 27 rotor, or by use of isopycnic CsCl gradients in the SW 65 rotor for 24 hr at 50.000 rpm (*R. R. Crichton*, unpublished work). Native apoferritin which constitutes about 20—25% of unfractionated ferritin (*63*) can also be prepared by differential centrifugation of ferritin solutions (*63*) and a fractionation according to iron content by ammonium sulphate precipitation has been described (*52, 64*) in which the iron rich fractions come down first.

IV. The Iron Core

A. Shape, Size and Composition

Ferritin consists of a shell of protein subunits surrounding a core of ferric hydroxyphosphate. For some time the inorganic component of ferritin was thought to be attached to the surface of the protein (*65*). However, the demonstration that ferritin and apoferritin have the same electrophoretic mobilities and are precipitated equally well with horse ferritin antibody (*66*) together with electron microscopic evidence (*67*) clearly showed that the iron is concentrated in the middle of the apoferritin protein shell. The iron micelle has a diameter of 70—75 Å whilst the protein shell has a diameter of the order of 120 Å (*67—70, 62*). Somewhat lower values are found in dried preparations (*71, 72*). The micelle contains ferric iron, predominantly as (FeO.OH) but also with some 1—1.5% phosphate (*4, 50*) and it seems that the iron: phosphorus ratio is constant for ferritins of different iron content (*52*). The composition $(FeO.OH)_8.(FeO.OPO_3H_2)$ has been suggested (*73, 74*). The percentage of iron in the micelle on this basis is 57%.

There would appear to be some discrepancy in the values calculated for the number of iron atoms/molecule of full ferritin *i.e.* the maximum iron content that can be obtained. *Fischbach* and *Anderegg* (*62*) isolated full ferritin by density gradient centrifugation and determined its molecular weight to be 900,000 by small angle X-ray scattering. They further showed that the full ferritin fraction had an iron: protein ratio of 0.52, and hence assuming that the iron was 57% of the micelle, and the molecular weight of the protein was 465,000 (*63*), they arrived at a molecular weight of 418,000 for the micelle and concluded that it contained 4,300 iron atoms. This would appear to agree well with the value that can be derived from molecular weight determination of full ferritin by low angle X-ray scattering (*62*) referred to above, and to the value calculated by *Harrison* (*75*) from the electron microscopic results of *van Bruggen* (*76*) of 860,000. Assuming the molecular weight of apoferritin to be 440,000 (as discussed in Section V. A below) these lead to values for the micellar molecular weight of 460,000 and 420,000 respectively; using the composition reported above, this corresponds to 4,700 and 4,300 iron atoms/molecules respectively. Further support for this value comes from calculations of the maximum content of (FeO.OH) that could be contained inside a sphere of diameter 74 Å. The total volume is 160,000 Å^3, and if one assumes (*77*) that the volume occupied by an (FeO.OH) molecule is 35—43 Å^3 (value based on packing volume in mineral iron hydroxides) and that packing is such that little or no water is present, then one could accomodate 3,800—4,500 iron atoms per ferritin molecule.

The data of *van Bruggen* (*76*) leads however to another and rather different conclusion: the iron: protein ratio of full ferritin can be calculated from his data as being 0.46, which for a protein of molecular weight

Table 2. *Iron and phosphorous contents of ferritin from different species*

Species and tissue	mg Fe:mg protein	Fe atoms/ molecule	Molar ratio P:Fe
Horse, spleen unfractionated			
(*4*)	0.32	2,550	0.084
(*50*)	0.35	2,730	0.130
(*52*)	0.30	2,370	0.117
full (*62*)[a]	0.52 0.53	4,120 4,200	
non crystalline fraction (*49*)	0.35	2,800	0.139
Rat, liver (*55*)	0.38	3,000	
kidney (*55*)	0.30	2,350	
spleen (*55*)	0.41	3,240	
heart (*55*)	0.40	3,170	
thigh muscle (*55*)	0.38	3,000	
hepatomas[b] (*55*)	0.10—0.44	800—3,500	
Rabbit, liver (*58*)	0.19	1,520	0.148
Dolphin, spleen (*34*)	0.21	1,660	0.061
Tuna fish (*35*)	0.16	1,160	0.058
Rana pipiens, egg (*36*)	0.13	1.030	
Phycomyces unfractionated (*48*)	0.10	800	
from CsCl gradient (*48*)	0.20	1,600	
Mortiarella alpina (*49*)	0.20	1,600	

[a] *R. R. Crichton*, unpublished observations.
[b] The iron content is a function of the growth rate of the tumour.

440,000 corresponds to an iron content of 3,600 atoms/molecule. And when one searches the literature the highest value for iron:protein ratio one finds (apart from that of *Fischbach* and *Anderegg* (*62*) for full ferritin) is 0.40 (*52, 64*), which corresponds to 3,200 atoms/molecule. We have determined the iron:protein ratio of a full ferritin fraction prepared according to *Fischbach* and *Anderegg* (*62*), and find a value of 0.53. This corresponds to an iron content of 4.200 atoms Fe/molecule of apoferritin.

Unfractionated horse spleen ferritin typically contains 20% of the dry weight as Fe. The iron:protein ratio is about 0.32, corresponding to about 2,500 iron atoms/molecule. In this calculation I have assumed that Fe is 57% of the micelle; hence the micelle contributes 35% of the dry weight, the protein 65%, and the Fe:protein ratio is 0.32. For comparison, on this basis full ferritin with 4,300 atoms/molecule would contain 27% of its dry weight as Fe. In Table 2 are collected together iron:protein ratios for ferritins from various species. The iron content of mammalian ferritins seems to be consistently higher than that found in any of the other species. In rat, the amount of iron/molecule of protein is fairly constant in different organs with the exception of kidney, where the value is somewhat lower, and in liver tumours (hepatomas) where the iron content is a function of tumour growth rate (*78*).

B. Electron Microscopic Studies

The iron oxide core of ferritin is of sufficient density to be seen in the electron microscope without staining. The cores do not appear to be homogenous, and it was suggested originally that the iron was distributed in four electron dense regions (*67*). Subsequently it was proposed that the core patterns seen in electron micrographs can be accounted for by assuming that they represent projections in various orientations of molecules containing six electron dense regions arranged either at the vertices of an octahedron (*79, 80*) or at the vertices of a trilateral prism (*76*). Although the "classical" maximum contrast electron micrographs of ferritin often show such an arrangement of four dense regions, often in a square or diamond formation and this arrangements is often thought to be typical of ferritin, the advent of electron microscopes of higher resolving power has led to the conclusion that such micrographs result from artifacts (*77, 81*). *Haggis* summarized the problem by pointing out (*77*) that such "classical" pictures of the ferritin core can only be obtained by gross underfocusing, and concluded that no particular number of electron dense groupings is common. More recently the problem has been reinvestigated by a high-resolution through focus study of ferritin (*82*). At high resolution (near 5 Å) there seems to be no evidence for substructure in the cores. There remains however, some controversy from electron microscopic studies as to whether the majority of cores contain only one crystallite, or whether more than one crystallite can grow within one molecule at a number of different sites (*83*, *84*). *Haggis* (*77*) reported variation in the appearance of the cores in different molecules. In some molecules the core was of uniform density and polyhedral, as has been reported for the in focus micrographs of the ferritin core made by *H. E. Huxley* (*75*). In others a number of different core configurations were seen. The shapes seen by *Haggis* approximated to irregular triangles, diamonds, squares, pentagons and hexagons. In other molecules two or more dense regions were seen divided from each

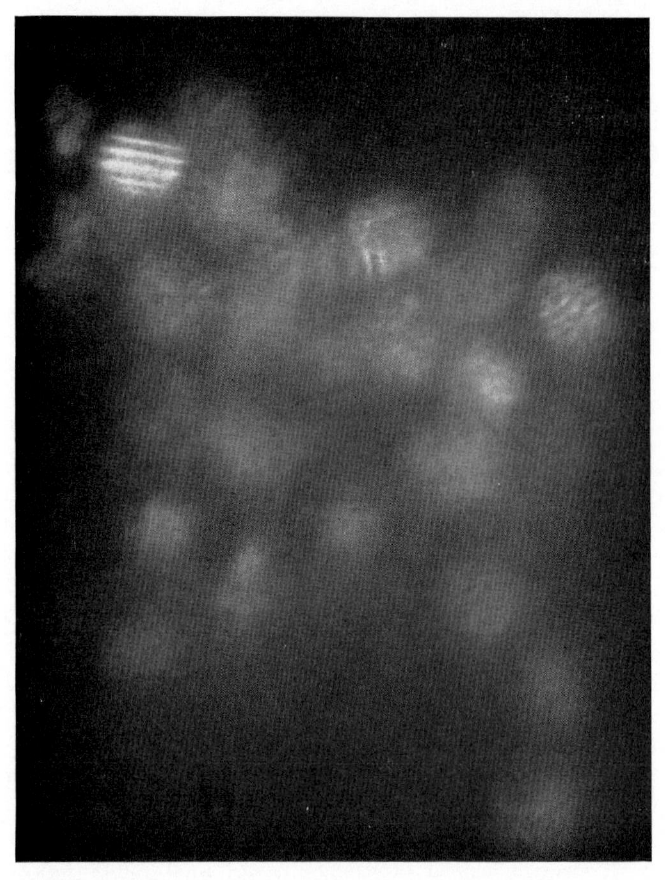

Plate 1. Dark field (hollow cone illumination) electron micrograph of the ferritin mineral core (*A. K. Kleinschmidt* and *A. Tanaka*, in preparation) at a magnification of 230,000, enlarged ×15. The cores appear with line spacings of around 10 Å, mostly unidirectional

other (even in true focus pictures) by narrow cleavage lines, or else cleavage lines which ran partly across the core to form U shapes. However, as discussed below evidence has been inferred from X-ray diffraction data that the cores have an ordered crystalline structure (*85*). *Haggis* (*77*) attempted to obtain an electron diffraction pattern from ferritin and succeeded in obtaining a ring pattern with strong spacings at 2.55 Å and 1.47 Å and weaker bands at a number of other spacings between 2.0 Å and 1.0 Å. Bands above 3.5 Å would not have been detected. The ferritin spacings could not be readily identified with those of any of the common ferric oxides or hydroxides.

Two more recent studies on the core of ferritin by electron microscopy should be mentioned here. *Massover* (*86*) using the ultrahigh voltage electron microscope at Toulouse has visualised ferritin cores with a low signal to noise ratio in brightfield images at both 1 MeV and 2 MeV. In dark field images, using exposure times of only a few seconds and beam currents of less than 1 μAmp. the cores are seen to have substantial elements appearing as smaller and larger granules, rod-like particles and varying amounts of homogenous material. These substructures have been visualised in one and the same core at 3 MeV under conditions of precise focus as confirmed by optical diffraction (*87*). Dark field micrographs of the mineral core have also been obtained (Plate 1) by *Drs. Kleinschmidt* and *A. Tanako* in New York and in these high resolution studies line spacings of close to 10 Å (between 9.6 and 10.3 Å) have been measured by optical diffractrometry. The lines (Plate 1) are mostly unidirectional. It is not yet clear if these are actually line spacings of microcrystals, or an indication of microcrystalline sheets. The multiple configuration that can now be seen in dark field microscopy (of Plate 1) suggests that model building will go through a 10 Å "unit structure" (*A. K. Kleinschmidt,* personal communication).

C. X-ray Diffraction Studies

Low angle X-ray scattering from ferritin solutions (*62*) shows that to a resolution of 20 Å the core approximates to a single uniform density object of average diameter 74 Å. An examination of ferritin cores under a variety of conditions (*85*) both wet and dry, by X-ray diffraction and electron diffraction resulted in essentially the same diffraction pattern both in intensity and in line breadth in each case. The spacings and approximate intensities correspond to those found by electron diffraction (section B, above): Thus the strongest intensities are at 2.5 Å (approx.), 2.23 Å and 1.47 Å.

It is possible to separate the cores from apoferritin by treatment with 1 N NaOH for several minutes at room temperature and this treatment removes about 80% of the original phosphate from the cores (*50*). Such cores gave the same diffraction pattern as intact ferritin (*85*). Thus the phosphate does not seem to contribute significantly to the diffraction pattern nor does it appear to be an essential structural component of the core. Confirmation of this view comes from studies on ferritins that were reconstituted from apoferritin by three different procedures (Section VI) in the absence of phosphate: these gave electron diffraction patterns that were similar to those of native ferritin although the crystallites were smaller in some cases (*85*).

Using single large wet crystals of ferritin, powder patterns typical of ferritin core patterns could be obtained. From a stationary crystal both sharp spots, from the protein crystal lattice, and very weak diffuse rings, from the iron cores can be seen simultanously. There seems to be no indication that the diffraction pattern of the core has any preferred orientation with respect to the crystal lattice of the protein, supporting earlier work on single crystals of ferritin and apoferritin (*88*) which had led to the conclusion that the atomic arrangement of the cores was not specifically related to the structure of the protein (see below). The X-ray diffraction patterns of ferritin cores give broad lines which are relatively weak, and show some similarity to that of δ-ferric oxyhydroxide. This material has a hexagonal cell with $a = 2.94$ Å and $c = 4.51$ Å (*89, 90*). Strong lines are present in the ferritin pattern corresponding to the (1010) and (1120) reflections of this cell, but other lines do not fit this cell. The ferritin pattern can however be indexed on a hexagonal cell with $a = 2.94$ Å and $c = 9.40$ Å. The spacings calculated for this cell agree quite well with those observed and it is interesting that in the more recent EM studies of Kleinschmidt and Tanaka referred to above line spacings of about 10 Å were observed.

The oxyhydroxides of iron consist essentially of close packed layers of oxygen atoms with iron atoms situated in the interstitial holes. Thus the iron atoms, depending on whether they are surrounded by 6 or 4 oxygen atoms have either octahedral or tetrahedral coordination. A crystalline model of the ferritin core has been proposed by *Harrison et al.* (*85*) which fits the X-ray and electron diffraction data involving close packed oxygen layers with iron randomly distributed among the eight tetrahedral and four octahedral sites in the unit cell.

A diffraction pattern somewhat similar to that of ferritin cores was found in a hydrolysate of ferric nitrate, and a different structure was proposed (*91*) which has however a closely related unit cell to that of *Harrison et al.* (*85*).

Another X-ray and electron diffraction study of ferritin (*92*) also led to the conclusion that the structure was not similar to that of natural oxides or hydroxides of iron. The diffraction pattern could also be indexed on a hexagonal cell with $a = 11.79$ Å and $c = 9.90$ Å. An examination of the intensity of the diffraction lines did not permit any conclusion to be reached on the position of the iron atoms between the layers of oxygen atoms and it was tentatively concluded that the Fe^{3+} was distributed between tetrahedral and octahedral sites. The conclusion is reached in a subsequent communication (*93*) that there are 32 molecules of FeO.OH per unit cell, and that for full ferritin containing 4,000 iron atoms there are 140 unit cells per core, corresponding to a volume of 170.000 Å^3. Thus the iron core occupies almost all of the cavity inside the apoferritin protein shell (about 160.000 Å^3).

D. Magnetic and Spectroscopic Properties

The magnetic susceptibility of ferritin was determined by *Michaelis et al.* (*73*) in the temperature range of 275°—301 °K and a value of 3.81 B.M. for the μ_{eff} was found. *Bayer* and *Hauser* (*94*) later reported an effective magnetic moment of 3.84 B.M. and noted a marked deviation of the temperature dependence of the magnetic susceptibility from the Curie laws at 239 °K and 90 °K. Measurements in the range 4.2—300 °K showed that ferritin is antiferromagnetic with a Néel temperature of 20° ± 3 °K (*95*). The theory of *Smart* (*96*) for antiferromagnetic exchange between iron atoms clustered in groups of two gave the best agreement between theoretical and experimental values. The antiferromagnetic exchange constant was $J/k = -4.8$ (°K) and the magnetic moment of $\mu_{eff} = 3.85$ B.M. The ground state of free Fe^{3+} ions is ^6S with $S = 5/2$. However, the experimentally derived magnetic moment of 3.85 B.M. is almost exactly what would be expected from spin quantum number $S = 3/2$. If we are dealing in ferritin with a normal Fe^{3+}-complex $S = 3/2$ would be theoretically impossible. Accordingly, *Schoffa* sought to explain the reduced magnetic moment for ferritin iron by the transfer of 2 electrons from ferritin atoms to Fe^{3+} on the cation-anion-cation super-exchange and further suggested that antiferromagnetism and super-exchange were possibly caused on the cubic magnetic structure of iron-oxygen micelles in ferritin.

Blaise et al. (*97*) made a thorough field dependence study and found a marked dependence of the magnetic susceptibility of ferritin on field strength. At very high fields they found a value of $\mu_{eff} = 5.08$ B.M. (*97*). The Curie point situated at −200 °K proves the existence of a strong negative coupling between the atoms. The measurements of the Mössbauer effect showed the progressive establishment of a magnetic order below 30 °K. The Mössbauer spectrum at liquid nitrogen temperatures and above (Fig. 4) has an isomer shift and quadropole splitting

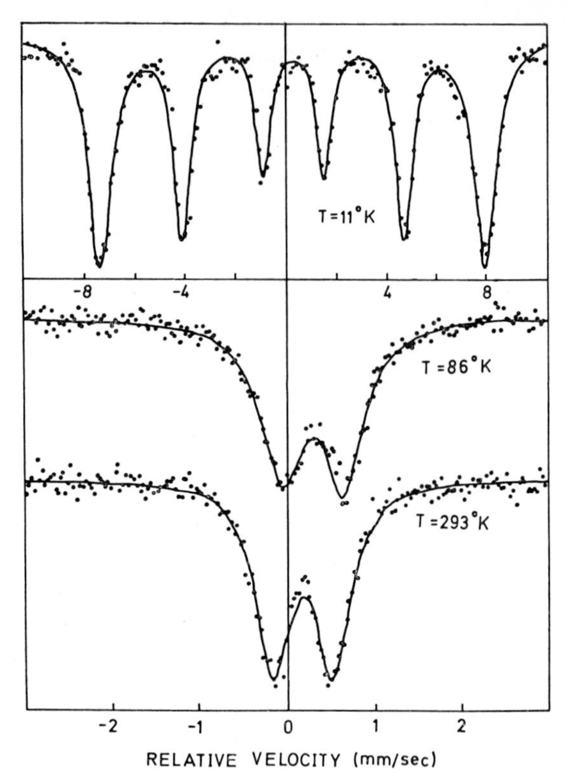

Fig. 4. Mössbauer absorption spectra obtained using a ^{57}Co in Pd source at room temperature and ferritin as absorber at 11 °K, 86 °K and 293 °K (reproduced from (*103*))

of 0.47—0.50 and 0.6—0.74 respectively (*97, 98*). At low temperatures, a six line hyperfine spectrum is observed (see Fig. 4), characteristic of magnetic ordering. As the temperature is raised, so the hyperfine spectrum is replaced by the paramagnetic quadropole split doublet spectrum. This supermagnetism was explained by *Blaise et al.* (*97*) as being due to an antiferromagnetism of fine particles (superantiferromagnetism) analogous to that seen in certain metal oxides reduced to fine powders. Girardet has summarized the quantitative results of these experiments (*99*): the magnetic moment of a grain is of the order of 10^{-18} μ.e.m., which, for 4000 ferric ions in the state $^6S_{5/2}$ in a molecule of full ferritin, is in accord with the theory of Néel of the antiferromagnetism of fine grains.

A study of the ESR and Mössbauer spectra of iron in ferritin (*100*) over the range 4.2—295 °K confirms that the temperature dependence of the spectra can be explained by the superparamagnetic properties of the micelles. A weak ESR signal is also observed near $g = 4$ at 77 °K and above, and this has been attributed to a small number of Fe^{3+} which are not bound to the micelle. Distinct differences are reported in the Mössbauer spectra of fractions of varying iron content which can be explained by differences in the size and shape of the micelles.

The similarity between the Mössbauer parameters of ferritin and a synthetic polymer derived from hydrolysed ferric nitrate have been pointed out previously (*101*): the polymer has a similar room temperature Mössbauer spectrum with isomer shift and quadropole splitting of 0.48 and 0.67 mm/sec. respectively. At very low temperatures the spectrum of the polymer is also closely similar to ferritin.

V. Structural Studies on Apoferritin

A. Molecular Weight and Subunit-Structure

Apoferritin was shown to have a sedimentation constant S_{20w} of 17.65 (*63*). A more recent study (*102*) has shown that the S value is markedly dependent on protein concentration, and a value of 17.12 S was obtained by extrapolation to zero concentration (Fig. 5). A comparison of the

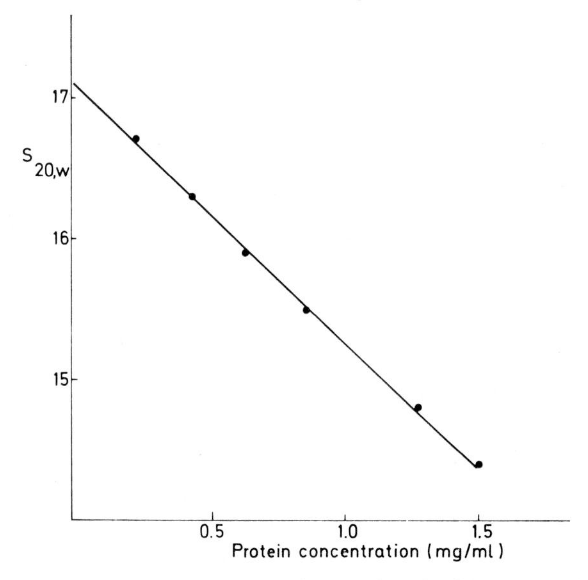

Fig. 5. Dependence of sedimentation coefficient of apoferritin on protein concentration

Table 3. *Molecular weights and sedimentation coefficients of apoferritins*

	Sedimentation Coefficient	Oligomer	Molecular Weight Subunit
Horse spleen	17.12S (*102*)	443,000 (*102*) 430,000 (*103*)	18,300—18,800 (*107*)
	17.6S (*63*)	465,000[a] (*63*) 480,000[a] (*88*)	25,000—27,000[b] (*109*)
		462,000 (*69*) 440,000 — 465,000 (*105*)	18,000—19,000 (*105*)
Horse liver	16.9S (*57*)		18,400—18,800 (*57*)
Human liver	17.3S (*57*)		18,000—18,700 (*57*)
Human spleen	16.8 S (*57*)		18,300—18,500 (*57*)
Rat liver	~17S[c]		19,000[c]
Guinea pig	17.7S (*113*)		
Phycomyces	18S (*47*) ~17S[e]		25,000[d] (*47*) 18,000[e]
Mortiarella alpina			19,300 (*48*)

The numbers in parenthesis refer to original references.

[a] The original values quoted here used a value of 0.747 ser $\bar{v}$. When the value of 0.731 (from amino acid analysis) is used these reduce respectively to 440,000 and 454,000.

[b] It it is assumed that 1.4 g of detergent are bond/g of protein (see text) this value reduces to 18,000—19,000.

[c] *R. R. Crichton* and *H. Hübers*, unpublished observations.

[d] A value of 24,000 was assumed for the subunit molecular weight of horse spleen apoferritin.

[e] *R. R. Crichton*, in preparation.

published sedimentation coefficients (Table 3) shows that for the species examined so far, they are closely similar.

The molecular weight of apoferritin was first determined by sedimentation diffusion (*63*) in 1944 and a value of 465,000 reported. Subsequent investigations have employed X-ray diffraction (*88*), low angle X-ray scattering (*69*), light scattering (*104*), approach to sedimentation equilibrium (*105*) (the meniscus depletion technique of Yphantis) and true sedimentation equilibrium (*102*). The values are given in Table 3, and range from 430,000—480,000. However, as has been pointed out (*106*) the value of 0.747 used in two of these determinations (*63, 88*) for $\bar{v}$, the partial specific volume of the protein, is rather higher than that which can be calculated from the amino acid composition of the protein (*61, 107*). When this latter value of 0.732 is used for $\bar{v}$ in the previous calculations values of 440,000 and 454,000 are obtained. Thus we may assume that the molecular weight of apoferritin is about 440,000 daltons.

That apoferritin consists of subunits, and not of one long polypeptide chain was conclusively demonstrated by *Harrison* and *Hoffmann* (*108, 109*). From chemical studies (discussed in Section V. B below) a value of 23,000—24,000 for the subunit molecular weight was calculated (*109, 110, 111*). Degradation with dilute alkali led to partial splitting of the protein into 2.1 S component (*109*). Treatment of native apoferritin with sodium dodecyl sulphate, urea or a number of other reagents did not result in any splitting into subunits. However, with frozen, lyophilised or alcohol dehydrated apoferritin, treatment with SDS resulted in complete splitting of the molecule into subunits. All of the apoferritin was in the form of an SDS-protein complex which sedimented at 2.5 S at SDS-protein ratios higher than 1:3. The molecular weight was determined for the SDS-apoferritin complex from measurements of the sedimentation and diffusion coefficients and by the approach to equilibrium method and found to have a molecular weight of 38,000 to 41,000. After allowing for the SDS bound, by equilibrium dialysis, the protein subunit molecular weight was calculated to be 25,000—27,000. Thus it was concluded that apoferritin consisted of 20 subunits of molecular weight 23—25,000 (*109*).

A reinvestigation of the subunit molecular weight was carried out by *Crichton* and *Bryce* (*112*) using the method developed by *Schapiro* and *Maizel* (*114*) of electrophoresis in polyacrylamide gels in the presence of SDS and a value of 18,200 $\pm$ 1200 was found for apoferritin subunits. Since this was in such considerable disagreement with the previously published value, independent confirmation of the subunit molecular weight was sought by other methods. Molecular weight determinations on calibrated columns of agarose gel in 6M guanidine hydrochloride, by sedimentation equilibrium centrifugation of subunits dissociated in 6 M guanidine hydrochloride and of subunits dissociated by treatment with acetic acid followed by dialysis into dilute buffer at *p*H 3.0, all gave values of 18,300—18,800 (*107*). The conclusion seemed inescapable that the polypeptide molecular weight was lower than had been previously thought and that there were most probably 24 subunits per apoferritin molecule. Shortly after these results were published confirmation of the subunit moleculer weight by sedimentation equilibrium and gel filtration in 6M guanidine hydrochloride and electrophoresis in SDS-polyacrylamide gels (*105*) was published by *Björk* and *Fish* (Table 2). As we have pointed out (*107*) if one assumes, as has been recently demonstrated (*115*), that 1.4 g SDS is bound per g of protein then *Hofmann* and *Harrison's* data (*109*) can be recalculated to give a molecular weight of 18,100—19,600. It would seem most likely that that the value reported by these authors was too high on account of underestimation of the amount of SDS bound to the protein. It has been shown that the removal of SDS from proteins is not easily achieved by dialysis (*116*).

Comparative data for the subunit molecular weights of a number of other apoferritins are given in Table 3. The human apoferritins and the horse liver and rat liver ferritin all have subunit molecular weights and sedimentation coefficients which are closely similar to horse spleen apoferritin, and they certainly must have the same subunit structure. In fact, the same is true for the Phycomyces apoferritin: although the first report of its subunit molecular weight gave a value of 25,000 (*48*) this was based on the assumption that horse spleen apoferritin had a subunit molecular weight of 24,000. Redetermination of the subunit molecular weight in our laboratory gave 18,600. The subunit molecular weight for the apoferritin of the soil fungus *Mortiarella* is also similar to that of horse, indicating that most probably the subunit structure is the same in this species.

B. Structural Studies on Horse Spleen Apoferritin

Apoferritin from horse spleen has an isoelectric point of 4,4 (*52*) and is clearly an acidic protein. The results of a number of determinations of the amino acid composition are listed in Table 4. Those of *Mazur et al.* were carried out using microbiological methods (*52*), and the remainder (*51, 61, 107, 110, 117*) by ion exchange chromatography using the amino acid analyser. There is poor agreement between the earlier analysis and those carried out using ion-exchange chromatography. However, when these latter are compared, the agreement is in general quite good with a number of notable exceptions. The content of cysteine, of proline and of tryptophan differs, consistently higher values being reported from our laboratory than from that of *Harrison* and *Tarver*. The question of whether there are 3 cysteine residues or only 2, and of whether these are present as a disulphide bridge and a free sulphydyl group, or as 2 or 3 cysteine residues is amenable to analysis by chemical modification experiments. Thus, the reaction of apoferritin, unfolded in 6 M guanidine hydrochloride with a number of SH reagents such as iodoacetic acid and N-ethyl maleimide both with and without prior reduction in mercapto-ethanol can be used to answer these questions. Reaction of apoferritin with N-ethyl maleimide leads to the incorporation of only one residue of reagent per subunit. The same result is found with the other reagent. However, when the protein is reduced with mercaptoethanol and unfolded in guanidine under conditions where no subsequent oxidation of SH groups can occur, the number of SH groups/subunit increases to 2: it thus seems that there are 2 cysteine residues per subunit in apoferritin,

Table 4. *Amino acid composition of horse spleen apoferritin. These results are expressed in residues of each amino acid per 18,500 daltons. Where data for one or more amino acid are lacking, those have been allowed for in calculating the number of residues for the remainder*

	(52)	*(110)*	*(51)*	*(117)*	*(61)*	*(107)*
Cysteine	2.5	1.8	1.4	2.0	2.8	2.8
Aspartic acid + asparagine	9.4	16.8	16.7	17.4	17.0	17.3
Threonine	10.2	5.4	5.2	5.5	5.4	5.5
Serine	n.d.	8.2; 6.9	8.4	9.5	8.8	9.0
Glutamic acid + glutamine	21.4	22.3	23.2	24.6	23.4	23.9
Proline	2.5	2.8; 2.2	2.1	2.1	2.8	2.8
Glycine	8.4	9.7	8.9	9.9	9.7	9.9
Alanine	4.0	13.4	12.4	14.2	13.7	14.0
Valine	6.6	6.7	5.6	7.1	6.8	6.9
Methionine	2.4	2.9; 2.6	3.1	3.0	2.7	2.8
Isoleucine	1.9	2.9; 3.5	2.4	3.8	3.4	3.5
Leucine	29.4	22.0; 23.9	23.5	25.2	24.5	25.0
Tyrosine	5.2	5.2	4.5	5.2	4.9	5.0
Phenylalanine	6.8	6.8	6.3	7.3	7.2	7.3
Histidine	5.6	4.6; 5.0	5.0	5.9	5.7	5.8
Lysine	9.9	7.5	7.3	9.0	8.6	8.7
Arginine	11.0	8.2	8.6	9.8	9.3	9.5
Tryptophan	1.1	0.8	n.d.	n.d.	n.d.	2.1

of which both are present as free SH groups (*K. Wetz* and *R. R. Crichton*, in preparation).

The variation in the proline content cannot be readily explained and it seems likely that whether there are 2 or 3 prolines per subunit must await final sequence determination. It is interesting that in the earlier report of *Harrison* (*110*) a value of 2.8 was found However, in the more recent report the proline value is the mean of 14 determinations, and should probably be regarded as more representative (*117*).

The tryptophan value reported by *Harrison et al.* (*110*) is less than half that found by us (*107*). The estimation by *Harrison et al.* (*110*) was by modification (*118*) of the method of *Spies* and *Chambers* (*119*) using the *p*-dimethylaminobenzaldehyde reagent with samples of apoferritin which had first been denatured. *Bryce* and *Crichton* (*107*) determined tryptophan both by the spectrophotometric method of *Edelhoch* (*120*) in 6M guanidine hydrochloride and using 2-nitrophenylsulphenyl chloride, a reagent which is rather specific for tryptophan (*121*) and has been widely used in the determination of tryptophan. The values obtained by both methods were in good agreement. A further check on the tryptophan content can be obtained by using the equation of *Edelhoch* (*120*)

$$E_{280} = N_{trp}5690 + M_{tyr}1280 + L_{cys}120 \qquad (1)$$

It is clear that if there were only 1 tryptophan residue/subunit, the molar extinction coefficient at 280 nm (E_{280}) would be 12,450, whereas if there were two it would be 18,140. The published values for $E_{1\,cm}^{1\,mg/ml}$ at 280 nm are 0.86—0.97 (*75*), 0.90 (*122*) and 0.982 (*123*): these correspond to E_{280} values of 16,080—18,000, 16,700 and 18,110 respectively. It thus seems clear that there are in fact 2 tryptophan residue/subunit.

The value reported for lysine and arginine by *Harrison et al.* (*110*) are lower than found in the later analyses and may reflect the fact that their overall recovery was only 94%. In fact this is true of almost all of the other amino acids as well. If the content of lysine and arginine is 19 per subunit, we would anticipate that there must be 25—26 acidic residues per subunit (assuming that the *p*K of Asp and Glu residues is about 4.0) *Harrison et al.* (*110*) report an amide content of 17—18 moles per 18,500 daltons and *Suran* and *Tarver* 15 (*51*). Since there are 41 residues of Glx and Asx per subunit this would imply that there are 23—27 residues of Gl*u* and Asp, and hence that the total number of carboxyl groups is 24—28 per subunit, which agrees well with the number calculated above. We have in fact coupled all of the free carboxyl groups in the protein to glycinamide using a water soluble carbodiimide after completely unfolding the protein to a random coil in 6M guanidine hydrochloride, and find a total of 22 additional glycine residues per subunit by amino acid analysis of the product (*K. Wetz* and *R. R. Crichton*, in preparation). This also agrees well with the data presented above.

Endgroup determination on apoferritin gave no significant amount of N-terminal group, and carboxypeptidase A did not release any C-terminal amino acid (*110*). The N-terminal amino acid was shown to be acetylated (*124*) and a ninhydrin-negative, arginine containing peptide isolated from tryptic digests of apoferritin was shown to have the sequence N-acetyl-SER-SER-GLU-ILE-ARG (*124*). The presence of N-terminal N-acetyl SER was confirmed and a value of 19.6 moles of bound acetyl group per mole of apoferritin reported (*111*). Carboxypeptidase B released 18—19 moles of arginine and 15 moles of lysine per mole of apoferritin. The C-terminal peptide was isolated from a peptic digest of apoferritin and its sequence reported as being SER-GLU-GLY-ASN-ALA-LEU-LYS-ARG (*111*).

Tryptic digestion of apoferritin was found to give fewer peptides than would have been expected had it been a single polypeptide chain of molecular weight 480,000 (*108*) thus reinforcing the view that the protein was made up of subunits. A total of 19 peptides were found together with 8 more which occurred in trace amounts. Varying amounts of an insoluble "core" were found ranging from 20% to 50% and this appeared to contain predominantly acidic peptides. Arginine was found in 8 of the major peptides, histidine in 6, tyrosine in 5 and tryptophan in only one. It is interesting to note that these numbers reflect rather well the amino acid composition in Table 4, with the exception of tryptophan: one of the arginines must be in a ninhydrin-negative peptide, and so 9 of the 9—10 arginines are accounted for. The other tryptophan might well be in the core, or else both tryptophan residues are in the same peptide. We have also carried out trypsin digestion and separated the peptides both by ion-exchange chromatography and by fingerprinting (*61*). The number of peptides that we find is 25, which is rather more than would be expected on the basis of the lysine plus arginine content. However, we know that several of these are the product of incomplete tryptic digestion (i.e. they contain more than basic amino acid) and in one case of chymotryptic digestion (*M. D. P. Boyle* and *R. R. Crichton,* unpublished work). Sequence analysis is presently in progress on these tryptic peptides in the author's laboratory.

From the amino acid analysis (Table 4) it is clear that treatment of apoferritin with CNBr, which cleaves peptide bonds adjacent to methionine residues, should result in 4 peptides. In fact, when such cleavage was carried out we were able to isolate 5 peptides (*125—130*). The molecular weight of these peptides can be determined by gel chromatography in 6M guanidine hydrochloride on columns of agarose (*130*), and they were found to be I-9,500; II-8,400; III-6,400; IV-3,400; and V-1,000 respectively. One of these peptides, I, contains an intact methionine residue, which can be cleaved in better yield following reduction with dithiothreitol (*130*). The four unique CNBr peptides II-V together account for the amino acid composition of the protein (*K. G. Bitar* and *R. R. Crichton,* in preparation). Determination of the new end groups released after CNBr treatment of apoferritin established that there were equimolar amounts of gly, glu (or gln) and ile, and these are the end groups found in 3 of the peptides (*128*). The order of the CNBr fragments has recently been established (Fig. 6): I is a mixture of III and IV, II is C-terminal (since it contains no homoserine) and III is N-terminal (since it has no free N-terminal amino group) — the order of peptides is thus III-IV-V-II (*K. G. Bitar* and *R. R. Crichton,* in preparation). The sequence of the CNBr peptides, which have been isolated in large amounts in our laboratory, is proceeding steadily at present.

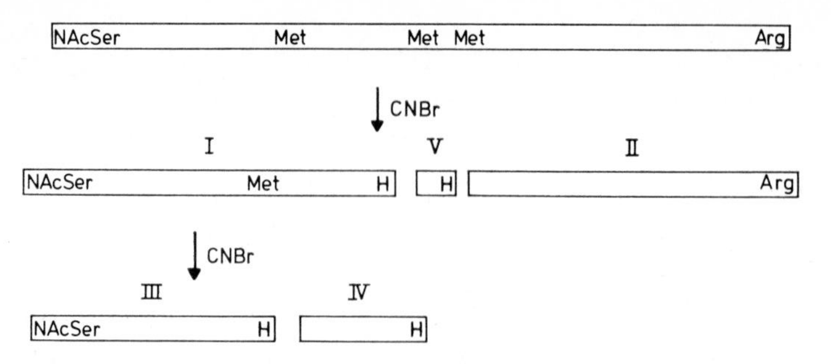

Fig. 6. Order of CNBr peptides in apoferritin: treatment of apoferritin with CNBr generates 3 peptides, one of which (I) can be further cleaved to yield the peptides III and IV. NAcSer — N-acetylserine; Met — methionine; H — homoserine

An analysis of apoferritin by isoelectric focussing (*131, 132*) led to the conclusion that there were several types of polypeptide chains present. Both ferritin and apoferritin were reported to give 7—9 bands in the *p*H range 4.1—4.6 which were grouped in three main components. As is clear from the discussion above there is no reason to anticipate that there is more than one polypeptide chain from the primary sequence studies, although at this stage the possibility that there is microheterogeneity such as is found for example in horse haemoglobins (*133*), cannot be excluded. Multiple bands on isoelectric focussing of ferritins have also been noted by *Lee* and *Richter* (*134*): it was found that when the same sample was refocussed, some of the bands were not consistenly present, and it was concluded that these minor bands represented ferritin-ampholyte complexes. We have applied isoelectric focussing to apoferritin from human and horse spleen and liver and found only one band in each case (*135*). We conclude (*135*) as do *Lee* and *Richter* (*134*) that the multiple bands seen on isoelectric focussing of apoferritin are an artifact of the method.

C. Comparative Studies of Apoferritin

Table 5 presents the amino acid compositions of a number of apoferritins together with their isoelectric point (taken from the literature) where available. There are a number of features whihc are of quite considerable interest. In the first place, although there are clear cut differences between the proteins of different species, a number of very considerable similarities exist. Thus, for all of the apoferritins for which data is available the content of non-polar amino acids (proline, glycine, alanine, valine, methionine, leucine, isoleucine, phenylalanine and tryptophan) remains constant at about 45%. What is even more remarkable is that the content of aspartic acid + asparagine and glutamic acid + glutamine for all species lies between 40—43 residues per subunit, whilst the content of basic amino acids lysine and arginine is also fairly constant at between 18—20 residues per subunit. The isoelectric point varies between 4.3—5.5, so that presumably the total number of glutamic acid and aspartic acid does vary from species to species. The content of lysine varies between 9 and 11 for all species except tuna, and the arginine content lies between 8 and 9.5 (probably 10 for horse) again with the exception of tuna, Tuna is quite unusual in that the lysine content is extremely high (12.5 residues) and the arginine is very low (6.2), and likewise the balance between Asp + Asn and Glu + Gln is also markedly different from that in all other species — 21 Asx and 19 Glx, compared with 17—19 Asx and 22—26 Glx in all other species. None the less the total Glx + Asx and lysine + arginine content remains within the limits even in tuna fish. When we consider potentially functional amino acids such as cysteine, histidine and tyrosine, there are some interesting variations. Thus if cysteine is required for the function of apoferritin in the Fe^{2+} oxidation (Section VII), guinea pig would seem an interesting species to examine, since it has only one cysteine. The histidine content never falls below 4 residues/subunit, so that if histidine is involved in the catalytic role of apoferritin (Section VII), it seems likely that any essential histidine or histidines are preserved. The tyrosine residues, for which an interesting role in subunit interactions seems likely in horse (Section VI) fall to a value of 3 per subunit in Phycomyces, which may turn out to have interesting consequences for the stability of subunit structure in this species. However, it seems from the data in Table 5 that the characteristic amino acid composition of apoferritin, with 25% Gly + Asx, 12% lysine + arginine, 45% non-polar amino acids and 12% of hydroxyl amino acids is strikingly conserved in the species included in Table 5.

D. Organ Specificity of Apoferritin

It has been claimed, on the basis of differences in electrophoretic mobilities, that different ferritins occur in a number of different organs within the same species (*136*, *137*). Thus, the ferritin of human bone marrow and reticulocytes was reported to differ electrophoretically from that of spleen or liver (*136*). Three electrophoretically distinct forms of ferritin were identified in spleen, liver and reticulocytes. In bone marrow (where there is a mixed cell population of erythroid cells and reticuloendothelial cells involved in salvaging iron from erythrocyte breakdown) two types of ferritin were found, corresponding in their electrophoretic mobilities to spleen and reticulocyte ferritin (*136*). Two types of ferritin had also

Table 5. *Amino acid compositions of apoferritin from various organisms. Values are expressed as moles of amino acid per 18,500 daltons*

	Horse spleen (*57*)	Horse liver (*57*)	Human spleen (*57*)	Human liver (*57*)	Guinea pig (*113*)	Dolphin spleen (*34*)	Tuna fish (*35*)	Phycomyces [b]
Cys	2.8	2.6	1.7	1.5	1.1	n.d.	n,d.	n.d.
Asx	17.3	17.9	19.3	19.2	19.5	18.8	21.0	17.1
Thr	5.5	5.6	6.1	6.2	6.2	7.6	4.6	8.2
Ser	9.0	8.9	7.7	9.3	7.8	11.8	11.0	11.0
Glx	23.9	25.3	22.3	23.9	24.0	21.2	19.4	25.9
Pro	2.8	3.1	3.8	2.9	3.5	4.6	4.6	3.5
Gly	9.9	10.2	10.8	10.1	10.4	18.2	15.7	12.5
Ala	14.0	13.4	13.8	13.7	13.4	15.2	18.9	15.1
Val	6.9	6.9	6.0	6.3	7.0	8.8	7.4	8.3
Met	2.8	2.5	2.7	2.9	2.8	3.3	3.4	1.4
Ile	3.5	3.6	3.8	2.5	3.9	4.9	6.0	7.9
Leu	25.0	24.2	23.3	23.4	23.0	17.9	17.2	18.8
Tyr	5.0	4.1	4.4	6.0	6.2	5.1	4.1	2.7
Phe	7.3	7.7	7.0	6.9	6.5	6.3	7.6	3.1
His	5.8	6.5	6.9	5.4	7.0	5.4	5.8	4.0
Lys	8.7	8.8	10.4	10.4	9.7	10.9	12.5	9.8
Trp	2.1	2.1	2.2	2.2	1.9	n.d.	n.d.	n.d.
Arg	9.5	9.0	9.3	8.4	9.2	9.3	6.2	7.8
I.P.	4.4[a]	5.5[a]			4.8	4.9	4.3	n.d.

[a] (*13*).
[b] *Crichton, R. R.*, in preparation.

been observed in human bone marrow by other workers (*137*), who found that the form corresponding electrophoretically to reticulocyte ferritin was the main incorporator of transferrin-bound ^{59}Fe. *Gabuzda* and *Pearson* (*138*) were able to prepare doubly labelled ferritin in bone marrow of rabbits. The administration of iron as ^{59}Fe bound to transferrin led to incorporation mostly into the reticulocyte-type ferritin; this form of bone marrow ferritin was termed anabolic, since it was presumed to be used for haemoglobin synthesis (*139*). Heat-damaged erythrocytes, labelled with ^{55}Fe were used to label the spleen-like ferritin: very little of this labelled iron went into the anabolic ferritin, and this ferritin was classified as catabolic, since it seemed to be involved in storing iron derived from haemoglobin breakdown. The ferritins of rat liver and kidney have been reported to be distinctive in electrophoretic mobilities and the presence of two ferritins in rat heart was also noted

(*78*). Some of these claims have been disputed by *Lee* and *Richter* (*134*) who found the electrophoretic mobilities of ferritins from human spleen, liver and bone marrow to be equal and identical, and the same to be true for ferritins from adult, foetal and regenerating rat liver, from rat spleen and from rat kidney, as well as for the ferritins from liver, spleen and bone marrow of rabbit. The same finding for the ferritins of human and horse liver and spleen has been made by *Crichton et al.* (*57*). Immunodiffusion experiments showed reactions of identity between homologous ferritins, whereas heterologous ferritins showed only partial identity, as had been reported by others (*13, 140*). All four proteins had the same subunit molecular weight and sedimentation coefficient (see Table 2). However, the amino acid composition revealed distinct differences both between organs and between species (Table 5): these differences were confirmed by analysis of the tryptic peptide patterns, and it was concluded that the apoferritins present in human and horse liver and spleen are both organ and species specific. The tryptic peptide results for human liver and horse spleen apoferritins differed markedly from those reported by *Richter et al.* (*141*) who found only one peptide in human apoferritin that was absent in horse and also one peptide in horse that was absent in human, whereas *Crichton et al.* (*57*) found five peptides present in human liver that were absent from horse spleen, whilst horse spleen had five peptides that were not present in human apoferritin. These differences may reflect different conditions of tryptic digestion and of fingerprinting.

The human cell lines HeLa and KB produce ferritins that are electrophoretically distinct from normal human ferritins (*18, 142*), although they have close similarities to human ferritins in their immunological properties (*18*). Similar observations have been made with two strains of embryonic human skin cells and with HEP-2 carcinoma cells (*17*). Since these early reports the presence of new molecular forms of ferritin in neoplastic cells has been widely observed (*143, 144*).

In five hepatomas studied by *Linder et al.* (*144*) the tumour ferritin migrated with an electrophoretic mobility similar to kidney ferritin from the same rat strain (*139*) and had other similarities to kidney ferritin. It is of interest that ferritin prepared from the livers of rats from 4 days to three weeks after birth differed from that of adult liver ferritin and was indistinguishable in its electrophoretic properties with adult kidney and spleen ferritin (*145*). Treatment of 3-week old rats with iron induced formation in liver of a ferritin with similar electrophoretic properties to adult liver ferritin. It is thus tempting to speculate that iron treatment at this stage of development may activate the genetic cistron for adult liver ferritin (*145*), and that in some liver tumours, deactivation of the adult liver ferritin cistron is associated with activation of the perinatal liver ferritin cistron.

E. Aggregation Behaviour

Both ferritin and apoferritin were found to give a number of stable fractions on gel electrophoresis (*18, 51, 142, 146—151*). Although heterogeneities were also demonstrated by chromatography of ferritin on DEAE cellulose, no significant differences were found in the amino acid compositions or in the fingerprints of tryptic peptides of these fractions (*51*). Apoferritin was eluted as a single peak from DEAE cellulose (*51, 152*). Three main fractions were observed by ultracentrifugation of horse spleen apoferritin (*51, 151*) and these were shown to correspond to the α, β and γ fractions that had been reported from gel electrophoresis. The components had sedimentation coefficients of 17—18S, 24—27S and 33—34S, which suggested that they were related as monomer, dimer and trimer. Two further components were reported in human ferritin (*17*) and a subsequent electron microscopic study showed that these correspond to tetramer and pentamer (*117*). Indeed, as we have shown (*57*) if the concentration of ferritin applied to polyacrylamide gels is sufficiently high, up to eight bands can be seen, presumably indicating that further aggregation to even higher oligomers can occur.

The possibility that disulphide bonds are involved in aggregation seems unlikely, since treatment of both ferritin and apoferritin with either mercaptoethanol or thioglycollate produces no changes in either sedimentation or gel electrophoretic patterns. The sedimentation pattern of an apoferritin fraction enriched in aggregates was unchanged by 24 hr in 1M NaCl (*51*). *Harrison* and *Gregory* suggested that higher aggregates of apoferritin were covalently linked, since the relative concentrations of the different components remained the same at different concentrations and after incubation with 8M urea or 2—6M guanidine hydrochloride, or after treatment with 1% SDS for 48 hr at 33 °C: nor was splitting of oligomer-oligomer bonds achieved by mercaptoethanol or sulphite, either in the presence or absence of urea (*152*). An analysis of the aggregation by light scattering (*104*) established that the association constants (25°) were: k_2 (monomer to dimer) $= 1{,}67 \times 10^6$ and k_3 (dimer to timer) $= 3.6 \times 10^5$. The corresponding free energy changes are $\Delta F_2^\circ =$ -7.72 kcal/mole and $\Delta F_3^\circ = -7.59$ kcal/mole. These values are consistent with reversible association involving non-covalent bonds. The possibility the dissociable Fe^{2+} located on the outside of the apoferritin monomer is involved in association was suggested.

Sephadex G-200 can be used to fractionate apoferritin aggregates (*117, 152*). Thus, although the apoferritin emerges as a symmetrical peak, the leading edge is enriched in β and γ components and the trailing edge contains predominantly α-component. In a subsequent report it was shown that ferritin oligomers did not dissociate on dilution or on storage, that apoferritin dimers were stable down to 0.19 mg/ml in dilute buffers, and that the intermolecular bonds involved in holding such stable dimers, trimers etc. together were resistant to a variety of reagents that would be expected to attack disulphide, peptide or ester linkages (*117*). The earlier view of *Harrison* and *Gregory* (*152*) that covalent bonds were involved was abandoned and in view of the failure of the urea, guanidine hydrochloride and various salts to cause splitting it seemed that predominantly polar interactions were unlikely. The stability of the aggregated forms in high salt concentrations, in which hydrophobic interactions would be stabilized (*153*) led to the suggestion that such interactions may be important in apoferritin aggregation. Conditions that lead to dissociation of the aggregates (high or low *p*H at low ionic strength) also leads to some dissociation of the apoferritin monomer into subunits (discussed in Section VI) which may imply that interactions between monomers are similar to those between subunits. If the interactions between monomers were predominantly hydrophobic, it is somewhat surprising that treatment of the aggregated forms with 40—50% dioxan did not cause dissociation (*117*). As will be discussed below, such organic solvents have little effect on the conformational properties of apoferritin monomers (Section VI).

It should be emphasised that in most ferritin preparations the content of oligomers is never more than 10—15%. Monomeric horse spleen ferritin can be prepared by gel filtration on Sepharose 4B columns and such monomers are reported to be stable during storage over several weeks at 4 °C (*154*). These monomer-enriched fractions are of considerable use in studies on protein penetration into cells and pinocytosis and facilitate quantitative determination of antigenic sites on cell surfaces with ferritin-labelled antibodies. Methods for the coupling of ferritin to antibody molecules, and the use of these ferritin-labelled antibodies in biology and medicine have been reviewed (*155*). The proportion of dimers and timers can be increased by treatment of ferritin with cross-linking reagents (*156*), *K. Wetz* and *R. R. Crichton*, unpublished work). Ferritin monomers can also be isolated from acrylamide gels: such monomers were also found to show no tendency to aggregate over a period of a few weeks (*157*). However, trimers readily dissociated to give predominantly monomer and dimer on subsequent electrophoresis suggesting that oligomer stability is not high.

F. Conformational Studies on Ferritin and Apoferritin in Solution

Mazur (*52*) showed in an early study that the iron in ferritin protected the protein against denaturation by acid, and also noted that at low *p*H ferritin was less rapidly and less extensively digested by pepsin than apoferritin. This already suggested that the protein conformation in ferritin might well be different from that in apoferritin. Further evidence pointing in this direction came from ORD studies (*158*) in which it was shown that the ORD spectrum of ferritin was independent of the iron content of the protein (which would imply that the iron core did not affect the determination of the α-helix content from the spectrum) and that in contrast to ferritin, apoferritin prepared by chemical reduction of ferritin had an ORD spectrum that was conistent with additional secondary structure. These ORD results were confirmed and similar results obtained from the CD spectrum of ferritin and apoferritin (*159*). From these studies it was apparent that the α-helix content of the protein in ferritin was around 35—45% whereas in apoferritin it was in excess of 60% (*159*).

The use of proteolysis as a probe of protein conformation in solution can detect changes in conformation of proteins, although it cannot be used to define any conformations. Apoferritin is digested more rapidly and to a greater extent than ferritin by trypsin, chymotrypsin and subtilisin, although with the latter enzyme this effect is only seen at low enzyme concentrations. With all three enzymes an examination of the peptides produced reveals that there are peptides present in apoferritin digests which are absent from the corresponding ferritin digests (*61, 160, 161* and *R. R. Crichton*, unpublished observations). With pepsin, or cathepsin D, ferritin and apoferritin were both digested at the same rate at *p*H 3.0 and the peptide patterns of the two enzymic digests of apoferritin were found to be closely similar to the corresponding ferritin digest. At lower *p*H values in agreement with the earlier results (*52*) apoferritin was digested more rapidly by pepsin than ferritin, although no qualitative difference in the peptide pattern was observed (*161*). It was proposed that these results could be explained by assuming that at *p*H 3.0 the broad specificity of pepsin and cathepsin D led to a rapid stripping of the protein from the iron core so that no difference was seen in the rates and extent of digestion. However, it was suggested that below *p*H 3.0 apoferritin underwent a conformational change that rendered it more susceptible to proteolysis (*161*).

Evidence for such a conformational change in apoferritin at low *p*H was obtained from ORD and CD experiments. Both proteins undergo a conformational change in the *p*H range 3.0—2.2 and the absolute magnitude of the change is twice as much for apoferritin as for ferritin: the changes for both proteins are consistent with the loss of about 10% of their ordered secondary structure (*159*). Near-ultraviolet CD of apoferritin shows marked sidechain optical activity which was tentatively attributed to tyrosine side-chains. Marked changes are seen in the near UV CD spectrum as the *p*H is lowered, especially between 270 nm and 295 nm (*159*). These results are further discussed in section VI B below. The effect of a number of organic solvents on the ORD spectra of ferritin and apoferritin were studied (*158*) and it was found that no additional helix formation took place in high concentrations of organic solvent. Since the presumed disruption of hydrophobic interactions does not produce substantial changes in the rotatory properties, it was concluded (*158*) that the molecule does not have a large degree of conformational flexibility. It is clear from the ORD studies that the binding of iron to ferritin does not induce an extrinsic Cotton effect in the region of absorption of the optically inactive chromophene which is related to the iron content. However, a subsequent CD analysis of native apoferritin isolated by density gradient centrifugation gave a CD spectrum identical with that of ferritin (*157*). The Fe:N ratio quoted for this fraction was <0.01. (this corresponds to <12.6 atoms of Fe per molecule[1])). In contrast the apoferritin preparation used in the ORD and CD study which was discussed above (*159*) and was prepared by chemical reduction contained <0.02 atoms of iron per subunit *i.e.* <0.5 atoms of iron per molecule of apoferritin. It seems clear that in order to resolve the question of whether ferritin really does have a quite different conformation from apoferritin one must either use native apoferritin that contains much less than 1 atom of iron per molecule (which from experience in our laboratory is virtually unattainable by centrifugal techniques), or else resort to an alternative strategy. This is to prepare ferritin from apoferritin which has been reduced by chemical methods and contains *no* iron, and observe whether this reconstituted ferritin has a CD spectrum which is typical of ferritin or of apoferritin. Methods for the reconstitution of ferritin from apoferritin are detailed in Section VII. If the reconstituted ferritin has a ferritin-like CD spectrum, then it is clear that the introduction of iron into apoferritin leads to changes in the conformation of the protein which are compatible with a reduction in the ordered secondary structure, and particularly in the α-helix content.

VI. Subunit Structure of Apoferritin

A. Introduction

The reasons for dealing with subunit structure in a separate section are as follows. The principal function of ferritin is to act as a soluble storage form for ferric iron, in which iron can be deposited and from which it

---

1) For a molecule containing one iron atom the Fe:N ratio would be 0.0008.

can be mobilized. For the maintenance of the ferric hydroxyphosphate micelle in a soluble form, the stability of the subunit structure of the protein is of obvious importance. However, as will be discussed in Section VII and VIII the various models proposed for iron uptake by and iron release from ferritin make a number of assumptions, as well as a number of predictions about the stability of the subunit structure of apoferritin, and of the protein in ferritin. Further, to answer the question of how the process of subunit assembly takes place within the cell during the synthesis of apoferritin (Section X) requires a knowledge of subunit-subunit interactions derived from studies on the assembly process outside the cell.

In this section the following terms will be used: apoferritin monomer — the molecule containing 24 subunits with S_{20w} value of 17S: subunit — the polypeptide chain of molecular weight 18,500 and sedimentation coefficient of 2.3S: Subunit dimer, subunit tetramer — molecules consisting respectively of 2 or 4 subunit polypeptide chains, of molecular weight 37,000 and 74,000.

B. Stability of Subunit Structure to Dissociation

The first demonstration that apoferritin could be converted to subunits was made by *Hofmann* and *Harrison* in 1962 (*109*), when they demonstrated that the anionic detergent SDS could split denatured apoferritin to a 2.5 S component. This observation was confirmed for ferritin (*162*), although in this study it was shown that the dissociation of ferritin into subunits was only 50% complete after 30 min at 60 °C in 0.1% detergent. Such dissociated material could be reconstituted by removal of the detergent using dialysis over a period of days. An apparent molecular weight of 12,000 was quoted for these subunits which was certainly in error on account of the method of molecular weight determination used. Subsequent studies showed that both ferritin and apoferritin, without any prior denaturation. could be dissociated completely into subunits of molecular weight 18,200 by treatment with 1% SDS, in the presence of mercaptoethanol for 4—6 hr at 37 °C followed by overnight dialysis into buffer containing 0.1% SDS, 0.1% mercaptoethanol (*112*). We have subsequently established that such SDS dissociated subunits can reassociate to apoferritin monomer on prolonged dialysis against buffer to remove most of the SDS — whether all of the SDS is removed on dialysis seems unlikely (*R. R. Crichton*, unpublished observations).

The effect of other protein denaturants on subunit dissociation has also been studied. Thus 10M urea did not dissociate apoferritin (*109*) and produced no change in the optical rotatory properties of ferritin and apoferritin (*158*). However, 10M urea did produce a considerable reduction in the depth of the trough at 233 nm for ferritin that had been lyophilised. Concentrations of guanidine hydrochloride of up to 3M had no effect on the ORD spectra of both ferritin and apoferritin. However, 6M guanidine hydrochloride caused changes in the optical rotatory properties of both ferritin and apoferritin consistent with considerable loss of helix: the *p*H of the solution was 6 (*158*). In a subsequent report it was shown that below *p*H 5.0 both ferritin and apoferritin underwent considerable decrease in secondary structure as determined by CD in 10M urea. Sedimentation velocity studies showed that ferritin was dissociated into subunits in 7M guanidine hydrochloride at *p*H 7.5 but apoferritin remained in an aggregated state under these conditions. CD studies indicated that extensive disruption of the ordered secondary structure accompanied subunit formation. At *p*H 4.5 guanidine hydrochloride did dissociate apoferritin into subunits. Reassociation of the protein to apoferritin monomer was achieved by removal of the guanidine in the presence of dithiothreitol and the monomer had a similar content of α-helix to that of native apoferritin. However, the β-pleated sheet structure observed in the native protein was not restored (*157*). In an earlier study (*107*) it was shown by sedimentation equilibrium studies that apoferritin could be completely dissociated into subunits of molecular weight 18,500 by treatment with 6M guanidine hydrochloride at *p*H 7.0. However, the dissociation procedure employed here used 4 hr at 37 °C followed by dialysis for 16 hr at room temperature in the guanidine hydrochloride solution in contrast to the 30 min-2 hr used in the later investigation (*157*). A complete time and temperature dependence study of the denaturation of ferritin and apoferritin in protein denaturatants would seem to be well worth carrying out.

By far the most useful procedure for the reversible dissociation of apoferritin is that of *Harrison* and *Gregory* (*152*). Using a procedure originally derived by *Fraenkel-Conrat* (*163*) for the dissociation of viral coat proteins from their nucleic acid, it was found that treatment of apoferritin with 67% acetic acid in the cold followed by dialysis into dilute buffer at *p*H 3.0 generated apoferritin solvents of sedimentation coefficient 2.7 S. These subunits could be reassociated by dialysis into buffer of *p*H 5.0 and the extent of reassembly was found to be favoured by the addition of mercaptoethanol or dithiothreitol. An extension of these studies on the reversible dissociation of apoferritin at low *p*H are discussed in Section VIC.

The stability of the apoferritin molecule to dissociation at low concentration has recently been examined (*102, 164*). Using apoferritin that had been acetylated (*164*) or N-methylated (*102*) with radioactive reagents no evidence for dissociation into subunits was found at concentrations of 0.1 μg/ml and 0.5 μg/ml respectively.

Richter has shown that by injecting SDS-dissociated subunits into rabbits antibodies are produced which are capable of precipitating subunits: such antibody populations also precipitate ferritin or apoferritin. When antibodies directed against ferritin are absorbed out, there remain antibodies capable of precipitating subunits. It is presumed that antibodies which are precipitated only by subunits are directed against antigenic determinants in the subunits which are hidden in apoferritin monomers (*165*).

C. Subunit Dissociation at Extremes of *p*H

A study of the dissociation of apoferritin in subunits by extremes of *p*H was initiated for a number of reasons. The observation (*152*) that treatment with acetic acid produces stable subunits at *p*H 3.0 together with evidence of a *p*H dependent conformation change in apoferritin below *p*H 3.0 (*159*) and the observation that at *p*H 3 apoferritin is present only as 17.6 S monomer (*R. R. Crichton* and *C. F. A. Bryce*, unpublished work) suggested that at low *p*H apoferritin can be reversibly dissociated, and that a significant hysteresis occurs in reassociation of the subunits to apoferritin monomers. In the course of studies on subunit dissociation (*109*) and on oligomer formation from apoferritin monomers (*152*) it was shown that partial dissociation of apoferritin into 2—3 S subunits occurs at high *p*H.

The dissociation of apoferritin at extremes of *p*H was examined by sedimentation velocity techniques (*166*). It was established that between *p*H 2.8—10.6 the apoferritin monomer (~17S) was the only species that could be detected. Between *p*H 2.8—1.6 and 10.6—13.0 both monomer (17S) and a low molecular weight component (2—3S), presumed to be subunit, were detected. The dissociation follows a smooth sigmoidal curve (Fig. 7a) in both cases with mid points, corresponding to equal amounts of 17S and 2—3S component at 2.2—2.4 for the acid dissociation and 11.8—12.2 for the alkaline dissociation (the exact values are dependent on the buffer used). When apoferritin is completely dissociated into subunits at low *p*H (either by exposure to buffer of *p*H 1.6, or by treatment with 67% acetic acid) and is then dialysed into buffer of higher *p*H, reassociation does not take place in dilute glycine buffers until *p*H values in excess of that required to induce subunit dissociation (Fig. 7a). The reassociation then follows a sigmoidal curve until complete reassociation to a 17S monomer is attained at *p*H 4.3. Although we have not as yet been able to follow reassociation completely in a more concentrated buffer (200 mM cf 10 mM) it is clear (Fig. 7a) that subunit reassociation occurs at a much lower value and that the hysteresis observed between dissociation and reassociation is much less in the more concentrated

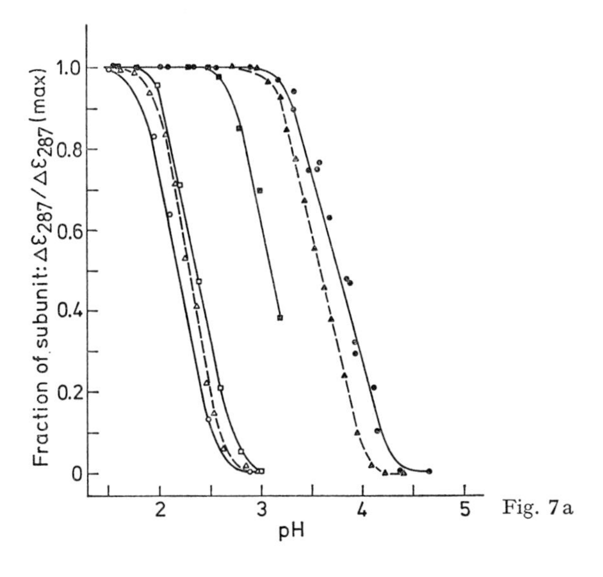

Fig. 7a

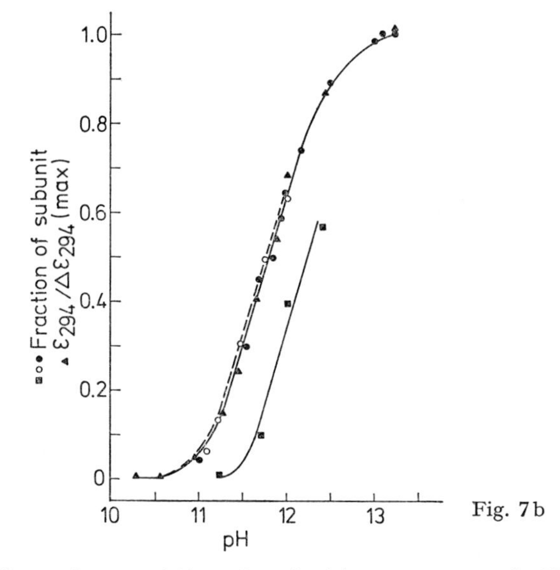

Fig. 7b

Fig. 7. Dissociation and reassociation of apoferritin at extremes of *p*H. (a) □ dissociation in 0.2 M buffer: ■ reassociation from *p*H 1.5 in 0.2 M buffer: ○ dissociation in 10 mM buffer: ● reassociation from *p*H 1.5 in 10 mM buffer: △ $\Delta\varepsilon_{287}/\Delta\varepsilon_{287max}$ for dissociation in 10 mM buffer: ▲ $\Delta\varepsilon_{287}/\Delta\varepsilon_{287max}$ for reassociation from *p*H 1.5 in 10 mM buffer. (b) ● dissociation: ○ reassociation from *p*H 12: ▲ spectrophotometric titration of tyrosine residues: ■ dissociation reported in (*117*). All other data from (*166*)

buffer than in the dilute glycine buffer. Reassociation of the protein to monomer from alkaline *p*H cannot be achieved (since the exposure to *p*H 13.0 causes irreversible denaturation): however, when the process of reassociation is carried out from *p*H 12.0 (where 70% dissociation has occurred) no hysteresis is observed (Fig. 7b).

Since it is apparent that 100% monomer and 100% subunit can be obtained at *p*H 3.0 in dilute glycine buffers depending on whether the protein has been exposed to low *p*H or not, it is possible to use difference spectroscopy to investigate conformational changes in the process of subunit dissociation. The acid difference spectrum can be accounted for by the transfer of 4—5 tyrosine per subunit from the interior of the protein into the solvent. This process is reversed on reassociation, but shows the same hysteresis as found by sedimentation techniques. The difference spectrum in alkali is more complex, but is consistent with the deprotonation of tyrosine residues which have a rather high *p*K (11.8). The acid and alkaline difference spectra follow the same profile as the fraction of subunit (Fig. 7). In addition to the involvement of tyrosine residues in the conformational change at low *p*H, spectral evidence was obtained that one tryptophan residue per subunit is transferred from a hydrophobic environment to the solvent just before subunit dissociation begins between *p*H 3.5—3.0 and returns to a hydrophobic environment subsequent to reassociation (*p*H 4.0—4.5). The dissociation and reassociation of apoferritin at low *p*H is a cooperative process involving the protonation and deprotonation of at least two carboxyl groups of rather low *p*K. In contrast the alkaline dissociation is not cooperative. The transfer of the tryptophan residue to the solvent just prior to dissociation is very sharp with a *p*K of 3.25, and may also reflect the involvement of a carboxyl group or groups in this process.

The reversible dissociation and reassociation has also been studied at low *p*H by gel filtration, and it is clear from these results (*167*) that during both dissociation and reassociation both subunit dimers and subunit tetramers were observed in addition to monomer and subunit. The ratio of monomer to subunit + subunit dimer + subunit tetramer observed by gel chromatography were quite different to that found by sedimentation techniques; this is illlustrated for subunit dissociation in Fig. 8, where at *p*H 2.16 the ratio of fast/slow component in sedimentation velocity is 1:1, the ratio of monomer to subunit + subunit oligomers is 1:20. In contrast, the ratio of monomer to subunit + subunit oligomers at *p*H 3.8 is 20:1 from gel filtration and 1:1 by sedimentation velocity. The imposition of a slow transport process on the equilibrium in gel filtration may affect the equilibrium be mass action so that rate limiting steps are affected. On this assumption we might assume that the rate limiting step in dissociation and reassociation is the conversion of subunit tetramer to subunit dimer (*R. R. Crichton*, in preparation).

An earlier report (*168*) that lyophilised apoferritin could be resolved by gel filtration in 10% acetic acid (*p*H 2.1) into several components gave essentially similar results to those found in the present study, but the correlation with subunits and oligomers of subunit was not established.

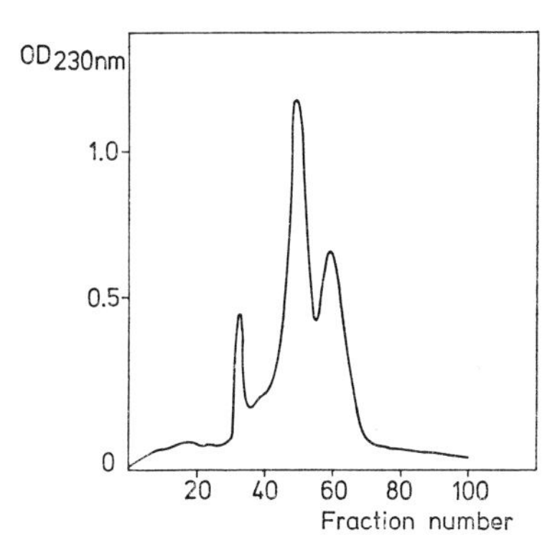

Fig. 8. Gel filtration of apoferritin (10 mg) at *p*H 2.16 in 10 mM glycine/HCl buffer on a column (100 cm × 1 cm) of Sephadex G-100. The peaks, in order of elution, correspond to oligomer, subunit-tetramer, subunit-dimer and subunit (*R. R. Crichton*, in preparation)

D. Towards a Model for Subunit Interactions

From the evidence presented in Section VI C above we may conclude that most likely all 5 tyrosine residues per subunit are at the subunit — subunit interface; that a tryptophan residue in the vicinity of the interface acts as a signal, changing its position in the interior of the protein for the solvent just before dissociation occurs; and that the subunit - subunit interface contains at least 3 carboxyl groups of rather low *p*K (2 of *p*K 2.16 and 1 of *p*K 3.25). The *p*K of the β- and γ-carboxyl group of aspartic and glutamic acid are 3.85 and 4.25 respectively; one reason for the low *p*K values associated with subunit dissociation may be the presence of salt linkages between carboxyl groups and amino groups at the subunit—subunit interface. It has been reported (unpublished work of *W. I. P. Mainwaring* cited in (75)) that succinylation of the lysine residues of apoferritin leads to its dissociation into subunits of molecular weight 25,000. In a detailed study of the effects of chemical modification on the stability of the apoferritin monomer (*166*) it was found that conversion of seven out of the nine lysine residues per subunit to homoarginine did not cause dissociation. It was also shown that the modification of nine out of the ten arginine residues per subunit with cyclohexanedione, of both tryptophan residues per subunit with 2-nitrophenyl chloride, nitration of one the five tyrosine residues per subunit or carboxymethylation of 2 cysteine residues per subunit and of 1 histidine residue per subunit did not cause subunit dissociation of apoferritin monomers (*166, 123*). More recent studies have established that following modification of 22 carboxyl groups per subunit with glycineamide in 6M guanidine hydrochloride the protein is dissociated in subunits and cannot reassociate to apoferritin monomer (*K. Wetz* and *R. R. Crichton*, unpublished work). This represents a rather drastic modification and further work will be required to establish the role of carboxyl groups and of basic residues in the subunit—subunit interactions. Fig. 9 presents a model of the subunit—subunit interactions in apoferritin, which is based on the spectroscopic and chemical studies discussed above.

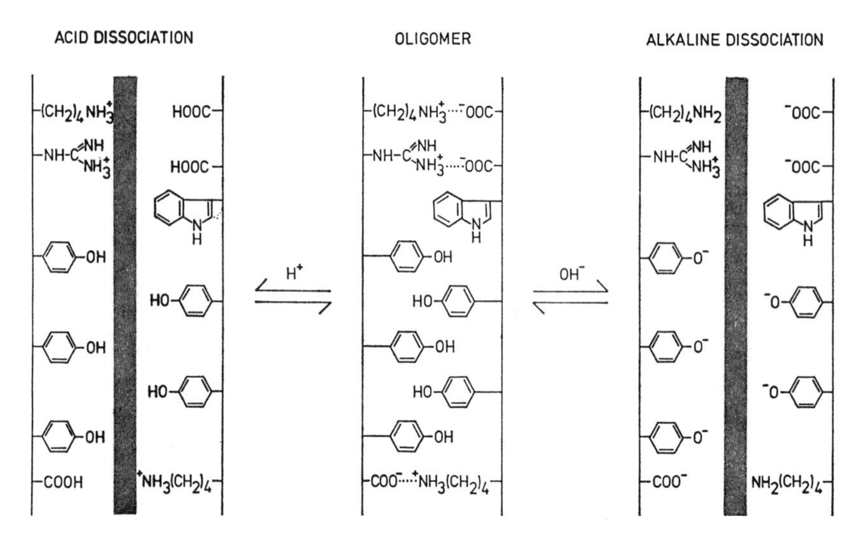

Fig. 9. A model for the subunit-subunit interface of apoferritin and for the effect of acid and alkaline dissociation. The contributions of one subunit to the interface have been represented schematically. However, it is clearly not possible to establish whether one subunit contributes all **5** tyrosine residues to one interface, whist the other contributes none, or any of the several other possibilities! An explanation of the various interactions can be found in the text and in (*166*)

The evidence from studies with protein denaturants (Section VIB) would imply that hydrophobic interactions may be of importance in stabilizing the subunit—subunit interactions, and the spectroscopic evidence supports the view that the five tyrosine residues per subunit are indeed in a predominantly hydrophobic environment. However, the *p*H-dependence of dissociation in urea and in guanidine, the failure of solvents such as dioxan to affect the secondary structure (Section V), and the fact that two lysine residues and one arginine residue per subunit are not modified in the monomer suggests that salt bridges may also be of importance. This is borne out by the finding that several carboxyl groups of rather low *p*K appear to be involved in the dissociation at low *p*H. However, to take the model outlined in Fig. 9 seriously would certainly be a mistake. An analysis of the forces responsible for the stability of the subunit interactions in haemoglobin (*169, 170*) shows that many different types of weak interaction are responsible for the stability of the tetrameric structure, and while further analysis of such interactions in apoferritin may throw light on possible amino acid residues that are at the subunit—subunit interface, a detailed picture must await the results of X-ray analysis.

E. Quaternary Structure of Apoferritin

Ferritin and apoferritin appeared to be cubic and isomorphous from X-ray powder photographs (*171*). An analysis of single crystals of ferritin showed that they were orthorhombic (*172, 173*). The space group $P2_12_12_1$ was derived but it was concluded that this was not the true space group. *Harrison* (*88*) examined crystals of apoferritin and of two crystalline forms of ferritin and reported that wet crystals of apoferritin and of one form of ferritin were face-centred cubic with $a = 184$ Å. The other form of ferritin was pseudocubic and truly face-centered on only one face. The lattice form could be described in terms of an orthorhombic cell of unit cell dimensions $a = 130$ Å, $b = 130$ Å and $c = 184$ Å. The space groups could not be determined unambigiously from the X-ray data, but it was concluded that for apoferritin the space group was possibly F 432 and that there were four molecules per unit cell. The molecule would thus have point group symmetry 432 and would be composed of 24 *n* subunits. A possible molecular structure consisting of 24 subunits situated at the vertices of a snub cube was proposed (Fig. 10a). A subsequent investigation of the same cubic form of apoferritin (*68*) led to the conclusion that the molecule contained "ordered disorder", and the apparent 432 symmetry together with a diffuse background was explained by assuming that molecules with a lower point group symmetry were lying on the lattice points, but in several different orientations such that the crystal symmetry was produced statistically. Evidence for pseudo-eicosohedral symmetry was found in the X-ray diffraction pattern, and this together with the occurrence of rather strong intensities near

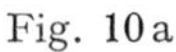

Fig. 10a

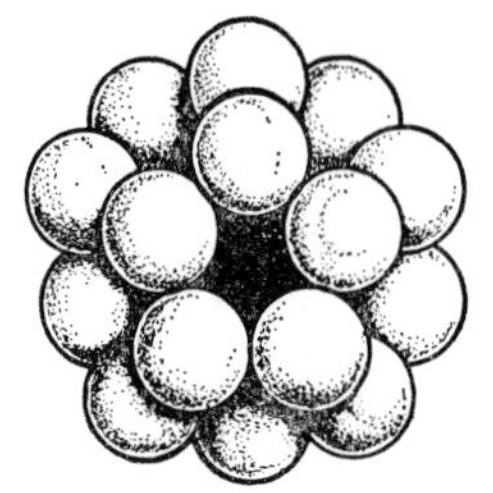

Fig. 10b

Fig. 10. Two possible models for apoferritin. On the right is a 20 subunit model with the subunits disposed at the vertices of a pentagonal dodecahedron. The model on the left consists of 24 subunits arranged at the vertices of a snub cube

the directions of the non crystallographic five-fold axes, and a body of chemical and physico-chemical data suggested that the most likely structure for the apoferritin molecule consisted of a pentagonal dodecahedron (Fig. 10b). Two more recent electron microscopic analyses of the arrangement of subunits in apoferritin by negative staining conclude that the results are compatible with a 20 subunit structure in which the subunits are situated at the vertices of a dodecahedron (*174*, *175*). As was subsequently pointed out one problem of such an eicosameric structure is that all of the subunits cannot occupy equivalent positions in the structure (*68*, *106*). There must therefore either be two kinds of subunits, for example 10 of one kind and 10 of another (*68*), or else the same subunit must exist in two distinct and different conformations in the same quaternary structure (*176*). The evidence for the existence of eicosameric structures is rather weak (*106*) and the chemical evidence available to date does not suggest that there are two kinds of subunit. The prospect of a structure in which all subunits are equal but some are more equal than others is not particularly appealing. As has been pointed out in Section V the physico-chemical evidence points overwhelmingly to a quaternary structure composed of 24 equal and identical polypeptide chains. A final decision on the quaternary structure must await further crystallographic investigations.

VII. Formation of Ferritin from Apoferritin

A. Introduction

As will become clear from the discussions below two quite different models have been advanced to explain the formation of ferritin. Historically the earliest demonstrations of ferritin formation by guinea pig liver homogenates from added apoferritin used ferric ammonium citrate (*74*). It was subsequently pointed out that the aqueous chemistry of iron (III) is dominated by its tendency to hydrolyse and polymerise, and that complexing agents which maintain ferric iron in solution at physiological *p*H do not necessarily prevent polymer formation (*101*). Thus, one model for ferritin formation assumes that the form of iron utilized for ferritin formation is polymerised ferric chelates (*72*). The alternative model, proposed first in 1955 (*177*) is a simple redox process in which ferrous iron undergoes conversion to ferric in the presence of apoferritin and a suitable electron acceptor with concomitant formation of ferritin. As we will see this latter model has been considerably supported in recent investigations.

B. The Iron Micelle Model

The formation of ferritin from liver homogenates, apoferritin and ferric iron was reported (*74, 178*). However, the incorporation of iron obtained was only 15% of the level found in native ferritin, whereas when ferrous iron was incubated with rat or horse apoferritin iron was rapidly incorporated up to 80% of the iron content of native ferritin (*177, 178*). The possibility that the ferric oxyhydroxide core of ferritin was first synthesised, and that protein subunits then formed a shell around the core to stabilize it was first considered by *Saltman* and his colleagues (*72*). It had been reported by *Granick* (*50*) that when ferritin was crystallised with $CdSO_4$ there always remained a small amount of ferritin-like material that resembled ferritin except in its ability to crystallise from $CdSO_4$. When this noncrystalline ferritin fraction was incubated with apoferritin crystalline ferritin could be recovered from the solution (*50*). The time course of this process was studied (*72*) and it was established that in the course of a 45 min incubation at 25 °C and *p*H 7.4 all of the iron present in the original protein could be recovered as crystallisable ferritin[2]). From electron microscopic studies it was concluded that the non-crystalline ferritin lacked some protein subunits, although this view has been challenged (*179*) and is certainly at odds with earlier results of *Farrant* (*67*). A previous article in this series (*101*) by *Spiro* and *Saltman* has emphasised the point that the chemistry of ferric chelates such as ferric fructose, ferric citrate and ferric hydroxynitrate in aqueous solution is dominated by the tendency to polymerise giving water soluble micellar cores, which in several cases have approximately the size of the cores of full ferritin (70 Å diameter).

---

[2]) The ratio of apoferritin to non-crystalline ferritin quoted (*72*) must be in error: a figure of 600 mg Fe/ml corresponding to 2 g ferritin protein/ml seems unlikely!

X-ray scattering, Mössbauer spectroscopy, infrared spectroscopy and magnetic susceptibility studies on the ferric hydroxynitrate polymer (approximate composition $Fe_4O_3(OH)_4 \cdot (NO_3)_2 \cdot 1.5\, H_2O$) shows that it is closely similar in those properties to the core of ferritin (*180*). It was also reported (*72*) that ferritin prepared from Fe^{2+} and apoferritin under oxidising conditions had a different morphology from native ferritin: the iron micelle had a diameter of 48 Å instead of 70 Å and the overall diameter was 102 Å instead of 120 Å.

These observations led *Saltman et al.* to propose that there is an equilibrium between ferritin and the non-crystalline form and apoferritin subunits (Eq. 2);

$$\text{Ferritin} \rightleftharpoons \text{non-crystalline ferritin} + \text{apoferritin subunits} \qquad (2)$$

which can be perturbed by addition of large amounts of apoferritin thus forwarding the formation of ferritin. A model was advanced for ferritin formation in the absence of any oxidative processes (Fig. 11).

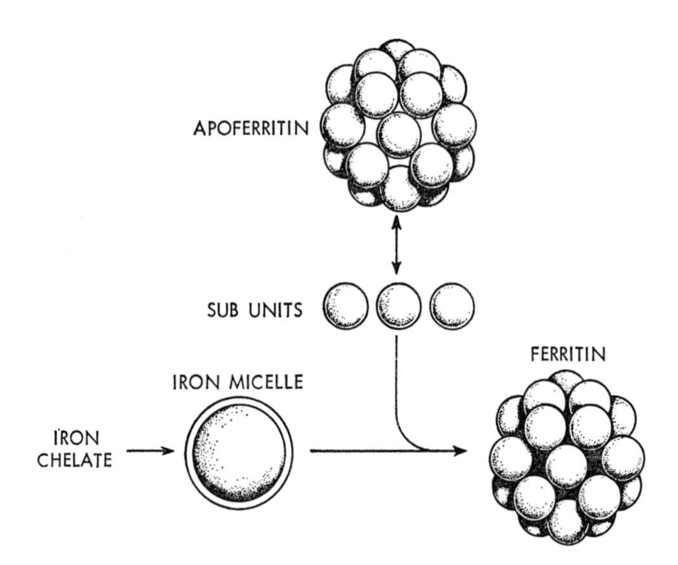

Fig. 11. Chelate model for ferritin formation: see text for details

Spontanous formation of ferric micelles is followed by their encapsulation by apoferritin subunits and results in formation of ferritin. A subsequent communication noted tht addition of apoferritin subunits, prepared from ferritin by treatment of ferritin in 10 M acetic acid at

0 °C followed by centrifugal separation of the core, to natural polymeric iron cores such as ferric nitrate, ferric citrate, ferric fructose and ferric ATP leads on rapid neutralization to formation of a synthetic ferritin that closely resembles ferritin (*181*).

This model has a number of attractive features. Not only is it compatible with the fact that Fe^{3+} is the preferred form of iron under physiological conditions, where the rate of autoxidation of Fe^{2+} to insoluble hydroxide under mildly oxidising conditions is extremely rapid but it could also account for the stimulation of ferritin synthesis by iron at the level of translation of a stable *m*RNA (Section X), since it seems reasonable to assume that, whereas the apoferritin monomer shows no tendency to dissociate under physiological conditions (Section VIB), the molecule is presumably synthesised as apoferritin subunits, which could then associate around performed micellar cores.

There are however a number of objections to this model, which have been discussed by several authors (*179, 182*). As pointed out above the observation that non-crystalline ferritin is deficient in apoferritin subunits is at variance with the electron micrographs of *Farrant* (*67*): further the Fe:N ratio quoted for non-crystalline ferritin was the same as that of the crystalline ferritin from which it was prepared (*72*), which would suggest that in addition to incomplete ferritin protein shells it also contained free subunits. Indeed one major criticism of the model is that it assumes that apoferritin monomer exists in equilibrium with subunits. There is at present no evidence for this at physiological *p*H. The model is also at variance with studies on the synthesis of apoferritin following iron administration *in vivo* (*183, 184*). It was shown in such experiments that the first ferritin to appear after iron administration was iron-poor ferritin and full ferritin only appeared some considerable time after iron administration. The model clearly requires that full ferritin should be the initial product of ferritin synthesis. If one assumes that subunit dissociation is very slow from iron rich ferritin compared with iron poor ferritin or apoferritin, then the newly synthesised apoferritin subunits would be incorporated preferentially into ferritin of low iron content, which would rescue the model (*185*). One argument which is at present difficult to assess is that in all of the synthetic ferric micelles studied by *Saltman et al.* the chelating agent is incorporated into the micelle. To date there is no satisfactory explanation for the formation of cores containing ferric oxyhydroxide and oxyphosphate.

C. Apoferritin—Catalyst for Ferritin Formation

Bielig and *Bayer* (*177*) demonstrated that Fe^{2+} in the presence of apoferritin at *p*H 7.4 was converted at 0 °C by bubbling of O_2 through the solution to a product that crystallised with $CdSO_4$ and had an Fe:N ratio of 2.4. Subsequently it was found that under identical conditions a ferritin-like material could be obtained from rat liver apoferritin (*178*). Thus the simple equation (Eq. 3) could be written to describe ferritin formation:

$$Fe^{2+} + O_2 + \text{apoferritin} \longrightarrow \text{ferritin} \qquad (3)$$

The formation of such ferritin-like material under several conditions was reported, although the crystallites were not in all cases as large as in native ferritin (*85*). Apoferritin and ferrous ammonium sulphate, or ferrous sulphate gave such products on a) oxidation with O_2 in bicarbonate buffer, *p*H 6.8—8.0 (*177*): in the absence of apoferritin the product was α-ferric oxyhydroxide. b) oxidation by KIO_3 in $Na_2S_2O_3$, *p*H 5.4—7.4: in the absence of apoferritin γ-ferric oxyhydroxide is formed. c) oxidation by O_2 in imidazole buffer, *p*H 6.7—7.0: this gave a third unidentified product in the absence of apoferritin. The fact that products which did not have the typical characteristics of the ferritin core were formed in the absence of apoferritin demonstrated that apoferritin had an effect on the way in which the ferric oxyhydroxide crystallised inside the protein shell. Apoferritin was reported to affect the rate of oxidation of Fe^{2+}, as well as influencing the product formed (*85*). *Niederer* (*185*) showed by following the disappearance of Fe^{2+} using the quantitative bipyridyl reaction, that apoferritin appeared to catalyse the oxidation of Fe^{2+} in the presence of suitable oxidising agents. In contrast, the rate of oxidation of Fe^{2+} in the presence of bovine serum albumin or lysozyme was slower than that with apoferritin: however, the rate in the absence of any protein, whilst slower than with apoferritin, was still quite rapid. From these observations he postulated that ferritin formation took place by penetration of Fe^{2+} ions into the apoferritin shell followed by oxidation at the inner surface of the protein catalysed by a catalytic active site. The Fe^{3+} was presumed to form an intramolecular precipitate which soon became too large to escape from the protein shell.

The absorption of the micelle at 420 nm can be used as a means of following ferritin formation directly (*122*); values quoted for the $E^{1\%}_{1cm}$ of the micellar core of ferritin range from 100 (*122*) to 61.4 (*123*). *Macara et al.* (*122*) have extended the previous studies of *Harrison et al.* (*85*) using this more direct assay and found that progress curves for iron uptake by apoferritin were sigmoidal, whereas those for ferritins of low iron content were hyperbolic. In these investigations, the reaction was started by addition of ferrous ammonium sulphate to a mixture of apoferritin and a large excess of oxidant (KIO_3 and $Na_2S_2O_3$) in imidazole buffer of *p*H 6.2—7.4. The rate of uptake of iron was dependent on the amount of iron already present in the molecule and the distribution of the iron content of reconstituted ferritins was inhomogenous. The reconstituted ferritin was found to be similar to native ferritin by a number of physical methods (electron microscopy, sedimentation velocity and X-ray diffraction). The effect of varying the iron concentration on the rate of ferritin formation gave a curve that was thought probably to be S-shaped, although the formation of FeO.OH outside of the protein at high iron concentration made it difficult to estimate reaction velocity at high iron concentration. The rate constant for the linear part of the curve was $k = 0.71\ \text{min}^{-1}$. An unusual dependence of the reaction rate on apoferritin concentration was observed: the rate rose to a maximum at about 1 mg/ml apoferritin and then fell off at higher concentrations. The rate of reaction increased with pH in the range 6.2—7.4, and it was noted that when stepwise additions of Fe^{2+} were made, although the initial uptake of iron followed a sigmoidal progress curve, subsequent additions did not.

These findings were interpreted in terms of a crystal growth model in which the surface area of the crystallites forming inside the protein increases until the molecule is half full and then declines, the surface of the crystallite controlling the rate at which new material is deposited. A two step mechanism was proposed to explain the results, an initial nucleation step which is zero order with respect to $[Fe^{2+}]$, followed by a more rapid growth stage. The apoferritin appears to increase the rate of nucleation, binding ferrous iron at sites where oxidation subsequently occurs. Three alternative models for the oxidation and growth of the

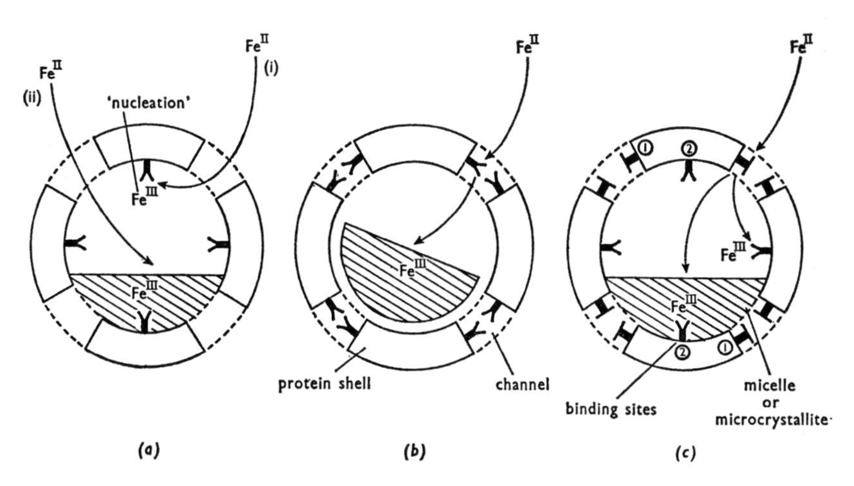

Fig. 12. Models for the formation of ferritin (reproduced from 122). For details of the three models see the text

core were considered (Fig. 12): a) a model in which the oxidation site was also the site at which the core starts to grow; here oxidation and nucleation can be regarded as a single step b) a model in which the oxidation sites are situated in the channel between subunits so that, in contrast to a) they remain accessible as the molecule is filled: crystal growth is presumed to occur by the growing ferric oxyhydrate inside the molecule binding Fe^{3+} more strongly than the oxidation sites, and thus causing their transfer to the interior of the molecule c) the last model is similar to b) except that in c) specific nucleation sites which bind Fe^{3+} are present inside the molecule in addition to the oxidation sites: the contrast is that in b) the interior of the protein simply provides a favourable environment for the crystal growth. *Macara et al.* (*122*) thus considered ferritin to be an enzyme in which the product remains firmly bound to the protein.

A subsequent analysis of this process by *Bryce* and *Crichton* (*123*) led to quite different conclusions. As had been shown in earlier studies apoferritin catalysed the oxidation of Fe^{2+} to Fe^{3+}: molecular O_2 was used as electron acceptor and since the reaction was carried out in cacodylate/borate buffer of *p*H 4.3—6.0 using a double beam spectrophotometer and the reaction initiated by addition of protein to ferrous ammonium sulphate already dissolved in the buffer, the progress of the apoferritin-catalysed reaction could be monitored directly at 420 nm. The reference cell effectively eliminated the rate of autoxidation, which at the *p*H values used, was slow. A number of other proteins were found to have no effect on Fe^{2+} oxidation under these conditions. The product of the reaction was similar to ferritin in its behaviour on electrophoresis and gel filtration and appeared to be similar to ferritin on electron microscopy and in the analytical ultracentrifuge. The progress curve was hyperbolic and the increase in initial velocity was linear with increasing apoferritin concentration in the range 0.1—4.0 mg/ml. With respect to Fe^{2+} the reaction followed *Michaelis-Menten* kinetics over the range 0.5—5 *m*M. The Km was 2.8 *m*M and k_{cat} 1.11 sec^{-1}. The corresponding k_{cat} reported by *Macara et al.* (*122*) was 4.50 sec^{-1}. The *p*H dependence study showed that at *p*H 6.0 the reaction rate was still increasing i.e. the *p*H optimum was well in excess of 6. The essential difference between this study (*123*) and that of *Macara et al.* (*122*) was that the reaction conditions were selected so as to ensure that the rate of autoxidation was slow, and could be continuously compensated for by the use of a double beam instrument, and that the buffer had sufficient buffering capacity to cope with the generation of protons resulting from hydration of Fe^{3+}. It was further demonstrated that identical progress curves were obtained when the loss of Fe^{2+} was followed and that Fe^{3+} was not taken up at all by apoferritin (*123*).

There are a number of questions concerning the catalytic activity of apoferritin that remain to be answered by further investigations. The form in which Fe^{2+} is presented to apoferritin in the cell itself is not known. Nor is it clear that molecular oxygen serves as electron acceptor, and if it is what the other products of the reaction are. The mechanism by which apoferritin can catalyse oxidation of Fe^{2+} remains to be clarified: some results from chemical modification and other approaches towards delineating the active site are discussed in Section VIID below. At any rate, we may conclude that the catalytic role of apoferritin in the formation of ferritin seems to be well established and can be envisaged in its simplest form as outlined in Fig. 12. Fe^{2+} diffuses into the apoferritin molecule through channels between subunits, and is there oxidised to Fe^{3+} in a reaction that involves a catalytic active site in the apoferritin molecule. Whether this is in the channels between subunits or in the interior of the molecules remains to be established. What happens thereafter is that the ferric oxyhydroxide forms a crystalline micelle which grows in size until it can no longer escape from the interior of the protein. Whether the apoferritin affects the process of micellar formation and whether deposition of the core after oxidation is also regulated by the protein must await further investigation.

D. The Active Site of Apoferritin

The term active site is now somewhat passé in a world of molecular mechanisms for enzyme reactions (*186*). However, we will use the term here in its original sense to define an area of the apoferritin molecule which is responsible for the process of oxidation and possibly also of core nucleation.

Mazur (*52*) noted that iron incorporation into ferritin could be suppressed by iodoacetamide. The system used, namely liver slices perfused with a ferric citrate solution, was too complex to enable any definite conclusion to be drawn, since a great many factors were involved in the iron uptake process. *Niederer* (*185*) found that the histidine (and also other amino acids) alkylating reagent diazonium-H-tetrazole does not affect iron uptake by apoferritin. In contrast bromoacetate, a weak alkylating reagent, formaldehyde and β-propiolactone, both of which react with amino groups and histidine residues, all inhibited iron uptake to varying degrees. Iodoacetamide had no effect when used in the *p*H range 4.5—8.0. The effect of Zn^{2+}, both *in vivo* (*187*) and *in vitro* (*185, 188*) suggests that histidine may be involved in the catalytic activity. In a study of the effects of chemical modification (*123*) it was found that using the assay system described above the modification of both tryptophan residues per subunit with 2-nitrophenyl sulphenylchloride does not affect either k_{cat} or Km for the oxidation. Nor does guanidination of seven out of nine lysine residues/subunit, the nitration of one tyrosine residue out of five/subunit, nor the modification of nine out of ten arginine residues/subunit with cyclohexanedione. The carboxymethylation of both cysteine residues/subunit and of one histidine residue/subunit out of six with iodoacetic acid at *p*H 8.0 results in a product which is catalytically completely inactive in Fe^{2+} oxidation. We cannot therefore decide at present whether histidine or cysteine or both are involved in the oxidation. It has been recently found (*K. Wetz* and *R. R. Crichton*, unpublished work) that the modification of one cysteine residue/subunit with N-ethylmaleimide does not affect the catalytic activity of apoferritin.

E. Mechanism of Apoferritin Catalysis

It is clearly premature to discuss the mechanism of a reaction for which data is only now beginning to accumulate. On the other hand the nature of this catalysis by a protein which lacks haem or copper in contrast to most other oxidases is clearly of interest and a few remarks about possible mechanisms would seem to be in order. One model suggests itself immediately, namely that Fe^{2+} can be bound, presumably on adjacent subunits, at two sites which are sufficiently close to one another to generate superoxide which could then bring about the oxidation to Fe^{3+} with concomitant formation of water (*B. G. Malmstrom*, personal communication). It has been recently suggested (*189*) that transferrin, which binds two molecules of iron per molecule or protein, can catalyse the oxidation of Fe^{2+} by oxygen. The formation of Fe^{3+}-transferrin from Fe^{2+} salts in the presence of dissolved oxygen was previously studied (*190*) and two mechanisms were advanced: a) the oxidation of Fe^{2+} to Fe^{3+} followed by rapid reaction of the Fe^{3+} with apoferritin; (b) the formation of an Fe^{2+}-transferrin complex followed by oxidation of the iron to Fe^{3+}. It has been recently found (*B. Sarkar*, personal communication) that Fe^{2+} appears to bind to transferrin in anaerobic conditions and that rapid oxidation of the Fe^{2+}-transferrin to Fe^{3+}-transferrin occurs on introduction of oxygen. *Gaber* and *Aisen* had earlier found no evidence for the formation of an Fe^{2+}-transferrin complex (*191*). If it is assumed that Fe^{2+} binding precedes oxidation, and that juxtaposition of two Fe^{2+} is required for oxidation, it can be further proposed that the catalytic site of apoferritin must be in the subunit—subunit interface since the apoferritin subunits appear to approximate to compact spherical molecules of roughly 25 Å diameter, and the alternative of having the active site internally would place the iron atoms too far apart. The location of the active site in the channels between subunits together with the postulated requirement for two adjacent Fe^{2+}-binding sites leads to the prediction that the apoferritin subunit should be catalytically inactive, although possibly still able to bind one molecule of Fe^{2+}. In contrast, subunit dimers might well retain full catalytic activity in Fe^{2+} oxidation.

A further unique feature of ferritin, as has been emphasised by *Macara et al.* (*122*), is that the enzyme retains the product and ceases to function once its maximum iron binding capacity has been attained. The present model would predict that diffusion of the Fe^{3+} into the interior of the molecule is the next step in ferritin formation, and that micellar growth occurs without any blocking of catalytic sites. This should also be capable of confirmation by detailed kinetic studies. One final advantage of the present model is that it suggests the possibility that Fe^{2+}, in the process of mobilization from ferritin (Section VIII) could use the same Fe^{2+} binding sites; this might explain the observation that ferritin contains very small amounts of Fe^{2+} 'at the surface' of the protein (*192*) and is not incompatible with recent electrochemical data on iron mobilization (*193, 194*).

VIII. Mobilization of Iron from Ferritin

A. Mobilization by Low Molecular Weight Reductants

Granick and *Michaelis* established that iron could be mobilized from ferritin as Fe^{2+} (*59*). However, sodium dithionite is clearly not likely to serve such a role physiologically. An analysis of the effectiveness of a number of biological reducing agents was conducted both by *Mazur et al.* (*192*) and by *Bielig* and *Bayer* (*195*). The former study established that crystalline ferritin contained small amounts of its iron in the Fe^{2+} state together with free sulphydyl groups. Treatment of the ferritin with anaerobic liver slices, glutathione, cysteine or ascorbic acid led to an increase in ferritin-Fe^{2+}. Thus following a 30 min incubation at *p*H 7.4 glutathione had converted about 0.9% of the ferritin iron to Fe^{2+}, ascorbate had converted 2.8% and cysteine 10.5%. The concentration of the reductants used, (67 mM) was, however, rather high for these results to be of physiological significance. *Bielig* and *Bayer* (*195*) obtained much less convincing evidence for such a pathway for iron mobilization. Whereas the halftime for ferritin iron release as Fe^{2+} with dithionite in the presence of bipyridyl was 0.025 h, the best of the biological reductants used (NADH, ascorbic acid and glutathione were tried), namely ascorbate, mobilized ferritin iron with a halftime of 96 h at *p*H 5.5.

In a more recent study the effect of NADH, ascorbate, glutathione and cysteine were examined under a variety of conditions (*R. R. Crichton* and *J. Dognin,* in preparation). Cysteine was by far the most effective in iron mobilization, but concentrations in excess of 50 mM were required to attain rates of physiological significance. Ascorbic acid at concentrations of 100 mM and above, and *p*H values of below 5 was also effective, although whether such levels of ascorbate occur even in the most dedicated of those seeking to avoid the common cold seems unlikely! The rates of mobilization were not significantly increased by the presence of biological chelating agents such as citrate or fructose. Release of iron from α-ferritin with sodium dithionite was found to be much slower than from β-ferritin (*196*).

B. Mobilization by Low Molecular Weight Chelators

The release of iron from ferritin without prior reduction was reported by *Saltman* and his colleagues (*197*). Both nitrilotriacetate and EDTA could mobilize iron from both ferritin and non-crystalline ferritin without any requirements for reduction. However, although 40% of ferritin iron could be mobilized in 120 h by nitrilotriacetate (at a concentration of 100 mM) the only physiological chelating agent used, sodium citrate, at the same concentration could only release 1% of the ferritin iron in this time. The release of iron by chelating agents was always more rapid from non-crystalline ferritin than from crystalline ferritin.

1.10-phenathroline released 12% of ferritin iron in 60 days at room temperature (*198*). The activation energy for the process was 13 ± 2 kcal/mole and the frequency factor of 10^1 sec^{-1}. From the data the mechanism was advanced (Eq. 4):

$$\text{iron (III) in ferritin micelles} \xrightarrow[\text{slow}]{} \text{iron (II, III) at surface of ferritin} \xrightarrow[\text{fast}]{\text{1.10-phenanthroline}} \text{iron (II) complex with 1.10-phenanthroline} \qquad (4)$$

The rate of release was greatly increased by ascorbic acid. A polarographic study (*194*) showed that ferritin iron can be released by EDTA, nitrilotriacetate and desferrioxamine B: all three chelates mobilize 20—25% of ferritin iron in 200 hr. From these results it was proposed that an equilibrium exists between the iron core and small quantities of iron outside of the protein bound to electron acceptor sites which contain sulphydyl groups (although no evidence for this was presented). The number of sites is dependent on the iron content of the ferritin. It was suggested that iron could be removed from these electron acceptor sites by the action of reducing agents and then chelated by complexing agents.

In recent studies in this laboratory the effects of a number of biological chelating agents on ferritin iron mobilization have been examined (*R. R. Crichton* and *J. Dognin*, in preparation). Glucose, fructose, glycine, succinate, citrate and AMP release very little iron from ferritin: the best of these, fructose can mobilize 15% of ferritin iron in 7 days. It seems unlikely that such a low molecular weight chelating agent plays a significant role in ferritin mobilization *in vivo*.

C. Enzymic Mechanisms for the Mobilization of Ferritin Iron

Mazur and his colleagues were the first to consider the possibility that an enzyme might be responsible for ferritin iron mobilization (*199, 200*). The reduction and release of ferritin iron during anaerobic incubation of ferritin with rat liver slices was accompanied by the accumulation of uric acid in the tissue and in the medium. Xanthine oxidase was reported to reduce ferritin iron under anaerobic conditions in the presence of hypoxanthine or xanthine. Reduction was increased in the presence of oxygen and by addition of catalase (*199*). However, it must be emphasized that the effect was not particularly great — at best a little less than 4umole Fe^{2+}/mmole of total ferritin iron/hr (*i.e.* about 0.4% of total iron mobilized/hr). *Tanaka* (*201*) had reported earlier that xanthine oxidase reduced ferritin iron anaerobically in the presence of hypoxanthine. The method used to detect this was to follow changes in the magnetic susceptibility of ferritin. *Green* and *Mazur* (*199*) also noted changes on specific susceptibility of ferritin iron on incubation with xanthine oxidase. Further studies showed that administration of xanthine or hypoxanthine lead to an increase in plasma iron and from inhibition studies it was suggested that the site of ferritin iron reduction was associated with the non haem iron of the enzyme and takes place by direct electron transfer (*200*).

Xanthinuria was observed in a patient with idiopathic haemochromatosis, an inheritable disease in which very large amounts of iron are found in liver and other tissues as ferritin and haemosiderin, together with a marked reduction of the activity of xanthine oxidase in the patient's liver (*202*). This would certainly be compatible with a role for xanthine oxidase in ferritin iron mobilization, as is the finding of high levels of the enzyme in the intestinal mucosa (*203, 204*). In most of the experiments of Mazur's group, the physiological stimulus for release of iron was tissue hypoxia, which stimulates formation of red blood cells and increases the level of xanthine oxidase substrates. The effect of this, it was argued (*200*) was to increase the reduction of ferritin and hence tc increase the level of plasma iron. Further support for a possible role for xanthine oxidase in ferritin iron mobilization comes from the observation that ferritin catalyses the oxidation of sulphite to thiosulphate and that addition of hypoxanthine and xanthine oxidase in the presence of ferritin stimulates the rate of thiosulphate production by 35% over that produced by ferritin alone (*205*).

We have recently reinvestigated the possible role of xanthine oxidase in ferritin iron mobilization and can find no evidence for its involvement. The incubation of xanthine oxidase with ferritin under conditions where the xanthine oxidase is metabolizing hypoxanthine extremely rapidly, releases virtually no Fe^{2+} from ferritin under anaerobic conditions (*R. R. Crichton* and *J. Dognin*, in preparation).

Niederer (*185*) has suggested that the reduction of ferritin involves the same active site in apoferritin as iron uptake, since treatment of apoferritin with formaldehyde, β-propiolactone and several divalent cations delayed iron release by dithionite. However, a report by *Osaki* and *Sirivich* (*206*) describing the isolation and characterization of a ferritin reducing enzyme in liver homogenates of a number of vertebrates including cow, pig, dog, rat, chicken and tadpole, and the partial purification from beef liver was briefly described. NADH and FMN are required for the enzymic reduction of ferritin iron to Fe^{2+} and the reaction is inhibited by >2 μM oxygen, but not by the xanthine oxidase inhibitor allopurinal: hypoxanthine, succinate, ethanol and glucose are not effective as electron donors. The enzyme activity for which the name ferriductase is proposed, seems to be located mainly in the soluble fraction of the cell: none is found in the mitochondrial fraction. The isolation and characterization of this enzyme will be awaited with great interest.

IX. Iron Exchange between Ferritin and Transferrin

Transfer of iron between ferritin and transferrin was first demonstrated by *Mazur et al.* (*192*), who found that the process required rat liver slices, and that only the ferrous iron of ferritin was transferred. *Bielig* and *Bayer* (*195*) in the same year demonstrated transfer of iron from ferritin to transferrin across a semipermeable membrane. This process required ascorbic acid. It was later reported from *Mazur*'s laboratory that the transfer of iron from transferrin to ferritin by liver homogenates requires ATP (*207*). A more detailed analysis of the transfer of iron from transferrin to ferritin across a semipermeable membrane using radioactive tracers was undertaken by *Perkins* (*208*). Transfer was found to be a passive process which was unaffected by the iron moving against a concentration gradient. A reducing agent was required for the transfer: of those used ascorbate was the best. This requirement was in order to remove iron from transferrin, to form a ferrous chelate as an intermediate and to ensure that iron was presented in the ferrous form for uptake by ferritin. The transfer was enhanced by chelates such as AMP, ADP, ATP, lactate, glucose, fructose, riboflavin and glycine. The chelate formed on removal of iron from transferrin must be stable enough to effectively remove the iron from transferrin, but not so stable as to prevent its uptake by ferritin. Thus EDTA and nitrilotriacetate, whilst extremely effective in chelating the Fe^{2+} released from transferrin by ascorbate, could not effectively transfer the iron to ferritin. The transfer reaction is *p*H dependent with a *p*H optimum at 7.3, and transfer is dependent on the iron content of the transferrin and ferritin. Thus transferrin of 50% or less iron saturation does not transfer iron: ferritin takes up most iron when it is about 50% saturated. The α fraction of ferritin (monomer) takes up iron much more effectively than β and γ-fractions. These results could be explained by the scheme presented in Fig. 13. Reduction of transferrin iron to form a stable Fe^{2+} chelate is

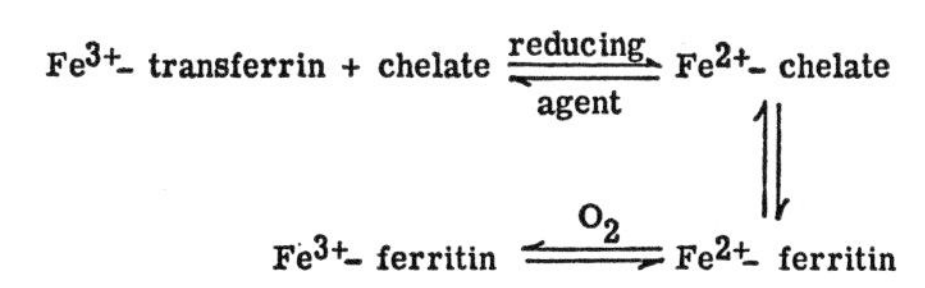

Fig. 13. Iron exchange between transferrin and ferritin. For details see text

followed by diffusion of this chelate through the semipermeable membrane and uptake of the Fe^{2+} by ferritin: the apoferritin then catalyses the oxidation of Fe^{2+} as discussed in Section VII. It was also reported (*209*) that when apotransferrin was dialysed against ferritin in the presence of reducing and chelating agents, the reverse transfer continues until transferrin is 40% saturated: the same mechanism is presumed to operate in reverse, namely reduction of ferritin iron, chelation of the Fe^{2+} and its transfer across the membrane followed by its oxidation and uptake by transferrin (again the question of whether the Fe^{2+} is taken up and then oxidised, or whether Fe^{3+} is taken up remains unclear).

When we consider the transfer of iron between ferritin and transferrin *in vivo*, there is much less known. Any mechanism that seeks to explain this process must take account of the following facts: iron must be presented to ferritin in the Fe^{2+} form and must be mobilized in this form. This necessitates a reduction step to convert transferrin iron to the divalent form, and its transport into the cell chelated to a suitable carrier. It has been reported that transferrin iron release to a membrane carrier is dependent on an NADH-linked electron transfer (*210*). Whether it is necessary to involve an enzyme to mediate oxidation of the Fe^{2+} prior to its binding by transferrin remains to be established. It has been suggested that ceruloplasmin might fulfil such a role (*211*): alternatively the transferrin might catalyse Fe^{2+} oxidation itself (*189*). The weak points in the scheme (Fig. 13) are clearly 1. the mechanism whereby iron is released from transferrin: it is known that there are receptor sites for transferrin on cell membranes (*212, 213*) but there is little information regarding the mobilization of transferrin iron and its transfer across the cell membrane. 2. the nature of the Fe^{2+}-chelator within the cell, and perhaps within the membrane is not established. 3. the precise mechanism by which transferrin takes up iron from the cell membrane after binding of apotransferrin to the appropriate receptor site remains unknown.

It seems that further progress in this area must await advances in our knowledge of the transport of iron across cell membranes and the interaction of transferrin, and perhaps also ferritin with receptor molecules on the outer and inner surface of the cell membrane.

X. Ferritin Biosynthesis and its Induction by Iron

Administration of iron to guinea pigs in large doses leads to the accumulation of large amounts of ferritin in the intestinal mucosal cells (*8*). It was suggested that iron would combine with pre-existing apoferritin to form ferritin and that the presence of the iron micelle would protect the protein against breakdown (*214*). The initial rate of synthesis of liver ferritin is increased several fold in iron-treated as compared to normal guinea pigs (*215*) and there is no comparable effect on the rate of synthesis of mixed liver proteins. This would suggest that iron has no effect on protein precursor pools and leads to the conclusion that iron administration accelerates the *de novo* synthesis of the protein moiety of ferritin. In experiments in which liver ferritin was fractionated by ultracentrifugation it was shown that the apoferritin rich fraction was more active than ferritin at short times after administration of tracer doses of ^{14}C-glycine to iron-stimulated guinea pigs and rats (*183, 184*). Stimulation of ferritin synthesis in liver slices was observed following iron treatment, and the uptake of labelled amino acids in slices from rats given actinomycin D was found to be 85% less than controls without this antibiotic[3]), which suggested that the control of ferritin synthesis was

[3]) Actinomycin D is a potent inhibitor of mammalian *m*RNA synthesis.

at the level of RNA synthesis (*216*). However, *Drysdale* and *Munro* (*184*) found the action of iron in stimulation of ferritin synthesis to be insensitive to actinomycin D. They revived the hypothesis of *Granick* that iron causes an apparent induction by stabilizing an unstable precursor of ferritin (presumably apoferritin) which would otherwise be rapidly degraded. Thus, the loss of radioactive molecules was considerably retarded by repeated injection of iron, suggesting that iron-rich ferritin molecules were less susceptible to degradation than iron-poor molecules. Further evidence that the effect of iron was not on *m*RNA synthesis came from experiments in which it was found that rats which had lost RNA from liver by protein deprivation had not lost their capacity to respond to iron (*217*). The explanation advanced for these findings was that *m*RNA for ferritin was stable during protein depletion or else that a reduced amount of template was being utilized more efficiently. That the stimulation of ferritin synthesis was not restricted to liver, but could be demonstrated in slices of spleen, testis, intestine and kidney was demonstrated by *Yoshino et al.* (*218*). These workers reported evidence favouring control of the transcription of RNA into *m*RNA as the site of iron action: actinomycin D inhibited the response to iron: iron treatment stimulated incorporation of radioactive precursors into a rapidly labelled fraction of liver nuclear RNA (presumed to be messenger RNA): direct addition of iron to liver slices *in vitro* did not stimulate ferritin synthesis, although the iron penetrated the liver cells. This last effect has been recently confirmed (*219*) and it has been shown that addition of iron in whole serum accelerates ferritin synthesis 2.5 fold. This effect of serum was attributed to its attenuation of the toxic effects of iron on protein synthesis: the serum also reduced iron uptake by the liver cells. Transferrin was not able to replace serum in this attenuation whereas albumin completely free of transferrin, was able to limit the uptake of excess iron by the liver cells and permit the specific induction of apoferritin biosynthesis by iron.

Chu and *Fineberg* (*220*) have studied the regulation of induction of ferritin synthesis in Hela cells, which originate from a human uterine carcinoma, are able to synthesise ferritin (*221*), and respond to iron administration by increasing their ferritin content (*222*). Acceleration of apoferritin synthesis by incubation of Hela cells with 10—100 μg of iron per ml was immediate and was not blocked by actinomycin D. Bipyridyl prevented the stimulatory effect of iron, whereas desferrioxamine prevented cellular iron uptake and removed much of the cellular ferritin iron. No measurable breakdown of radioactive ferritin was observed in a 4hr incubation of desferrioxamine-treated cells which argues against the stabilization hypothesis. More recently it has been reported that the half life of ferritin in young rats is 1.9 days and in old rats 4 days: the view is advanced that, since the rate of synthesis is the same, the higher levels of ferritin iron found in older rats, about 4 times as much iron, reflects the decreased rate of degradation (*223*).

The results of *Chu* and *Fineberg* (*220*) have been confirmed by *in vivo* studies on ferritin synthesis in rat liver (*224*). In this latter study cyclohexamide, which inhibits protein synthesis in rat liver (*225*) abolished the effects of iron; this supports the view that iron acts solely on the rate of *de novo* ferritin synthesis and not by stabilization. Actinomycin D treatment enhanced the effect of iron in Hela cells (*220*) and in rat liver (*224*). It was suggested that the induction of apoferritin by iron could be explained in one of two ways: 1. that binding of iron to nascent apoferritin chains on the polysomes relieves an inhibition of synthesis according to the general mechanism of *Cline* and *Bock* (*226*) 2. in view of the synergistic effect of actinomycin D, that there is an actinomycin-sensitive inhibitor at the level of translation for apoferritin synthesis, such as has been proposed in the case of tryptophan pyrrolase and tyrosine transamninase induction by glucorticoids (*227*). Iron would inactivate the inhibitor, and actinomycin could enhance the effect of iron by inactivation or blocking the regeneration of the inhibitor (*220*).

A different effect of iron-induced ferritin synthesis is obtained by excess dietary Zn^{2+} (*187*). This results in formation of iron-poor ferritin: the treated animals have an iron:protein ratio in liver ferritin that is only one-third of control animals. However, the Zn toxicity does not appear to interfere with the incorporation of radioactive acids into apoferritin and the Zn-treated animals can still synthesise apoferritin *de novo* in response to iron administration. The turnover rate of ferritin iron and ferritin protein from Zn-fed rats seem to be more rapid than the controls, which would be consistent with the view of *Drysdale* and *Munro* (*184*) that ferritin of low iron content is more rapidly degraded than ferritin of high iron content. The effect of Zn on the catalytic activity of ferritin has been discussed earlier.

The site of ferritin synthesis within the cells seems to be on the "free" polysomes (*228—230*). In contrast proteins such as albumin which are secreted from the cell appear to be made on "bound" polysomes (those which are attached to the endoplasmic reticulum). However, the regulation of ferritin synthesis has not been studied in such *in vitro* protein synthesising systems.

The synthesis of apoferritin by reticulocytes has been studied (*231*) and it has been established that administration of iron stimulates ferritin synthesis two fold over controls. The control of ferritin induction again seems to be at the level of translation. An analysis of apoferritin synthesis in human erythroid cells from normal and thalassemic subjects showed that thalassemic cells synthesised 37 times as much apoferritin as normal cells (*232*). The ratio, of ferritin to haemoglobin also increased in the thalassemic cells. Erythropoietin can induce erythrocyte formation in marrow cells in culture. When rat bone marrow cells are incubated with erythropoietin a system is found which accumulates extracellular iron within the interior of the committed cells: a part of the protein formed as a result of the action of erythropoietin is ferritin (*233*). The induced uptake of iron is inhibited by actinomycin D, puromycin and cyclohexamide, which suggests that both RNA synthesis and protein synthesis are required for iron accumulation and perhaps for ferritin synthesis.

Studies on ferritin synthesis in duodenal mucosal cells have been carried out by many workers. Oral and parenteral administration of iron causes a 3-fold and 10-fold increase in synthesis of mucosal apoferritin respectively (*234*). The doses of iron used were pharmacological *i.e.* greatly in excess of those encountered under physiological conditions. Similar doses were used to show that direct duodenal administration of iron stimulated the synthesis of ferritin by iron-deficient rats and that this newly synthesised ferritin did not prevent the uptake of more iron (*235*). Using an introduodenal dose as low as 10 μg induced synthesis of apoferritin by both iron-replete and iron-deficient rats and increasing amounts of apoferritin were made as the dose was increased (*236*). Mucosal ferritin synthesis in response to iron is also insensitive to actinomycin D but blocked by cyclohexamide, again suggesting that control is at the level of translation and does not involve gene activation (*237*).

The effect of iron administration on induction of ferritin in a number of tissues and in a series of hepatic tumours has been studied by *Linder* (*78, 144*) and the increase in synthesis of ferritin following hepatectomy of rats carrying transplantable tumours has been examined (*238*).

XI. Summary and Conclusion

Studies on the structure of apoferritin are advancing rapidly to the stage where the primary structure of the horse spleen molecule should soon be established. The quaternary structure of apoferritin from a variety of organisms seem to be similar; the protein is in all cases made up of 24 identical polypeptide chains of molecular weight 18,500 daltons. Studies on the subunit-subunit interactions are progressing, and a number of amino acid residues at the subunit interface have been identified. The process of oligomer formation from subunits is amenable to study at low *p*H. Apoferritin catalyses the oxidation of Fe^{2+} in the presence of oxidising agents with concomitant formation of ferritin. There is some evidence for the involvement of cysteine and histidine in this process. The mobilization of iron from ferritin may proceed via an NADH-dependent ferriductase which has been detected in liver cells from a number of animals. The regulation of iron uptake by ferritin and its mobilization from ferritin has also been studied in model systems, but the exact mechanisms involved *in vivo* for iron exchange between ferritin and transferrin remain to be established. Ferritin synthesis can be induced by administration of iron both *in vivo* and in cells growing in cell culture: the process appears to be regulated at the level of translation of a stable *m*RNA.

There remain a number of important questions concerning ferritin which are as yet unanswered. Information regarding the micellar iron core is not yet sufficiently advanced to allow a detailed molecular structure for the iron core to be determined unequivocally. Nor is it clear whether the structure of the core is determined by the protein, although this seems likely. Whether apoferritin undergoes a major conformation change on the uptake of iron is not yet established: there is evidence both for and against. Although the catalytic role of apoferritin in ferritin formation has been clearly shown, the precise mechanism for the formation of the iron core remains uncertain. The mobilization of iron from ferritin and the regulation of the balance between iron transfer from transferrin to ferritin and vice versa remain to be clearly determined. Although induction of ferritin synthesis by iron has been shown not to require DNA directed *m*RNA synthesis, the precise nature of the regulation involved is still a matter for speculation. Finally, the precise subunit structure, as well as the nature of subunit-subunit interactions, although under investigation, still remain to be clearly eatablished. In short the present review is, as must needs be, a progress report. However, that progress is being made, and will continue to be made, in the clarification of the structure and function of ferritin is clear from the results presented herein.

XII. Acknowledgement

I would like to thank a number of colleagues for their criticisms and suggestions during the course of the preparation of this review: *Drs. B. G. Malmström, J. L. Girardet, B. Sarkar, P. Aisen* and *P. M. Harrison,* and my research associates here in Berlin — *Drs. K. Wetz* and *K. Bitar, Miss M. Dognin* and *Mrs. C. Piekarek.* The generous facilities and interest of Professor *H. G. Wittmann* are gratefully acknowledged.

XIII. References

1. *Feeney, R. E., Komatsu, S. K.:* Struct. Bonding *1*, 149 (1966).
2. *Schmiedeberg, N.:* Arch. Exp. Pathol. Pharmakol. *33*, 101 (1894).
3. *Laufberger, V.:* Biol. Listy *19*, 73 (1934).
4. — Bull. Soc. Chim. Biol. *19*, 1575 (1937).
5. *Pollycove, M.:* In: The metabolic basis of inherited disease, 2nd edit., p. 780 (1966).
6. *Jandl, J. H., Katz, J. H.:* In: Iron metabolism ed. *Gross, F.* p. 103. Berlin-Göttingen-Heidelberg-New York: Springer 1964.
7. *Hahn, P. F., Bale, W. F., Ross, J. F., Balfour, W. M., Whipple, G. H.:* J. Exptl. Med. *78*, 169 (1943).
8. *Granick, S.:* J. Biol. Chem. *164*, 737 (1946).
9. *Crosby, W. H.:* Blood *22*, 441 (1963).
10. *Brown, E. B., Dubach, R., Moore, C. V.:* J. Lab. Clin. Med. *52*, 335 (1958).
11. *Stewart, W. B., Yuile, C. L., Claiburn, H. A., Snowman, R. T., Whipple, G. H.:* J. Exptl. Med. *92*, 375 (1950).
12. *Conrad, M. E., Crosby, W. H.:* Blood *22*, 406 (1963).
13. *Mazur, A., Shorr, E.:* J. Biol. Chem. *176*, 771 (1948).
14. *Baez, S., Mazur, A., Shorr, E.:* Am. J. Physiol. *162*, 198 (1950); *169*, 123, 134 (1952).
15. *Kuhn, R., Sörenson, N. A., Birkhofer, L.:* Ber. Chem. Ges. *73B*, 823 (1940).
16. *Granick, S.:* J. Biol. Chem. *149*, 157 (1943).
17. *Richter, G. W.:* Nature *207*, 616 (1965).
18. — Brit. J. Exptl. Pathol. *45*, 88 (1964).
19. *Arora, R. S., Lynch, E. C., Whitley, C. E., Allfrey, C. P.:* Texas Rep. Biol. Med. *28*, 189 (1970).
20. *Zweifach, B. W.:* In: Methods of medical research, p. 131. Chicago: Year Book Publishers Inc. 1948.
21. *Shorr, E.:* Harvey Lectures, p. 112 (1956).
22. *Skrikantia, S. G., Gopalan, C.:* J. Appl. Physiol. *14*, 829 (1959).
23. *Shorr, E., Baez, S., Zweifach, B. W., Payne, M. A., Mazur, A.:* Trans. Assoc. Am. Physicians *63*, 39 (1950).
24. *Reissman, K. R., Dietrich, M. R.:* J. Clin. Invest. *35*, 588 (1956).
25. *Wöhlers, F., Schonlau, F.:* Klin. Wochschr. *37*, 445 (1959).
26. *Aungst, C. W.:* J. Lab. Clin. Med. *71*, 517 (1968).
27. *Beamish, M. R., Addison, G. M., Llewellin, P., Hodgkins, M., Hales, C. N., Jacobs, A.:* Brit. J. Haematol. *22*, 637 (1972).
28. *Addison, G. M., Beamish, M. R., Hales, C. N., Hodgkins, M., Jacobs, A., Llewellin, P.:* J. Clin. Pathol. *25*, 326 (1972).
29. *Simson, J. V., Spicer, S. S.:* J. Cell. Biol. *52*, 536 (1972).
30. *Tanaka, Y., Cushard, J. A.:* J. Ultrastruct. Res. *33*, 436 (1970).
31. *Helgeland, L.:* FEBS Letters 1, 308 (1968).
32. *Björklid, E., Helgeland, L.:* Biochim. Biophys. Acta *221*, 583 (1970).
33. *van Wyck, C. P., Linder-Horowitz, M., Munro, H. N.:* J. Biol. Chem. *246*, 1025 (1971).
34. *Kato, T., Shimada, T.:* J. Biochem. *68*, 681 (1970).
35. — *Shinjo, S., Shimada, T.:* J. Biochem. *63*, 170 (1968).
36. *Brown, D. D., Caston, J. D.:* Develop. Biol. *5*, 445 (1962).
37. *Heneine, I. F., Gazzinelli, G., Tafuri, W. L.:* Comp. Biochem. Physiol. *28*, 391 (1969).

38. *Nardi, G., Muzii, E. O., Puca, M.:* Comp. Biochem. Physiol. *40*, 199 (1971).
39. *Roche, J., Bessis, M., Breton-Gorius, J., Stralin, H.:* Compt. Rend. Soc. Biol. *155*, 1790 (1961).
40. *Towe, K. M., Lowenstam, H. A., Nesson, M. H.:* Science *142*, 63 (1963).
41. *Hyde, B. B., Hodge, A. J., Kahn, A., Birnstiel, M. L.:* J. Ultrastruct. Res. *9*, 248 (1963).
42. *Robards, A. W., Humpherson, P. G.:* Planta *76*, 169 (1969).
43. *Seckbach, J.:* J. Ultrastruct. Res. *22*, 413 (1968).
44. *Robards, A. W., Robinson, C. L.:* Planta *82*, 179 (1968).
45. *Laulhere, J. P., Lambert, J., Berducon, J.:* Compt. Rend. *275*, 759 (1972).
46. *Craig, A. S., Williamson, K. I.:* Virology *39*, 616 (1969).
47. *Peat, A., Banbury, G. H.:* Planta *79*, 268 (1968).
48. *David, C. N., Easterbrook, K.:* J. Cell Biol. *48*, 15 (1971).
49. *Bozarth, R. F., Goenega, A.:* Can. J. Microbiol. *18*, 619 (1972).
50. *Granick, S.:* J. Biol. Chem. *146*, 451 (1942).
51. *Suran, A. A., Tarver, H.:* Arch. Biochem. Biophys. *111*, 399 (1965).
52. *Mazur, A., Litt, I., Shorr, E.:* J. Biol. Chem. *187*, 473 (1950).
53. *Drysdale, J. W., Ramsey, W. N., M.:* Biochem. J. *95*, 282 (1965).
54. — *Munro, H. N.:* Biochem. J. *95*, 851 (1965).
55. *Linder, M. C., Munro, H. N.:* Anal. Biochem. *48*, 266 (1972).
56. *Penders, T. J., de Rooij-Dijk, H. H., Leijnse, B.:* Biochim. Biophys. Acta *168*, 588 (1958).
57. *Crichton, R. R., Millar, J. A., Cumming, R. L. C., Bryce, C. F. A.:* Biochem. J. *131*, 51 (1973).
58. *Muraoka, N., Nakajima, S., Kato, T., Shimada, T.:* J. Biochem. *60*, 489 (1966).
59. *Granick, S., Michaelis, L.:* J. Biol. Chem. *147*, 91 (1943).
60. *Behrens, M., Taubert, M.:* Z. Physiol. Chem. *290*, 156 (1952).
61. *Crichton, R. R.:* Biochim. Biophys. Acta *194*, 34 (1969).
62. *Fischbach, F. A., Anderegg, J. W.:* J. Mol. Biol. *14*, 458 (1965).
63. *Rothen, A.:* J. Biol. Chem. *182*, 607 (1944).
64. *Philipott, J., de Bornier, B. M.:* Bull. Soc. Chim. Biol. *41*, 119 (1959).
65. *Granick, S.:* Physiol. Rev. 31, 489 (1951).
66. *Mazur, A., Shorr, E.:* J. Biol. Chem. *182*, 607 (1950).
67. *Farrant, J. L.:* Biochim. Biophys. Acta *13*, 569 (1954).
68. *Harrison, P. M.:* J. Mol. Biol. *6*, 404 (1963).
69. *Bielig, H. J., Kratky, O., Rohns, G., Wawra, H.:* Biochim. Biophys. Acta *112*, 110 (1966).
70. *Kleinwachter, V.:* Arch. Biochem. Biophys. *105*, 352 (1964).
71. *Richter, G. W.:* J. Biophys. Biochem. Cytol. *6*, 531 (1959).
72. *Pape, L., Multani, J., Stitt, C., Saltman, P.:* Biochemistry *7*, 606 (1968).
73. *Michaelis, L., Coryell, C. D., Granick, S.:* J. Biol. Chem. *148*, 463 (1943).
74. *Granick, S., Hahn, P. F.:* J. Biol. Chem. *155*, 661 (1944).
75. *Harrison, P. M.:* In: Iron metabolism (ed. *F. Gross*), p. 40. Berlin–Göttingen–Heidelberg–New York: Springer 1964.
76. *van Bruggen, E. F. J., Wiebenga, E. H., Gruber, M.:* J. Mol. Biol. *2*, 81 (1960).
77. *Haggis, G. H.:* J. Mol. Biol. *14*, 598 (1965).
78. *Linder-Horowitz, M., Reutinger, R. T., Munro, H. N.:* Biochim. Biophys. Acta *200*, 442 (1970).
79. *Muir, A. R.:* Quart. J. Exptl. Physiol *45*, 192 (1960).
80. *Bessis, M., Breton-Gorius, J.:* Compt. Rend. *250*, 1360 (1960).
81. *Chescoe, D., Agar, A. W.:* J. Microscopie *5*, 91 (1966).

82. *Haydon, G. B.:* J. Microsc. *89*, 251 (1969).
83. *Harrison, P. M., Hoy, T. G.:* J. Microsc. *91*, 61 (1970).
84. *Haydon, G. B.:* J. Microsc. *91*, 65 (1970).
85. *Harrison, P. M., Fischbach, F. A., Hoy, T. G., Haggis, G. H.:* Nature *216*, 1188 (1967).
86. *Massover, B.:* 30th Ann. Electron Microsc. Soc. America, p. 182 (1972).
87. — *Lacaze, J. C., Durrien, L.:* J. Ultrastruct. Res., in press (1973).
88. *Harrison, P. M.:* J. Mol. Biol. *1*, 69 (1959).
89. *Francombe, M. H., Rooksby, H. P.:* Clay Min. Bull. *4*, 1 (1959).
90. *Bernal, J. D., Dasgupta, D. R., Mackay, A. L.:* Clay Min. Bull. *4*, 15 (1959).
91. *Towe, K. M., Bradley, W. F.:* J. Colloid Interface Sci. *24*, 384 (1967).
92. *Girardet, J. L., Lawrence, J. J.:* Bull. Soc. Franc. Mineral. Crist. *91*, 440 (1968).
93. *Alix, D., Girardet, J. L., Lawrence, J. J., Mouriquand, C.:* J. Microscopie *12*, 33 (1971).
94. *Bayer, E., Hausser, K. H.:* Experientia *11*, 254 (1955).
95. *Schoffa, G.:* Z. Naturforsch. *20b*, 167 (1965).
96. *Smart, J. S.:* In: Magnetism, Vol. III, p. 69 (ed. *Rado, G. T.* and *Suhl, H.*). New York: Academic Press 1963.
97. *Blaise, A., Chappert, J., Girardet, J. L.:* Compt. Rend. *261*, 2310 (1965).
98. *Boas, J. F., Window, B.:* Australian J. Phys. *19*, 573 (1966).
99. *Girardet, J. L.:* 1st Europ. Congress Biophys., Baden, p. 119 (1971).
100. *Boas, J. F., Troup, G. J.:* Biochim. Biophys. Acta *229*, 68 (1971).
101. *Spiro, T. G., Saltman, P.:* Struct. Bonding *6*, 116 (1969).
102. *Crichton, R. R., Eason, R., Barclay, A., Bryce, C. F. A.:* Biochem. J. *131*, 855 (1973).
103. *Fischbach, F. A., Gregory, D. W., Harrison, P. M., Hoy, T. G., Williams, J. M.:* J. Ultrastruct. Res. *37*, 495 (1971).
104. *Richter, G. W., Walker, G. F.:* Biochemistry *6*, 2871 (1967).
105. *Björk, I., Fish, W. W.:* Biochemistry *10*, 2844 (1971).
106. *Crichton, R. R.:* Biochem. J. *126*, 761 (1972).
107. *Bryce, C. F. A., Crichton, R. R.:* J. Biol. Chem. *246*, 4198 (1971).
108. *Harrison, P. M., Hofmann, T.:* J. Mol. Biol. *4*, 239 (1962).
109. *Hofmann, T., Harrison, P. M.:* J. Mol. Biol. *6*, 256 (1963).
110. *Harrison, P. M., Hofmann, T., Mainwaring, W. I. P.:* J. Mol. Biol. *4*, 251 (1962).
111. *Mainwaring, W. I. P., Hofmann, T.:* Arch. Biochem. Biophys. *125*, 975 (1968).
112. *Crichton, R. R., Bryce, C. F. A.:* FEBS Letters *6*, 121 (1970).
113. *Friedberg, F.:* Can. J. Biochem. Physiol. *40*, 983 (1962).
114. *Shapiro, A. L., Maizel, Jr., J. V.:* Anal. Biochem. *29*, 505 (1969).
115. *Reynolds, J. A., Tanford, C.:* J. Biol. Chem. *245*, 5161 (1970).
116. *Weber, K., Kuter, D. J.:* J. Biol. Chem. *246*, 4504 (1971).
117. *Williams, M. A., Harrison, P. M.:* Biochem. J. *110*, 265 (1968).
118. *Harrison, P. M., Hofmann, T.:* Biochem. J. *80*, 38P (1961).
119. *Spies, J. R., Chambers, D. C.:* Anal. Chem. *21*, 1249 (1949).
120. *Edelhoch, H.:* Biochemistry *6*, 1948 (1967).
121. *Scoffone, E., Fontana, A., Rochi, R.:* Biochemistry *7*, 971 (1970).
122. *Macara, I. G., Hoy, T. G., Harrison, P. M.:* Biochem. J. *126*, 151 (1972).
123. *Bryce, C. F. A., Crichton, R. R.:* Biochem. J., *133*, 301 (1973).
124. *Suran, A. A.:* Arch. Biochem. Biophys. *113*, 1 (1966).
125. *Crichton, R. R.:* Technicon Symposium "Automation in Analytical Chemistry" London 1969, p. 167 (1971).
126. — Biochem. J. *117*, 34P (1970).

127. — *Barbirolli, V.:* FEBS Letters *6,* 134 (1970).
128. — Z. Physiol. Chem. *352,* 9 (1971).
129. *Bryce, C. F. A., Crichton, R. R.:* Z. Physiol. Chem. *352,* 9 (1971).
130. — — J. Chromatog. *63,* 267 (1971).
131. *Drysdale, J. W.:* Biochim. Biophys. Acta *207,* 256 (1970).
132. *Uroshizaki, I., Niitsu, Y., Ishitani, K., Matsuda, M., Fukada, M.:* Biochim. Biophys. Acta *243,* 187 (1971).
133. *Kilmartin, J. V., Clegg, J. B.:* Nature *213,* 269 (1967).
134. *Lee, J. C. K., Richter, G. W.:* Comp. Biochem. Physiol. *35,* 325 (1971).
135. *Bryce, C. F. A., Crichton, R. R.:* Z. Physiol. Chem. *354,* 344 (1973).
136. *Allfrey, C. P., Lynch, E. C., Whitley, C. E.:* J. Lab. Clin. Med. *70,* 419 (1967).
137. *Gabuzda, T. G., Gardner, F. H.:* Blood *29,* 770 (1967).
138. — *Pearson, J.:* Biochim. Biophys. Acta *194,* 50 (1969).
139. — *Silver, R. K.:* J. Cell. Physiol. *74,* 273 (1969).
140. *Richter, G. W.:* Exptl. Mol. Pathol. *6,* 96 (1967).
141. — *Moppert, G. A., Lee, J. C. K.:* Comp. Biochem. Physiol. *32,* 451 (1970).
142. — Lab. Invest. *16,* 187 (1967).
143. — *Lee, J. C. K.:* Cancer Res. *30,* 880 (1970).
144. *Linder, M., Munro, H. N., Morris, H. P.:* Cancer Res. *30,* 2231 (1970).
145. — *Moor, J. R., Scott, L. E., Munro, H. N.:* Biochem. J. *129,* 455 (1972).
146. *Saadi, R.:* Rev. Franc. Etudes Clin. Biol. *7,* 408 (1962).
147. *Theron, J. J., Hawtrey, A. O., Schirron, V.:* Clin. Chim. Acta *8,* 165 (1963).
148. *Kopp, R., Vogt, A., Maass, G.:* Nature *198,* 892 (1963).
149. *Fine, J. M., Harris, G.:* Clin. Chim. Acta *8,* 794 (1963).
150. *Carnevali, F., Tecce, G.:* Arch. Biochem. Biophys. *105,* 207 (1964).
151. *Kopp, R., Vogt, A., Maass, G.:* Nature *202,* 1211 (1964).
152. *Harrison, P. M., Gregory, D. W.:* J. Mol. Biol. *14,* 626 (1965).
153. *Kauzman, W.:* Advan. Prot. Chem. *14,* 1 (1959).
154. *Jones, W. A. K., Williams, M. A.:* Biochem. J. *126,* 17P (1972).
155. *Singer, S. J.:* Nature *183,* 1523 (1959).
156. *Gregory, D. W., Williams, M. A.:* Biochim. Biophys. Acta *133,* 319 (1967).
157. *Listowsky, I., Bauer, G., Englard, S., Betheil, J. J.:* Biochemistry *11,* 2176 (1972).
158. — *Englard, S., Betheil, J. J.:* Biochemistry *6,* 1341 (1967).
159. *Wood, G. C., Crichton, R. R.:* Biochim. Biophys. Acta *229,* 83 (1971).
160. *Crichton, R. R.:* Biochem. J. *119,* 40P (1970).
161. — Biochim. Biophys. Acta *229,* 75 (1971).
162. *Smith-Johannsen, H., Drysdale, J. W.:* Biochim. Biophys. Acta *194,* 43 (1969).
163. *Fraenkel-Conrat, H.:* Virology *4,* 1 (1957).
164. *Jaenicke, R., Bartman, P.:* Biochem. Biophys. Res. Commun. *49,* 884 (1972).
165. *Richter, G. W.:* Biochim. Biophys. Acta *257,* 471 (1972).
166. *Crichton, R. R., Bryce, C. F. A.:* Biochem. J., *133,* 289 (1973).
167. — Biochem. J. *130,* 35P (1972).
168. *Vanacek, J., Keil, B.:* Collection Czech. Chem. Commun. *34,* 1067 (1969).
169. *Perutz, M. F.:* Proc. Roy. Soc. (London), Ser. B *173,* 113 (1969).
170. *McLachlan, A. D., Perutz, M. F., Pulsinelli, P. D.:* 23rd Mosbach Colloquium der Gesellschaft für Biologische Chemie, p. 91 (1972).
171. *Fankuchen, I.:* J. Biol. Chem. *150,* 57 (1943).
172. *Crowfoot, D. C.:* Cold Spring Harbor Symp. Quant. Biol. *14,* 65 (1949).
173. *Cowan, P. M.:* D. Phil. thesis, Oxford (1953).
174. *Easterbrook, K. B.:* J. Ultrastruct. Res. *33,* 442 (1970).

175. *Hawkins, H., Mergner, W. J., Henkins, R., Kinney, T. D., Trump, B. F.:* 28th Ann. Meeting of the Electron Microscopic Soc. of America, p. 268 (1970).
176. *Hanson, K. R.:* J. Mol. Biol. *38,* 133 (1968).
177. *Bielig, H. J., Bayer, E.:* Naturwissenschaften *42,* 125 (1955).
178. *Loewus, M. W., Fineberg, R. A.:* Biochim. Biophys. Acta *26,* 441 (1957).
179. *Drysdale, J. W., Haggis, G. H., Harrison, P. M.:* Nature *219,* 1045 (1968).
180. *Brady, G. W., Kurjian, C. R., Lyden, E. F. X., Robin, M. B., Saltman, P.: Spiro, T., Terzis, A.:* Biochemistry *7,* 2185 (1968).
181. *Cepurneek, C. P., Hagenmauer, J. C., Meltzer, M. L., Winslow, J. C., Spiro, T. G., Saltman, P.:* Federation Proc. *30,* Abstr. 1469 (1971).
182. *Crichton, R. R.:* New Engl. J. Med. *284,* 1413 (1971).
183. *Fineberg, R. A., Greenberg, D. M.:* J. Biol. Chem. *214,* 107 (1955).
184. *Drysdale, J. W., Munro, H. N.:* J. Biol. Chem. *241,* 3630 (1966).
185. *Niederer, W.:* Experientia *26,* 218 (1970).
186. Cold Spring Harbor Symp. Quant. Biol. *36,* (1971).
187. *Coleman, C. B., Matrone, G.:* Biochim. Biophys. Acta *177,* 106 (1969).
188. *Macara, I. G., Hoy, T. G., Harrison, P. M.:* Trans. Biochem. Soc. *1,* 102 (1973).
189. *Bates, G. W., Workman, E. F., Schlabach, M. R.:* Biochem. Biophys. Res. Commun. *50,* 84 (1973).
190. *Ross, J., Kochwa, S., Wasserman, L. R.:* Biochim. Biophys. Acta *154,* 70 (1968).
191. *Gaber, B. P., Aisen, P.:* Biochim. Biophys. Acta *221,* 228 (1970).
192. *Mazur, A., Baez, S., Shorr, E.:* J. Biol. Chem. *213,* 147 (1955).
193. *Deron, S., Dognin, J., Girardet, J. L.:* Compt. Rend. *269,* 729 (1969).
194. *Dognin, J., Girardet J. L., Chapron, Y.:* Biochim. Biophys. Acta *292,* 276 (1973).
195. *Bielig, H. J., Bayer, E.:* Naturwissenschaften *42,* 466 (1955).
196. *Niederer, G.:* Experientia *25,* 804 (1969).
197. *Pape, L., Multani, J. S., Stitt, C., Saltman, P.:* Biochemistry *7,* 613 (1968).
198. *Jones, M. M., Johnston, D. O.:* Nature *214,* 509 (1967).
199. *Green, S., Mazur, A.:* J. Biol. Chem. *227,* 653 (1967).
200. *Mazur, A., Green, S., Saha, A., Carleton, A.:* J. Clin. Invest. *37,* 1809 (1958).
201. *Tanaka, S.:* J. Biochem. *37,* 129 (1956).
202. *Ayzvasian, J. H.:* New Engl. J. Med. *270,* 18 (1964).
203. *Stromeyer, G. W., Miller, S. A., Scarlata, R. W., Moore, E. W., Greenberg, M. S., Chalmers, T. C.:* Am. J. Physiol. *207,* 55 (1964).
204. *Mazur, A, Carleton, A.:* Blood *26,* 317 (1965).
205. *Baxter, C. F., van Reen, R.:* Biochim. Biophys. Acta *28,* 573 (1958).
206. *Osaki, S., Sirivech, S.:* Federation Proc. *30,* Abstr. 1394 (1971).
207. *Mazur, A., Green, S., Carleton, A.:* J. Biol. Chem. *235,* 595 (1960).
208. *Miller, J. P. G., Perkins, D. J.:* Europ. J. Biochem. *10,* 146 (1969).
209. — Ph. D. thesis, London University (1968).
210. *Morgan, E. H., Baker, E.:* Biochim. Biophys. Acta *184,* 442 (1969).
211. *Osaki, S., Johnson, D. A., Frieden, E.:* J. Biol. Chem. *241,* 2746 (1966).
212. *Jandl, J. H., Katz, J. H.:* J. Clin. Invest. *42,* 314 (1963).
213. *Morgan, E. H.:* Brit. J. Haematol. *10,* 442 (1964).
214. *Granick, S.:* Bull. N. Y. Acad. Med. *25,* 403 (1959).
215. *Fineberg, R. A., Greenberg, D. M.:* J. Biol. Chem. *214,* 97 (1955).
216. *Yu, F. L., Fineberg, R. A.:* J. Biol. Chem. *240,* 2083 (1965).
217. *Drysdale, J. W., Olafsdotter, E. Munro, H. N.:* J. Biol. Chem. *243,* 552 (1968).
218. *Yoshino, Y., Manis, J., Schachter, D.:* J. Biol. Chem. *243,* 2911 (1968).
219. *Smith, L., Fineberg, R. A.:* Blood *39,* 274 (1972).

220. *Chu, L. L. H., Fineberg, R. A.:* J. Biol. Chem. *244,* 3847 (1969).
221. *Radola, B. J., Kellner, G., Frimmel, J. S.:* Nature *207,* 206 (1965).
222. *Richter, G. W.:* Nature *190,* 413 (1961).
223. *Ove, P., Obenrader, M., Lansing, A.:* Biochim. Biophys. Acta *277,* 211 (1972).
224. *Millar, J. A., Cumming, R. L. C., Smith, J. A., Goldberg, A.:* Biochem. J. *119,* 643 (1970).
225. *Yeh, S. D. H., Shils, M. E.:* Biochem. Pharmacol. *18,* 1919 (1969).
226. *Cline, A. L., Bock, R. M.:* Cold Spring Harbor Symp. Quant. Biol. *31,* 321 (1966).
227. *Tomkins, G. M., Thompson, E. B., Hayashi, S., Gelehrte, T., Granner, D., Peterkovsky, B.:* Cold Spring Harbor Symp. Quant. Biol. *31,* 349 (1966).
228. *Hicks, S. J., Drysdale, J. W., Munro, H. N.:* Science *164,* 584 (1969).
229. *Redman, C. M.:* J. Biol. Chem. *244,* 4308 (1969).
230. *Puro, D. G., Richter, G. W.:* Proc. Soc. Exptl. Biol. Med. *138,* 399 (1971).
231. *Matioli, G. T., Eylar, E. H.:* Proc. Natl. Acad. Sci. *52,* 508 (1964).
232. *Eylar, E. H., Matioli, G. T.:* Nature *208,* 661 (1965).
233. *Hrinda, M. E., Goldwasser, E.:* Biochim. Biophys. Acta *195,* 165 (1969).
234. *Smith, J. A., Drysdale, J. W., Goldberg, A., Munro, H. N.:* Brit. J. Haematol. *14,* 79 (1968).
235. *Brittin, G. M., Raval, D.:* J. Lab. Clin. Med. *75,* 811 (1970).
236. — — J. Lab. Clin. Med. *77,* 54 (1971).
237. *Millar, J. A., Goldberg, A., Cumming, R. L. C.:* In: "Iron deficiency: Pathology, Clinical aspects, therapy, p. 121 (ed. *L. Hallberg, H. G, Hallwerth A. Vannotti*). London: Academic Press 1970.
238. *Lee, J. C. K.:* Am. J. Pathol. *65,* 347 (1971).

Received April 5, 1973

Metal-Polypeptide Interactions: The Conformational State of Iron Proteins

Miguel Llinás

Department of Biochemistry*, University of California
Berkeley, Calif. 94720, U.S.A.

Table of Contents

---

* A major portion of this article was written while the author was a member of the Laboratory of Chemical Biodynamics on the same campus.

I. Introduction

Molecular evolution of metalloproteins has had to incorporate and preserve structural features that allow a fit of the metal atom at the binding site (*1*). The resulting conformation had to allow for the ligands (certain amino acid side chain groups) to properly arrange in space so that the enclosed complex could be one of well defined (square planar, octahedral, etc.) symmetry. Such an ideal situation, yielding minimal ground state energy for the metal atom, is difficult to achieve because of conformational pressures introducing distortions which obstruct perfect symmetry at the metal center. A strained situation thus prevails which raises the ground state energy of the active site and enhances the catalytic reactivity of the molecule. This effect has been named '*entasis*' (from the Greek 'under tension') by *Vallée* and *Williams* (*2*) who have made use of the metal complex chromophore as a most sensitive spectrochemical label to probe the state of the protein (*2, 3*). As a concept, entasis is thus mainly concerned with the results of structural constraints on the state of the active site and is one of the various factors distinguishing the metal-binding properties of metalloenzymes from those of more simple model compounds (*4*).

There is, however, another aspect of the metal-polypeptide chain interaction which is of relevance if the structural-functional relationships in metalloproteins are to be understood. This is the effect of the metal-binding event on the protein conformation and it can be formulated as follows: the peculiar configuration of the ligands dictated by the chelate symmetry requirements somehow distort the conformation which, under the same conditions, the protein would assume in the absence of the metal ion. In other words: metal-binding can result in a significant conformational change of the polypeptide part of the molecule and the resulting structure has to compromise the strains and stresses generated on the polypeptide backbone by each single residue and by the metal complex site. This raises the conformational free energy of the molecule, the thermal balance of the

$$\text{Metal} + \text{Apoprotein} \rightarrow \text{Metalloprotein}$$

reaction being thus accounted for by the free energy of metal-binding. The final conformation of the protein may, however, be kinetically very stable: *i.e.* the metal is not necessarily easily released to yield the apo ("denatured") species. This is because the spatial arrangement of the ligands, "imperfect" as it may be, and the protein conformation as a whole are so coupled that the conformational fluctuations responsible for metal-exchange become an unlikely cooperative event. This "Multiple Juxtapositional Fixedness" effect has been recognized by *Bush et al.* (*5*) to be a major factor in determining the stability of macrocyclic ligands and explicitly formulated as an outlook complementary to the entatic aspects of the metalloprotein state. It is thus obvious that a static conformational description of the protein is somewhat insufficient in that the stability of the molecule is ignored. This might, indeed, be a most important feature of the protein as a whole in order to distinguish homologous polypeptides that are, otherwise, conformationally similar.

In this article we propose to survey the role of iron as a conformational determinant in polypeptides and in non-haem proteins. This aspect of iron coordination research has been either ignored or only sporadically dealt with in reviews concerned with metalloproteins. By stressing this often circumstantial aspect of iron biochemistry it is our hope that the relevance of metals in general as fundamental structural factors in biomolecules will be brought into proper perspective.

It is not our intention to review the entire field of iron proteins. Rather, we will focus on certain cases for which the experimental evidence concerning the conformational consequences of the iron-polypeptide interaction are well established. Stereochemical aspects of metalloporphyrins have recently been surveyed (*6*) and some conformational studies on haemoproteins were elegantly presented by *Wüthrich* in this series (*7*). We will also ignore ferritin, where iron is non-structural but is present as a micellar aggregate of mixed hydroxide-phosphate salt coated by *20—24* globular protein units (*8*) (See the review by *R. R. Crichton* in this issue).

The structural role of iron to be discussed here is, in general, somewhat broader than that already described in the literature for Ca^{2+}, Mg^{2+}, and the light alkali metal ions (*9—11*). These cations appear to function mainly as effectors of enzyme activity, either by modulating conformational changes of small amplitude that regulate the affinity of the substrate for the active site or, more directly, by bridging the substrate to the enzyme. These proteins possibly are already in a native state, even in the absence of the metal ion. By contrast, and with the possible exception of Mn^{2+}, in most cases heavy metal ions are necessary constituents to hold the structure of the metalloprotein in a conformation that is close, if not identical, to that of the active species. This role is in addition to whatever action the heavy ion may play in the catalytic event itself.

Another aspect of interest is how the oxidation state of the iron atom controls the conformational state of the polypeptide. In certain cases the electronic configuration of the ion is of fundamental importance since only one valence state can exist chelated by the polypeptide. Such appears to be the case in the siderochromes and in transferrin, where only Fe^{3+} is coordinated with significant affinity. Indeed, it was proposed that reduction of ferrichrome might be the mechanism of release of the metal inside the cell so as to make it available for biosynthetic purposes (*12*, *13*). With regard to non-haem electron transport iron proteins, a suggestion made by *Takano et al.* (*14*) for cytochrome c may be extended. By mediating an electron transfer between a preceding reductase and a subsequent oxidase, a redox protein may have to alter its shape slightly to fit the active sites of either one of the two connected enzymes.

We will also be concerned with the extent to which the polypeptide chain itself participates in determining the final conformation of the iron protein. That is, the extent to which the conformational state of these compounds is perturbed by residue substitutions that are not fundamental either for metal-binding or for (contact or covalent) steric interactions.

The compounds to be surveyed here comprise the siderochromes, the transferrins, hemerythrin, the rubredoxins and the iron-sulfur proteins. They differ widely in their size, function, oligomeric state, iron content and nature of the metal-binding site. Historically this has caused their study to follow different routes. Although the problem of the chemical and electronic nature of the iron chromophore has dominated their investigation, the peculiarities of their active sites have demanded different approaches in each case. Furthermore, the procedures which have been used to reveal differences among metal-free and iron-bound (oxidized and reduced) proteins have provided only partial outlooks of the conformational process. The 'lock and key' insights gained from very sensitive biological recognition phenomena, namely, proteolytic digestion, antigen-antibody reactivity, and membrane attachment (iron transport) experiments, are not too instructive in molecular terms. In this sense, although hydrodynamic properties are also of limited value, they yield information that is unequivocal. Optical spectropolarimetry, very sensitive to conformational features of the polypeptide backbone, is complicated by the side chain aromatic chromophores and the metal site itself; unless these side effects can be screened, interpretation of the data is ambiguous at best. Of all the spectroscopic methods, NMR[1)] is probably the one that offers the highest resolution but this is still far from sufficient for protein-sized molecules. For low molecular weight ($\lesssim$2,000 dalton) substances the technique is extremely useful if resonance broadening by

[1)] Abbreviations: NMR, nuclear magnetic resonance; PMR, proton magnetic resonance; ppm, parts per million; TMS, tetramethylsilane; MW, molecular weight; EPR, electron paramagnetic resonance; CD, circular dichroism; ORD, optical rotatory dispersion; HiPISP, high potential iron-sulfur portein.

a paramagnetic ion is not extreme. Contact shifts of the resonances and iron substitution by a similar ion can provide valuable information for solution studies. If atomic resolution is intended, X-ray crystallography closely approaches this goal and is probably the most satisfying. However, it gives data for conditions that are far removed from those prevailing in solution and provides no information on conformational dynamics. As discussed below, this last aspect is suitably probed by amide hydrogen exchange studies once the purely conformational contribution to the measured kinetics have been estimated by other techniques.

We intend to select from the literature only that information bearing on the conformational state of these molecules. It should be stressed, however, that because of the non-parallel evolution of their individual research, the treatment given to each of these iron compounds necessarily has to differ.

II. The Siderochromes

The siderochromes (from Greek: *sideros* = iron, *khroma* = color) comprise a group of ferric chelators, mostly peptides, which are involved in microbial iron transport (*13, 15, 16*). Their molecular weight ranges from about 500 to 1,100 daltons, which makes these compounds especially suited for conformational studies. Our concern with the siderochromes at this initial point is justified not only because of their relative simplicity but also because they most clearly lead to the concept of 'conformational state' which, in turn, will serve as a guideline in discussing larger molecules. Furthermore, extensive studies in our laboratory and elsewhere make these compounds among the best understood iron biomolecules in physical terms. They should serve as realistic models to foster understanding of more complex structures.

According to *Neilands* (*13*) the siderochromes may be classified into two broad classes; the hydroxamates and the phenolates, depending on which of these two groups provide the dentates for metal-binding. Our survey will be mainly concerned with the ferrichromes, a family of similar ferric trihydroxamate cyclohexapeptides. Mention will be made of results on ferrioxamine B, a linear ferric hydroxamate and some recent evidence on enterobactin will also be presented. The latter, a cyclic triester containing amide bonds, exemplifies phenolate iron-binding in general. A number of reviews have been published which deal with the chemical and biological properties of these compounds (*13, 16—21* a,b).

A. The Ferrichromes

The primary structure of the ferrichromes discussed here can be represented as

$$\lfloor \text{Res}^3\text{-Res}^2\text{-Gly}^1\text{-Orn}^3\text{-Orn}^2\text{-Orn}^1 \rfloor \;(\text{cyclic},\ \leftarrow)$$

where $\text{Res}^{2,3}$ represent glycyl (Gly) or L-seryl (Ser) and Orn represents δ-N-acyl-δ-N-hydroxy-L-ornithyl residues. The hydroxamic acid bidentates are thus furnished by the δ-N substituted ornithyl side chains.

In the ferrichromes the iron is high spin ferric, with a magnetic moment of 5.73 Bohr magnetons (*22*). The nature of the iron locus has been studied by Mössbauer (*23, 24*) and far infrared (*25*) techniques. The low temperature EPR spectrum of ferrichrome A exhibits characteristic $g = 4.3$ and $g = 9.6$ bands (*26*). These studies have indicated a strong crystal field of low symmetry.

All the ferrichromes whose conformations have been studied are described in Fig. 1. This structural model is based on the X-ray crystallographic study of ferrichrome A by *Zalkin et al.* (*27, 28*) and in most respects, it is confirmed by PMR solution studies (*29, 30*). The structure exhibited by the model in Fig. 1 is globular with the three substituted ornithyl side chains embracing the metal ion in octahedral coordination while the cyclohexapeptide backbone assumes a distorted antiparallel β-pleated sheet structure.

On the basis of chemical evidence suggesting cyclodimerization of linear tripeptides ('doubling reaction') *Schwyzer* has proposed (*31*) a conformational model for cyclohexapaptides involving two transannular hydrogen bonds which pair the carbonyl oxygen and amide nitrogen atoms of opposed residues, e.g., in ferrichrome it would be

$$\text{Res}^3 \begin{matrix} \diagup \text{C=O}\cdots\text{H—N} \diagdown \\ \diagdown \text{N—H}\cdots\text{O=C} \diagup \end{matrix} \text{Orn}^3 .$$

The X-ray data yields, however, a distance between the Res^3 N and the Orn^3 O atoms too great for an H bond; the 2.99 Å between the Res^3 O and the Orn^3 N is, in contrast, consistent with the presence of a moderately strong transannular H-bond between them. X-ray studies on cyclohexaglycine (*32*) and on cyclo Gly_4-(D-Ala)$_2$ (*33*) have detected the presence of *Schwyzer* conformers in the crystalline samples. Furthermore, a number of studies on synthetic cyclohexapeptides in solution (*34*) by PMR have yielded consistent evidence supporting the pleated sheet structure, especially in solvents of low polarity. This may not, however,

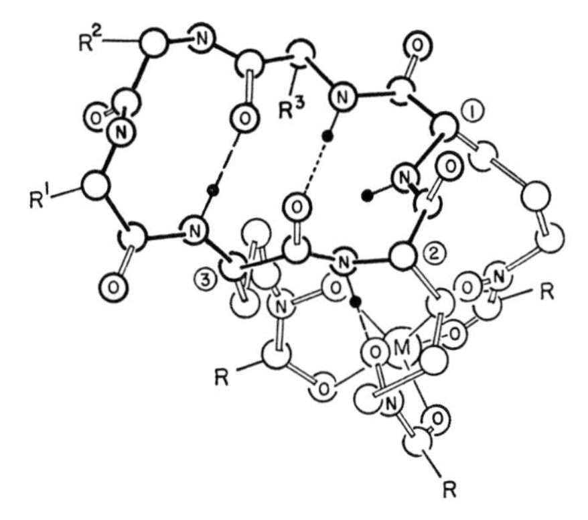

Fig. 1. The structure of the ferrichromes (*28, 30*). The backbone bonds and atoms are shown in heavier lines. H bonds detected in the crystallographic study (---) were confirmed in the PMR investigations which also suggested the presence of a second, weak, transannular H bond (· · ·) for the seryl analogues in solution (*59*). The compositional differences among the ferrichromes surveyed here are summarized below:

	R^2	R^3	R
ferrichrome	H	H	CH_3
ferricrocin	CH_2OH	H	CH_3
ferrichrysin	CH_2OH	CH_2OH	CH_3
ferrichrome A	CH_2OH	CH_2OH	$>C=C<$ (CH_3; H, CH_2COOH)

R^2 and R^3 represent side chains of those glycyl or seryl residues at sites 2 and 3 respectively, site 1 being occupied by a glycyl residue ($R^1 = H$). R represents the acyl group in the hydroxamate moiety (acetic acid or *trans*-*β*-methyl glutaconic acid). The prefix 'ferri' refers to the naturally occurring ferric chelates. When replaced by 'alumi' (as in 'alumicrocin'), or 'galli' (as in 'gallichrome'), Fe^{3+} has been substituted by Al^{3+} or Ga^{3+}, respectively

be a general feature as structural constraints may yield other conformations (*35*).

A *Pseudomonad* has been isolated that can grow on ferrichrome as sole carbon and nitrogen source. The first step in the degradation is catalyzed by a highly specific peptidase which hydrolyses the amide linkage between Orn^1 and Res^3 (*36*). The peptidase acts on the ferric complexes but not on the iron-free peptides. More recently, Emery has found that the ferrichrome-mediated iron transport in *Ustilago sphaerogena* is specific for this ionophore; other ferric chelators or natural ferric hydroxamates being inactive (*37*). The data suggest that the basis for recognition is conformation of the chelate rather than other physical properties of the carrier since both the *Pseudomonad* peptidase and the *Ustilago* transport system acted on the Al^{3+} analogue of ferrichrome ('alumichrome') as well (*37, 38*). Furthermore, it has been noted that while ferrichrome is readily crystallizable from methanol solutions the metal-free peptide has not been crystallized (*39*).

We have extensively studied the solution conformation of the metal-free ferrichromes, deferriferrichrome, deferriferricrocin and deferriferrichrysin (*29, 40*). By plotting the PMR chemical shift versus temperature for the different amides, an estimate of the relative extent of steric shielding or intramolecular H-bonding can be obtained from their linear slopes. Fig. 2, a and b, show this kind of plot for deferriferrichrome both in H_2O and in $(CD_3)_2SO$. It is evident from such graphs that while little distinction can be made between the extent of shielding of the amides in water, at least two amides, namely those denoted by G_3 (a glycyl

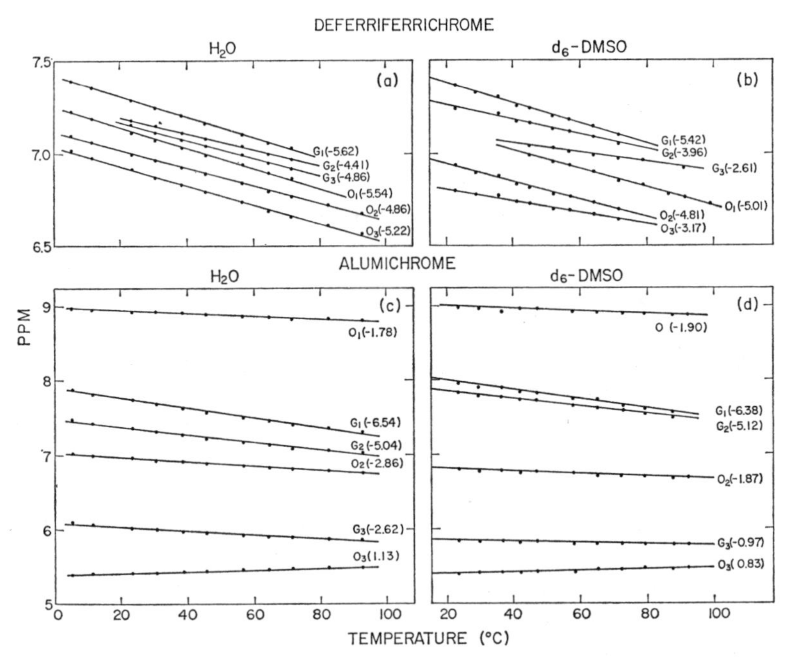

Fig. 2. The temperature dependence of the deferriferrichrome and alumichrome amide N*H* PMR chemical shifts, in water (*p*H 5.14) and in $(CD_3)_2SO$ (*29, 43*). O^i and G^i refer to the ith ornithyl and glycyl amide proton, respectively, in the order of scanning from low to high magnetic field (see Fig. 4). The resonances are assigned in Table 1. Values in parenthesis are (10^3 x) the slopes of the linear plots, the magnitude of these numbers reflecting the extent of exposure of the amide proton

resonance) and O_3 (an orthinyl resonance) could be internal in dimethylsulfoxide, a less polar solvent. It should be observed that glycyl and ornithyl N*H* resonances are readily distinguished since the first are triplets (spin-spin coupled to two α-protons) and the second doublets (coupled to a single α-proton). The relative shift of the resonances of the two 'internal' amides towards higher fields has been explained (*41*) on the basis that anisotropic shielding effects arising from the neighbor —CO—NH— π electron clouds at the two β-turns will oppose the deshielding resulting from intramolecular H-bonding, as, for example, in a *Schwyzer* type model. We favor this model for deferriferrichrome in dimethylsulfoxide, but not in water, and have proposed (*29*) that such a conformer is basically the same as in the chelate (Fig. 1). In the less polar solvent, the main effect of complexation is thus on the ornithyl side chains which, in the process of metal coordination, tend to acquire the "right" configuration to achieve the octahedral geometry at the ionic center. In water the conformational change is more drastic since the whole peptide backbone, in addition to the ornithyl residues, is driven into the new configuration. The ultraviolet CD of aqueous solutions of deferriferrichrome and of ferrichrome are shown in Fig. 3. Although a conformational interpretation of the CD of metalloproteins is unwarranted unless the optical rotation contribution of the metallic center itself is substracted (*42*), the spectra shown in Fig. 3 are consistent with a conformational change at the polypeptide backbone concomitant with the chelation event (*43*). Furthermore, while the CD spectrum of deferriferrichrome in water is similar to that of random-coil polypeptides, that of ferrichrome shows a negative peak at 217.5 nm and a positive one at 195 nm characteristic of β-pleated structures (*44*).

Hydrogen-tritium exchange experiments by the Englander two-column gel filtration technique had indeed suggested such a drastic conformational difference between ferrichrome or ferrichrome A and their metal-free derivatives (*45*). Thus while the time of exchange of the deferriferrichromes was too short (<3 min) to be detected by this technique, the iron complexes clearly showed a slowed kinetics, two to four hydrogens exchanging with half times of 14 to 426 minutes, depending on the *p*H and temperature. Inspection of the model (Fig. 1) shows that there are only two exposed amide hydrogens, namely those belonging to Gly^1 and Res^2, which should exchange with faster rates. The other four are protected by various extents of steric shielding and intramolecular H-bonding. Thus, while the Orn^1 NH is extensively buried in a pouch limited by the peptide backbone itself and the side chains of the ornithyl and site 3 residues, the Orn^2 amide is only partially shielded but participates in a strong H bond to its own side chain NO oxygen atom [2.80 Å, from the X-ray data (*28*)]. Res^3 and Orn^3, in turn, structure the

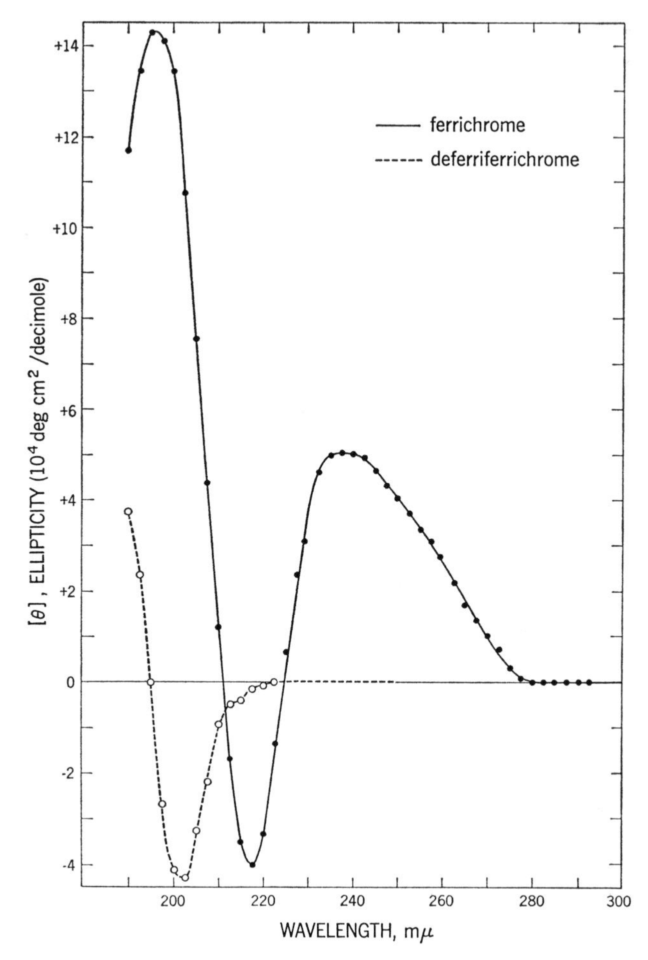

Fig. 3. Ultraviolet circular dichroism spectra of deferriferrichrome and ferrichrome dissolved in water (*43*).

β-sheet. This early interpretation of the exchange data on the basis of the X-ray model has been corroborated by PMR studies on the alumichromes, namely the Al^{3+} analogues of the ferrichromes. Substitution of the metal ion was necessary for the PMR work because of extensive resonance broadening by the paramagnetic Fe^{3+} ion.

The PMR spectra of deferriferrichrome and of alumichrome in $(CD_3)_2SO$ are shown in Fig. 4a and b respectively. The changes, apparent everywhere in the spectrum, are more dramatic at the amide resonances which appear clustered in the range of 7.5 to 8.5 ppm in deferriferri-

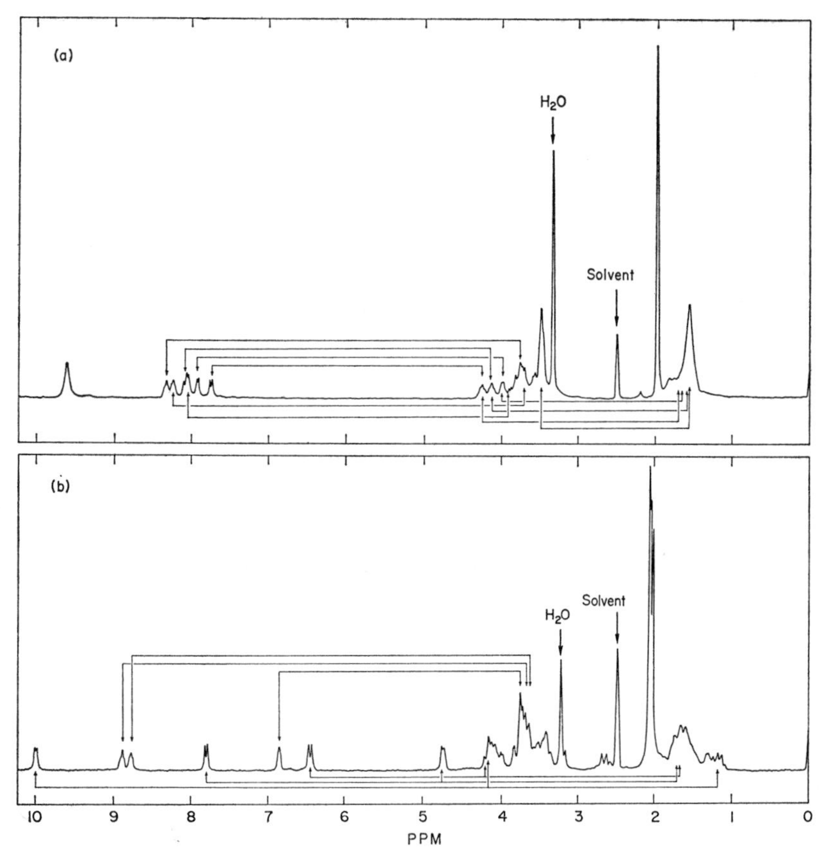

Fig. 4. The 220 MHz PMR spectra of deferriferrichrome (a) and alumichrome (b) dissolved in $(CD_3)_2SO$ at about 45 °C (*29*). The lone peak at ~9,60 ppm in (a) is assigned to the hydroxamic acid NO*H* resonance which is absent in (b). The rest of the peaks below and above 5 ppm are amide N*H* and aliphatic resonances, respectively. The sharp peak(s) at ~2 ppm arises from the acetyl methyl groups. Glycyl and ornithyl N*H* resonances are readily distinguished by their multiplet character as the former are triplets and the latter doublets. Arrows connect resonances showing vicinal spin-spin coupling interactions, which allow their assignment to the α, β, etc. protons. The resonances are referred to internal tetramethylsilane

chrome and expand to the range of 6.5 to 10 ppm in alumichrome. Since, due to their H-bonding capability, the amide spectrum is so sensitive to the N*H* environment, such a spread in their resonance region is strong indication that the conformational location of the individual protons is more constant with time in the chelate than in the deferripeptide. Indeed, no other oligopeptide examined by PMR to-date has exhibited such a wide range of NH chemical shifts. What is most interesting is that the temperature dependence of these resonance positions (Fig. 2c and d) do indeed show that two glycyls, namely Gly_1 and Gly_2, exhibit a more pronounced slope (similar to those in the random deferripeptide in water), suggesting the other four amides, one glycyl and three ornithyl, are more protected from interaction with the solvent as required by the model (Fig. 1).

On the basis of the single alumichrome spectrum shown in Fig. 4b little can be said regarding an absolute assignment of the resonances to specific residues in the hexapeptide, although relative assignments can be achieved by double irradiation spin decoupling experiments. The resonance identification could be solved, however, by a comparative study of glycyl- and L-seryl-containing analogues (*30*). The consistency of the model (Fig. 1) with the spectral observations and the resonance assignments was most illuminating and gave confidence that the model provided by the X-ray studies would hold in solution, slightly perturbed by side chain solvation effects. Thus the extreme low and high field shifts of the Orn_1 (at ~10 ppm) and Orn_3 (at ~6.5 ppm) resonances respectively result from the strong H-bonding of the first (Orn^2 in Fig. 1) to its own side chain and the lack of any H-bonding of the second (Orn^1 in Fig. 1) that is buried in a hydrophobic environment. The two glycyl resonances at ~8.8 ppm, *i.e.* those exhibiting the strongest temperature dependencies, were assigned to the glycyls at positions 1 and 2 in the sequence (see Fig. 1). Finally the ornithyl and glycyl resonances at ~7.8 and ~6.8 ppm respectively, which exhibit reduced temperature dependencies (Fig. 2) and that are assigned to Orn^3 and Gly^3 in the primary structure, show chemical shifts that indicate stronger H-bonding of the first relative to the second. The relative shift of this pair of resonances towards higher fields is consistent with the β-turn peptide bond anisotropies discussed above.

The correspondence between the PMR spectral characteristics of alumichrome and the X-ray picture for ferrichrome A justifies using the crystallographic model to describe the alumichrome conformation in solution; it would be desirable, however, to have some quantitative way of evaluating the extent of fitness between both sets of data. In 1969 *Bystrov et al.* (*46*) proposed a Karplus-type relationship that allowed a reasonable estimate of the conventional ϕ dihedral angle at the $N-C_\alpha$ bond (*47*) on the basis of the amide NH resonance splitting due to proton-proton spin coupling (J_{NC}). By applying Bystrov's formula to the alumichrome NH doublets a set of ϕ angles were obtained which, within the accuracy of the method, gave an excellent fit between both sets of data. Indeed, it was on the basis of this correspondence that a preliminary assignment of most PMR resonances of alumichrome (*29*) was achieved.

Later, comparative criteria proved this assignment to be correct and enabled its extension to all the resonances (*30*). More recently *Ramachandran et al.* (*48*) have derived an improved $J_{NC}(\phi)$ relationship by fitting the parameters involved to more extensive experimental data. Table 1 shows the ϕ dihedral angles calculated by this latter method, for a number of ferrichrome analogues in H_2O and in $(CD_3)_2SO$, together with the X-ray values.

Table 1

	Ser_1	Ser_2	Orn_1	Orn_2	Orn_3
(a) H_2O					
Gallichrome			285	210	257
Alumichrome			286	211	260
Alumicrocin	unresolved		287	205	253
Alumichrysin	unresolved	189	286	208	268
Alumichrome A	unresolved	189	287	208	270
(b) $(CD_3)_2SO$					
Alumichrome			288	209	259
Alumicrocin	307		287	206	252
Alumichrysin	315	180	285	205	260
(c) X-ray					
Ferrichrome A	303	197	283	215	256
	Ser^2	Ser^3	Orn^2	Orn^2	Orn^1

The ϕ dihedral angle between the HNC_α and the $NC_\alpha H$ planes. Values in (a) and (b) are calculated from the amide $HN\text{-}C_\alpha H$ doublet proton spin-spin splittings (J_{NC}) in water and in deuterodimethylsulfoxide, respectively, and on the basis of the data reported in (*30*) and the Karplus relationship of *Ramachandran et al.* (*48*). The labelling of the residues at the top refers to an arbitrary PMR classification of the resonances while the data in (c) corresponds to the angles determined by *Zalkin et al.* (*28*) for crystalline ferrichrome A, correctly labelled according to the criteria used in Fig. 1.

In comparing the structures of these peptides, a point of interest is the constancy of the glycyl residue at site 1. PMR studies on sake colorant A (*43*) and ferrichrome C (*49*), ferrichromes which contain alanine at site 2 and serine or glycine, respectively, at site 3, have shown unequivocally that site 1 is also occupied by a glycyl residue. It has been speculated that either a glycyl or a D-amino acid residue at a position equivalent to 1 in Fig. 1 are required if a stable β-turn (3_{10} helix) is to be formed (*41, 50*). Some statistical evidence supports this claim (*51* a,b) which is consistent with the X-ray crystallographic and PMR solution studies on ferrichrome in that the

$$Orn^3—NH\cdots O=C—Res^3$$

is a relatively stable hydrogen bond while the alternate

$$Orn^3—C=O\cdots HN—Res^3$$

is of much lesser strength. It is thus to be expected that if the conformation of the polypeptide is of any biological relevance, as seems to be the case, constancy of a glycyl at site 1 would be mandatory. It is hence rather surprising that albomycin, a ferrichrome-type antibiotic, has been reported to fill sites 1-2-3 with a triseryl sequence (*16*). A recent communication (*52*) indicates, however, that this peptide possesses only one seryl residue, which makes it more likely to fit the structural features common to other ferrichromes. Further research on albomycin would thus be of interest.

It should be noted that in spite of the excellent correspondence between the solution and crystallographic data exhibited by the ϕ angles, distinct differences occur among the analogues. These differences, although too small ($\lesssim 10$ degree) to modify the model shown in Fig. 1 to any significant extent, are such that they could affect the stability of the molecule as a whole. Thus, by slightly reorienting the peptide NH or CO groups, the strength of any bridging H bond could be drastically affected given its angular and distance dependence. In going from ferrichrome to ferricrocin to ferrichrysin, single L-seryl for glycyl substitutions are involved. The steric and solvation requirements of these two residues are different enough to result in noticeable conformational differences among the deferripeptides (*40*). Upon chelation, the free energy for metal-binding overcomes these forces so that formation of the *tri*ornithyl-metal complex dictates a uniform conformation for the different cyclohexapeptides. However, the strains and stresses that control the conformation of the metal-free peptides are still present in the chelate and although they are not of sufficient strength to control its final structure, they will contribute in determining its stabiltiy.

Following *Berger* and *Linderstrøm-Lang* (*53*), *Hvidt*, and *Nielsen* (*54*) have elaborated the idea that proteins are not rigid structures but exist in conformations which fluctuate in time about their native, folded state. In studying isotopic hydrogen exchange, those amides that are external, *i.e.* hydrogen-bonded to the solvent, will exchange faster than those that are buried or merely intramolecularly H-bonded. According to the *Linderstrøm-Lang* hypothesis, during the conformational 'breathing' process that proteins undergo in solution, amides that were protected, in their native, ground state conformation, may become exposed and hence able to undergo hydrogen exchange with the solvent. Although *sensu stricto* other factors such as the dielectric nature of the amide environment or the structuring of water at its proximity will also affect the kinetics of the exchange reaction (*55*), the hypothesis advanced by the *Carlsberg* group is most useful in interpreting the H exchange behavior of proteins in terms of their conformational stability.

Early magnetic susceptibility measurements by *Ehrenberg* (*22*) showed that iron-binding in ferrichrome is only weakly covalent. The ionic character of ferrichrome was clearly evidenced by *Lovenberg et al.* (*56*) by detecting a relatively fast exchange ($t_{\frac{1}{2}}$ ~10 min.) of its ion for $^{59}Fe^{3+}$ at *p*H 6.3. Furthermore, *Czech* workers reported evidence for a complex of ferrichrome with a second iron atom when the metal iron is in excess (*57*). These last two experiments indicate that there is enough conformational flexibility in the molecule to allow iron release or uptake in a fluctuating process that could randomly modulate the degree of exposure of amide hydrogens to the solvent. Although the rate of amide hydrogen exchange is minimal around *p*H 3, Emery noticed that in ferrichrome the exchange rate is accelerated as the *p*H drops from ~5 to ~3 (*45*). The converse was observed with ferrichrome A, however. On raising the *p*H to 7.0, the exchange rates of both peptides increase (but still are readily measurable) as would be expected from the known base catalysis of the exchange process. Because of the excellent spectral resolution of the N*H* resonances, it was found possible to study the ^1H—^2H exchange of the individual amides of alumichrome by PMR (*58*). The *p*H trends were consistent with those observed by *Emery* for the Fe^{3+} complex but, on the average, faster rates of exchange at low (~3) *p*H and slower rates at high (~7) *p*H were observed for the Al^{3+} relative to the ferric complexes. We have noticed that external Fe^{3+} does not exchange for Al^{3+} in alumichrome unless the solution is brought to acidic *p*H. Thus, although the binding constant of model hydroxamates is substantially higher for Fe^{3+} than for any other metal, the kinetics of the ion exchange process is slower in alumichrome than in ferrichrome. However, because of competition between the metal and hydrogen ions for the hydroxamate NO^- dentate, the metal exchange is acid-catalyzed. On the basis of the *Linderstrøm-Lang* hypothesis we have proposed that it is complex formation which confers the structural rigidity to the peptide backbone and that as the complex is loosened (as detected, *e.g.*, by the metal exchange kinetics) the concomitant relaxation of the molecule results in more exposure of the protected, internal amides to interact — hence exchange — with the solvent. That is: the conformational fluctuations are mainly those involved in the random metal exchange, which is itself acid-

Table 2

	ΔH†	ΔS†	ΔF†(25)	$t_{1/2}$(25)
(a) Gallichrome				
Gly[3]				fast
Orn[1]	15.9±1.3	−24.3±4.1	23.1±1.8	161.30
Orn[2]	16.0±0.9	−22.2±2.8	22.7±1.2	77.73
Orn[3]	17.9±1.7	−15.7±5.6	22.6±2.4	68.99
(b) Alumichrome				
Gly[3]	16.1±0.6	−17.1±1.9	21.2±0.8	6.25
Orn[1]	20.9±1.4	− 9.3±4.7	23.7±2.0	415.87
Orn[2]	21.9±0.6	− 4.7±2.1	23.3±0.9	220.05
Orn[3]	21.6±0.8	− 5.8±2.6	23.3±1.1	219.46
(c) Alumicrocin				
Gly[3]	19.3±5.0	− 6.5±16.8	21.2±7.1	6.80
Orn[1]	18.8±1.3	−18.6±4.0	24.4±1.7	1396.75
Orn[2]	19.4±0.8	−15.3±2.5	24.0±1.1	1335.57
Orn[3]	21.1±1.0	− 8.3±3.2	23.6±1.4	350.10
(d) Alumichrysin				
Ser[3]	21.0±1.7	− 4.2±5.7	22.2±2.4	38.08
Orn[1]	22.3±1.6	− 9.7±4.9	25.2±2.2	5491.57
Orn[2]	20.3±0.6	−14.8±1.9	24.7±0.8	2383.23
Orn[3]	25.9±1.5	3.2±4.5	24.9±2.0	3463.08

Kinetic parameters for the amide $^{1}H—^{2}H$ exchange for several ferrichrome analogues dissolved in D_2O, pD 5.14 (*59*). The data are for the Al^{3+} chelates except in (a) where it refers to the Ga^{3+} complex of deferriferrichrome ('gallichrome'). ΔH† and ΔS† are the activation enthalpy and entropy values, respectively, as estimated from Eyring plots. ΔF† (25) and $t_{1/2}$(25) are the activation free energy and the reaction half time calculated from the ΔH† and ΔS† data at 25 °C.

catalyzed. Indeed, analysis of the temperature dependence of the individual amide exchange rates indicates that at low *p*H the exchange mechanism proceeds with relatively large entropy barriers (ΔS†) suggesting significant conformational changes to reach the activated state. At neutral *p*H, the converse is observed, and this is especially sensed by the strictly buried Orn[1] amide hydrogen, namely, that the activation entropies are relatively small while the corresponding enthalpy values (ΔH†) are significantly larger (*58*). The extent to which the conformational stability of the chelate is influenced by the primary composition of the peptide was clearly exhibited by comparing the amide exchange rates of the alumichrome analogues (*59*). At *p*H 5.14 (Table 2) the exchange half times of all the ornithyl NH's increased on replacing a serine for a glycine at the Res^2 site (alumichrome → alumicrocin) and even more on a subsequent seryl-for-glycyl substitution at the Res^3 site (alumicrocin → alumichrysin). Thus a naive substitution of a glycyl for a seryl at site 3 more than doubles the exchange half times of all the slow amides and multiplies by 10 that of Orn^3 without inflicting any significant conformational change. Indeed, for this particular amide the effect has been mainly enthalpic, as if the residue substitution resulted in a structural stability gain.

The extent to which the conformational stability is coupled to the metal-binding affinity is shown by the fact that iron is retained ~10 times more firmly to ferrichrysin than to ferrichrome (*60*) (there are no reported data for ferricrocin). A similar argument may explain the relative trends observed by *Emery* between ferrichrome and ferrichrome A: the stability of ferrichrome A ($\sim 10^{32}$) is higher than that of ferrichrome ($\sim 10^{29}$) (*16, 39*) and the latter should exchange faster than the former. The observation that the rate of exchange of ferrichrome A is further retarded at *p*H 3 does in fact agree with the above interpretation as ferrichrome A is the only ferrichrome that, owing to the three free carboxylate groups on its ornithyl side chain acyl group (Fig. 1), should be conformationally stabilized at about this *p*H. Indeed, ferrichrome A can be easily crystallized out of aqueous solution in the range $2.5 < p\text{H} < 3.0$. The faster ^1H—^2H exchange of the Ga^{3+} complex of deferriferrichrome ('gallichrome') relative to alumichrome [Table 2 (a) and (b), notice the $\Delta S^\dagger$ differences] may reflect, again, different chelate stabilities.

In summary, what in the deferripeptides is manifested as conformational differences among the analogues (*40*), in the chelates is evidenced as stability differences dramatically expressed in their different hydrogen exchange kinetics and only slightly reflected in the PMR spectra.

B. *Ferrioxamine B*

Ferrioxamine B is a linear trihydroxamate with two peptide bonds (Fig. 5). Its affinity for Fe^{3+} (10^{31}) is comparable to that of the ferrichromes, the chelation not being stereo-specific as the complex lacks optical activity (*18, 61*). Mössbauer studies have indicated the ferric center is high spin with rhombic distortion (*62*). Although it is relatively small molecule (MW = 613), *Emery* (*45*) found that there are two hydrogens that exhibit retarded exchange rates, the kinetics being strongly *p*H dependent with minimal rates at *p*H ~4.5. The iron-free derivative exchanged with very fast rates, thus indicating the importance of metal-coordination in determining protection of the two amides. Space-filling models show, however, that it is possible to embrace the

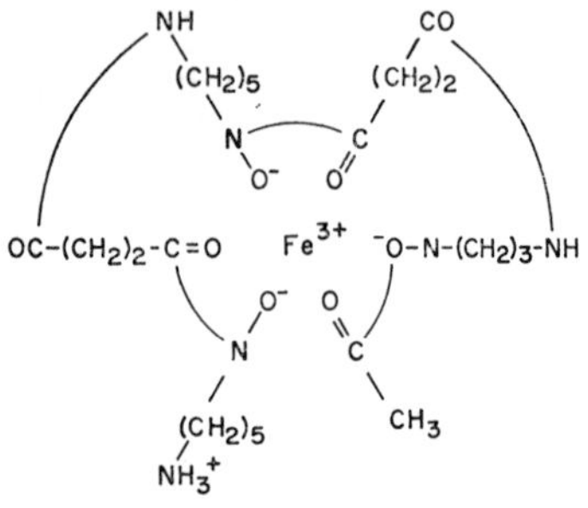

Ferrioxamine B

Fig. 5. Structure of ferrioxamine B. (Desferal = deferriferrioxamine B mesylate)

ferric ion by the linear molecule so that octahedral coordination is achieved while orienting the amides in such a way that they become H-bonded to the hydroxamate NO^-, as is the case with the Orn[2] N*H* in the ferrichromes.

C. Enterobactin

Enterobactin is a cyclic triester of 2,3-dihydroxy-N-benzoyl-L-serine (Fig. 6). It is accumulated by *E. coli* and other bacteria when grown under low iron conditions and mediates the anabolic utilization of iron by these microbes, as do the hydroxamate siderochromes mentioned above (*13, 21*). In enterobactin, however, the metal-binding ligands are provided by catechol groups and the sustaining backbone is held together by ester, rather than by amide, bonds.

The EPR of the iron complex exhibits the characteristic $g = 4.3$ signal of high spin ferric ion in rhombic fields (*63*). It should be noted that since binding of a 3+ cation releases two protons per catechol ligand, the chelate is a 3− anion, *i.e.* there is an excess of one electronic charge on each benzenoid ring.

Cell cultures producing enterobactin do not accumulate it indefinitely because an esterase (presumably involved in the iron release process) degrades the cyclic trimer to the linear tri-, di-, and monomeric species. It has been observed that the hydrolytic activity of the purified enzyme(s) depends on whether the substrate is free or complexed (*64, 65*), which suggests a conformational recognition of the siderochrome by the esterase.

The proton and carbon-13 NMR spectra of enterobactin and of its Ga^{3+} chelate are consistent with the free and coordinated forms having C_3 symmetry (*66*). The PMR spectra (Fig. 7) show dramatic shifts in

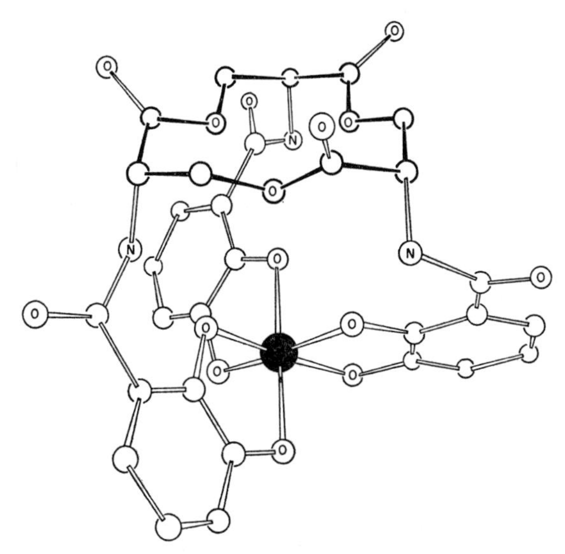

Fig. 6. Conformational model of chelated enterobactin (*66*). Bonds and atoms at the cycloester triseryl backbone are marked with heavier lines. Hydrogen atoms are not shown. A left-handed propeller configuration is shown for the ligands around the metal ion center, but this has not yet been established experimentally

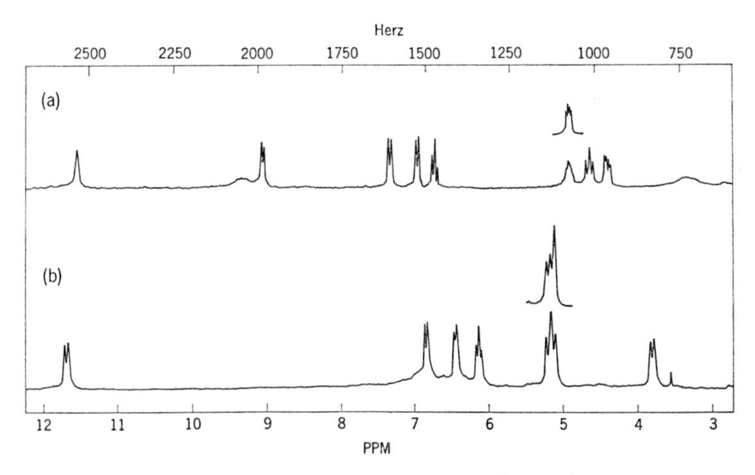

Fig. 7. The PMR spectra of enterobactin (a) and of Ga^{3+}-enterobactin (b) dissolved in $(CD_3)_2SO$ at ~45 °C (*66*). The frequency scales are referred to TMS as internal standard. The resonances below 8 ppm are catechol O*H* (singlets) or amide N*H* (doublets). The group of three peaks, between 6 and 8 ppm, are ascribed to the aryl C*H*. The other three protons, resonating towards higher fields, are the seryl $C_\alpha H$ and $C_\beta H$. Inserts show the α proton resonances after the amide hydrogen has been exchanged for deuterium

the resonances as well as changes in the magnitudes of the vicinal proton spin-spin couplings as the metal is chelated. These reflect changes in the electronic state of the catechol ligands as well as an overall conformational change of the molecule.

A conformational analysis of the PMR data was achieved on the basis of the J couplings between the $C_\alpha H - C_\beta H$ and $NH - C_\alpha H$ protons, which yield estimates of dihedral angles along these bonds. Gross consistency between the chemical shifts of the resonances and the particular location of the H atoms in the three dimensional structure was required. It was proposed that binding of the metal effectively enlarges the virtual 'pore' defined by the cycloester ring by rotating the $C_\alpha H - C_\beta H$ bond from $\theta \sim 42°$ to $\theta \sim -60°$ ($\theta = 0°$ for the eclipsed conformation), so that the ligands can embrace the metal. In the process the $NH - C_\alpha H$ bond rotates from $\phi \sim 60°$ to $\phi \sim -150°$ while the N*H* resonance shifts from 9.06 to 11.72 ppm from TMS (*66*). The latter has been attributed to a change in the hybridization state of the benzoyl-serine amide bond which could arise from electronic delocalization between the *ortho* phenoxide and the carbonyl groups, so that the ligands assume a quinoid character:

Such an interpretation is also supported by the ^{13}C-NMR data. An *ortho, para* electronic shift of this type has recently been proposed to explain the optical spectral differences between free and metal-bound phenoxides in the transferrins (*67*). In enterobactin, however, this tautomerism is of fundamental conformational importance as it lowers the energy barrier to rotate the amide bond to $\sim 47°$ from the planar *trans* configuration (Fig. 6), as is required in order to form a mononuclear octahedral complex. Upon reorientation, the amide proton becomes sterically more shielded from interaction with the solvent and this is reflected in its temperature coefficient which (in absolute value) drops from 4.7×10^{-3} ppm/deg in enterobactin to 1.3×10^{-3} ppm/deg in the Ga^{3+} chelate. Conversely, the driving force to form the mononuclear complex will favor the electronic delocalization implied in the above tautomerism.

The PMR data can be interpreted as suggesting that the ligands have the left-handed propeller configuration shown in the model (Fig. 7), although a right-handed configuration cannot be ruled out. Molecular models do not show either configuration to be any more strained than the other which might mean that, in actuality, both forms could exist in equilibrium. The net optical activity of the ferric chelate (*68*) could then be due to predominance of one form in solution because of some minor energy difference in its favor.

In showing how the electron density at the active site of an enzymic metallochromophore might be affected by conformational constraints, complexed enterobactin clearly exemplifies entasis. Furthermore, enterobactin is known to bind Fe^{3+} with much more affinity than its monomeric unit, 2,3-dihydroxy-N-benzoyl-L-serine ($k \cong 10^{25}$). Although a chelate effect should account for part of this difference, the conformational change involved in metal-complexation by the cyclic trimer will impose a significant barrier to metal release. Indeed, the final conformational state of the chelate results from a number of concerted bond twists both at the triester backbone and at the amide link, and release of one phenolate bidentate is somewhat hindered by the whole molecular conformation. This exemplifies the "Multiple Juxtapositional Fixedness" effect, already referred to in the Introduction, which helps account for the strong binding (hence, conformational stability) of metalloproteins in spite of the 'entatic' ligand misfit (hence, catalytic reactivity) of the metal.

In the case of enterobactin, the complete reduction of the complex in a 1:3 Fe_2SO_3:monomer solution, was achieved at a hydrogen (calomel) electrode potential of ~ -0.6 V, which should be contrasted with the irreducibility of the corresponding cyclic triester complex down to electrode potentials approaching hydrogen discharge (~ -1.8 V) (*63*). A plausible explanation is that the reducible species is a partially dissociated (*i.e.* hydrated) complex, present in equilibrium with the tricatechol chelate, which itself is practically irreducible. This is consistent with the observation that the potential for full reduction decreases as the concentration of monomeric ligand is increased beyond the stoichiometric requirement.

D. Discussion

The conformational studies on the ferrichromes and on ferrioxamine B indicate that a number of intramolecular hydrogen bonds are formed in the process of metal-chelation and that these contribute to the overall stability of the coordinated conformation relative to that of the free species. Consistent with this view, it should be mentioned that extensive hydrogen bonding has also been observed in the low molecular weight monovalent cation complexes of the antibiotics monensin, nigericin, dianemycin, the enniatins and valinomycin by NMR spectroscopy (*69, 70*), X-ray crystallography (*71, 72, 73*), or both. Like the siderochromes, these compounds act by mediating cation fluxes across membranes.

The ferrichromes have exemplified the relation between the stability factors in the metallopeptide conformation and the chelating tendency of the free ligand. As discussed below for small proteins, the metal is the gross conformational determinant while the primary polypeptide sequence specifies the unique stability pattern of each metal-protein complex. The need to think in terms of a dynamic *conformational state* rather than a static geometrical description has thus been clearly demonstrated. In the case of ferric enterobactin it is suggested that the low redox potential is related to extremely difficult metal exchange, which is a result of the structural rigidity imposed on the ligands by features of the molecular conformation.

III. The Transferrins

The transferrins are a class of iron-binding glycoproteins which have been found in the blood serum of a variety of vertebrates and are presumably also present in moth hemolymph [see (*74*) and references therein]. These proteins would appear to mediate the absorption and distribution of iron (*75*), and could play some role in the movement of carbon dioxide in the body (*76*). Proteins of similar characteristics, lactoferrin and conalbumin, have been isolated from milk and avian egg white respectively. It has been proposed (*77*) that conalbumin could prevent bacterial contamination of the egg yolk by removing free iron. A similar bacteriostatic action could be performed by lactoferrin in milk, and it is possible that this protein is involved in controlling the intestinal flora in infants (*78*).

There is some controversy regarding molecular weights of the transferrins, recent evidence suggesting a value of about 80,000 dalton for that from human serum. This figure varies over a wide range depending on the vertebrate species. The amino acid compositions for a number of these transferrins have been reported (*74*); little is known regarding their primary structures.

Besides Fe^{3+}, the transferrins are able to bind a variety of divalent and trivalent metal and rare earth ions (see *79—81* and references therein). They do not, however, bind Fe^{2+} (*82*). Metal displacement experiments (*79*) have shown that the highest binding affinity is for iron [dissociation constant of about 10^{-23} to 10^{-29} (*83—85*)].

Studies by equilibrium dialysis (*83*), free electrophoresis (*86*), isoelectric focusing (*87*), and DEAE-cellulose chromatography (*88*) have suggested that the two metal-binding sites on the protein molecule are independent and equivalent. These data, together with evidence obtained by chemical oxidation and alkylation (*89*) and tryptic digestion (*90*), have led to the suggestion that the protein is an aggregate of two identical subunits. This possibility has been considered especially for bovine transferrin from chemical modification, sedimentation and viscosity data (*91*). On the basis of the physical-chemical properties of reduced-alkylated (*92, 93*), sulphitolyzed (*93*) and succinylated (*94*) transferrins, *Bezkorovainy* and coworkers favor an opposing view, namely that the transferrins are not integrated as an aggregate of subunits joined by either physical (*e.g.*, hydrophobic or hydrogen) or disulfide bonds (*95*). Sequence determinations of cysteic acid peptides from conalbumin do not favor the presence of two independent subunits (*96*a). It has been suggested, however, that the present-day polypeptide is an evolutionary elongation of a shorter chain precursor by a process of partial gene duplication with mutations accumulated with time (*96*a,b). The claim (*74*) that hagfish, a most primitive vertebrate, contains a serum transferrin of about 44,000 dalton which binds a single atom of iron has recently been challenged (*97*). *Aasa* (*98*) has shown, however, that at about 5 °K two partly overlapping rhombic-type Fe(III) electron spin resonances can be detected in both human transferrin and in conalbumin while the hagfish protein exhibits no such sign of inhomogeneity.

The extent to which the two metal-binding sites are independent is not yet satisfactorily understood. According to *Davis et al.* (*85*) the dissociation constant of the monoferric complex

$$\mathrm{Tr{-}Fe} \longleftrightarrow \mathrm{Tr} + \mathrm{Fe^{3+}}$$

is approximately 100 fold larger than that of the diferric complex

$$\mathrm{Tr{-}Fe_2} \longleftrightarrow \mathrm{Tr{-}Fe} + \mathrm{Fe^{3+}}.$$

The electron spin resonance of the Cr^{3+} complexes showed differences between the Hamiltonians for the two binding sites (*99*). Furthermore, ultraviolet difference absorption and fluorescence spectroscopy studies on the binding of trivalent lanthanide ions to transferrin indicate that binding of a second ion is impaired when Nd^{3+} or Pr^{3+} have been bound first (*100*). Since other lanthanides, such as Tb^{3+}, Eu^{3+}, Er^{3+} and Ho^{3+}, which have smaller ionic radii, are bound in pairs, it was suggested that a nonequivalence between the two sites becomes manifest for large metal ions. On the basis of metal-binding kinetic studies *Woodworth* (*101*) has proposed a cooperative model wherein binding of a first metal ion drives a conformational change in the protein that facilitates rapid coordination of another ion at the second site. Although electrophoretic evidence has been provided by *Woodworth et al.* (*102*) suggesting that the claimed one-iron transferrin species (*83, 86—88, 103*) might in fact arise by dimer formation between the zero-iron and the two-iron species, recent sedimentation, electrophoretic, osmotic and water proton relaxation dispersion studies (*104, 105*a) tend to confirm the existence of three types of transferrins according to their iron content. Consistently, spectroscopic studies on mixed Ga^{3+}—Fe^{3+} complexes of conalbumin have also been interpreted to indicate a difference between the iron-binding sites of this protein (*105b*).

Warner and his collaborators (*80, 106*) first proposed involvement of tyrosyl residues as ligands for the ferric ion on the basis of potentiometric and spectrophotometric titration data. This has been further substantiated by chemical modification (*107, 108*) and other difference spectrophotometric (*67*) studies. Apparently two or three tyrosyl (*81, 108*), two histidyl (*83, 108—110*) and an equivocal number of tryptophanyl (*111* a, b) residues are present at each binding site (*112*). EPR evidence indicates that oxygen and as many as 4 nitrogen nuclei interact with the metal (*113, 114*). Furthermore, bicarbonate is bound to the protein, resulting in the usual salmon-pink colored form of the ferric complex which is otherwise colorless (*115* a, b). The minimal possible distance be-

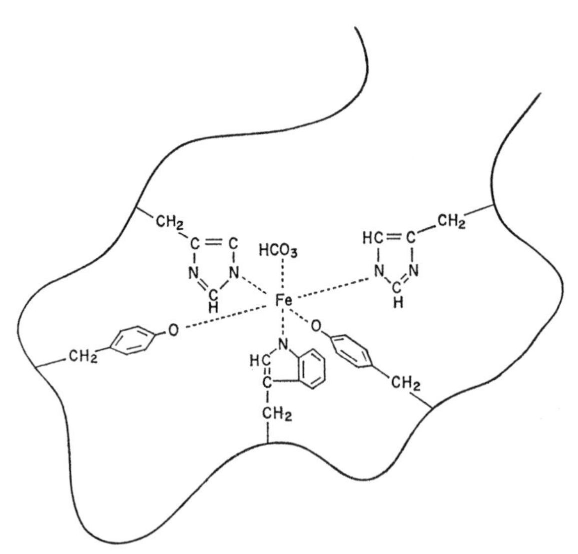

Fig. 8. Active center of the transferrins. The diagram summarizes a number of contradictory reports regarding the structure of the coordination site and does not pretend to satisfy all such evidence. The purpose is simply to hint how the active center of the complexed proteins might be composed. The continuous line represents the polypeptide backbone; two such centers are present per molecule

tween the metal-binding sites has been estimated by EPR (*83*) and fluorescence (*100*) experiments to be 9 and 43 Å respectively. Although the structure of the transferrin iron-binding sites is still a matter of extensive research and speculation, an octahedral complex as shown in Fig. 8 may be proposed as a useful model for the present discussion. In regard to the carbohydrate moiety (*116a,b*) and other aspects of this class of proteins, ample information is contained in a number of recent reviews (*117—120*).

A. The Conformational State

Since a number of denaturing treatments and chemical modifications of conalbumin resulted in an impaired ability to complex iron, *Fraenkel-Conrat* and *Feeney* (*121*) were led, in 1950, to suggest that metal coordination does not arise from a simple interaction of the ferric ion with some group in the protein. Rather, it was proposed, chelation is the result of a specific configuration assumed by a number of residues which generates a site of the proper geometry to receive the iron atom. A close correspondence between iron-binding and the protein conformation as a whole was thus implied, analogous to the lock-and-key theory of enzyme specificity and catalysis. The titration studies of *Wishnia, Weber* and *Warner* (*106*) referred to above, led these authors to propose that seven of the tyrosyl residues in conalbumin are buried (*i.e.* not in H^+ equilibrium) and a large portion of the remainder, including those involved in forming the iron complex, are hydrogen-bonded, thus contributing to the secondary and tertiary structure of the protein. Furthermore, amino acid analysis of fragment peptides derived from cyanogen bromide-cleaved transferrin, have shown that the tyrosyl residues that chelate the iron are indeed distributed over a relatively large region of the polypeptide chain rather than localized in a small section of the molecule (*95*).

Early studies showed that the metal-free and complexed transferrins differ in their solubilities in 40% ethanol, and also in their resistances toward proteolysis, alkali, thermal and urea denaturations, iodination and treatment with disulfide-breaking reagents (*122—126*). Under all conditions the ferric complex was found to be more stable than the apoprotein. The stabilization against urea and guanidine · HCL denaturation conferred by iron-binding has been corroborated recently by *Bezkorovainy* and collaborators (*92*), who also observed that the metal protects transferrin against the denaturation induced in the apoprotein by succinylation (*94*). In an investigation where reductive alkylation and trinitrophenylation procedures were combined with succinylation, the same team found that six to seven protected free amino groups are less reactive in the iron-complexed than in the metal-free protein (*127*). Although these groups were apparently not involved in binding the metal, their modification lead to denaturation of the protein which implies that in a metalloprotein of the size of transferrin the contribution of certain residues in determining the tertiary structure, vis-à-vis metal chelation, may be very important.

According to *Feeney* and collaborators (*109*, *125*) metal-binding simply stabilizes one of the several conformations which the apoprotein normally assumes and which happens to differ in its susceptibility towards a variety of degradative processes, such as denaturation by several agents (heat, solvent, etc.), chemical reactivity of certain groups, proteolytic digestion, etc.

a) Biological Probes

At this point it is interesting to note that while iron chelation labilizes the ferrichromes towards cleavage by the *Pseudomonads FC*-1 peptidase, the opposite effect happens with the transferrins towards tryptic or chymotryptic digestion. In both cases, however, the peptidases appear to discriminate between two distinct states of the polypeptide backbone and proteolytic digestion just exemplifies enzymatic recognition as a conformational probe. Immunology and transport are two other approaches to this problem also arising from biological specificity.

Antitransferrin sera from rabbits reacted differently with iron-saturated and unsaturated transferrin, as assayed by *Ouchterlony* plates and immunoelectrophoresis (*128*, *129*). The time course of light scattering for a reaction between bovine anti-conalbumin antibody and free and complexed conalbumin (*130*) indicates that the response of the anti-conalbumin antibody is higher towards conalbumin than towards its ferric complex.

A partially succinylated transferrin can be obtained by succinylation of Fe-transferrin and subsequent removal of the metal. It has been found that antibodies to native transferrin reacted with the partially succinylated protein but not with the fully succinylated derivative while partially succinylated transferrin antiserum precipitated the native apoprotein (*131*). It is unclear, however, whether this non-succinylated area includes determinants exposed by a metal-driven conformational transition or if it is simply a region masked by the Fe^{3+} ion itself. In general, immunochemical differences do not exclude other possibilities such as electrostatic charge effects in the free and chelated proteins or mere coverage of the antigenic determinant by the metal ion. However, the investigations of *Faust* and *Tengerdy* described below, which revealed no differences between the primary antibody association constants of the free and chelated conalbumins, argue against these alternatives.

The equilibrium and kinetics of the immunochemical reactions have been studied by means of fluorescence quenching and polarization, light scattering, and saturated $(NH_4)_2SO_4$ precipitation techniques (*132, 133*). The antigen-antibody (Ag—Ab) reaction may be divided into three sequential steps:

$$Ag + Ab \underset{k_{1'}}{\overset{k_1}{\rightleftarrows}} (AgAb) \underset{k_{2'}}{\overset{k_2}{\rightleftarrows}} (AgAb)_n \xrightarrow{k_3} \text{precipitate}$$

Neither $(NH_4)_2SO_4$ precipitation nor fluorescence quenching indicated differences between the rate of reaction of conalbumin or ferric conalbumin with the anticonalbumin antibody. It was hence proposed that binding of iron to conalbumin does not affect the primary phase (→AgAb) of the immunochemical reaction but rather the rate of lattice formation, which is what scatters light (*130, 133*). Fluorescence polarization competition studies with conalbumin and with iron-conalbumin (*134*) have provided evidence that the average size of the AgAb complex is smaller for the ferric conalbumin than for the apoprotein. This has been further substantiated by $(NH_4)_2SO_4$ precipitation, antigenic competition with labelled protein and AgAb ultracentrifugation experiments, all of which have consistently indicated that the antigenic reactivity difference between iron-conalbumin and the apoprotein manifests itself only in the relatively slow aggregation of the AgAb complexes (secondary phase of the Ag—Ab reaction).

The third line of biological evidence comes from studies on the effect of metal-binding on the affinity of transferrin for reticulocytes (immature red blood cells). *Jandl* and *Katz* (*135*) demonstrated that ferric transferrin binds with relatively higher affinity than apotransferrin to the reticulocyte receptors. Later research by *Fletcher* and *Huehns* (*75, 136*) showed a 10 fold faster rate of reticulocyte iron uptake from diferric transferrin than from the monoferric species. The authors indicated that '*the availability of bound iron to immature red blood cells seems to vary according to the way in which the iron is distributed in the molecule of transferrin*'. Such a recognition by the reticulocyte would imply that the iron-binding sites are nonequivalent. *Kornfeld* (*137*) found a five fold enhancement of the binding affinity of human [^{131}I]transferrin to rabbit reticulocytes when the protein is saturated with iron. The affinity of 43% and 100% Fe^{3+}-saturated transferrins for the reticulocyte were compared and the latter shown to attach more firmly to the cell (*86*). Controlled chemical modification experiments in which lysine and N-terminal free amino groups were blocked by reagents of different charge and size showed that binding to reticulocytes was more affected than the ability of the protein to bind iron (*138*). Furthermore, extensive removal of the carbohydrate moiety from the conjugated protein was not reflected in the biological activity (*138*). This would suggest that the protein backbone, rather than the heterosaccharide chains or the ferric chromophore, is responsible for the transferrin-reticulocyte interaction. In any event, binding the first iron atom causes a significantly greater effect on the biological action than does the addition of a second ion (*75, 136, 137*).

b) Physical Evidence

Gel filtration may be used as a means of detecting different conformations of a protein. *Kornfeld* (*137*) has found that the elution of Fe^{3+}-saturated transferrin through Sephadex G-100 is retarded relative to the apoprotein. Similar results have been reported by *Charlwood* (*139*) who appears to have been able to resolve the three species (0-Fe, 1-Fe and 2-Fe) of transferrin by this method, indicating a 0.7% decrease in the Stokes radius of the molecule per iron atom bound. Differential measurements of sedimentation velocity showed about 1.8% increase in $\mathring{S}_{20w}$ upon binding of 2 iron atoms per mole (*139*).

The hydrodynamic properties of iron-free and iron-complexed conalbumin have been reported by *Faust* and *Tengerdy* (*140*) (Table 3).

Table 3

Property	Conalbumin	Fe-Conalbumin
Molecular Weight	85,000	85,000
$D_{20 \cdot w}$	5.30	5.72
$S_{20 \cdot w}$	5.05 – 5.15	5.26 – 5.35
f/f_0	1.372	1.294
$\bar{v}$	0.732	0.732
Axial ratio a/b	4.5	3.0

Physicochemical properties of conalbumin and Fe-conalbumin (*140*). $D_{20 \cdot w}$ is the diffusion constant in water at 20 °C, $S_{20 \cdot w}$ is the sedimentation constant in water at 20 °C, f/f_0 is the frictional ratio (f = observed coefficient, f_0 = coefficient of an unhydrated sphere), and $\bar{v}$ is the partial specific volume in ml/g.

Fuller and *Briggs* (*141*) obtained the same molecular volumes for conalbumin and ferric conalbumin but found differences in the axial ratios of their equivalents ellipsoids. From the molecular volumes calculated on the basis of the Stokes radius of an unhydrated sphere ($\rightarrow \sim$102,000 Å^3) the lengths of the equivalent prolate ellipsoid axes as well as the surface area may be estimated for the metal-free and iron-complexed conalbumin (*140*):

	a (Å)	b (Å)	Area = $4\pi ab$ (Å^2)
conalbumin	79	17.5	17,600
ferric conalbumin	60	20	15,000

which means that the ferric complex has a more spherical and compact shape than the apoprotein.

A prolate shape also appears to explain better the hydrodynamic properties of iron-free (*142*) and of iron-saturated (*143*) transferrin. Ferric transferrin (a/b = 3) would, however, be more elongated than the iron-free form (a/b = 2) while the effective hydrodynamic volume (V_e) would be higher for the iron complex than for the apoprotein. These results not only differ from those given in Table 3 for conalbumin but are also in partial disagreement with dielectric dispersion and viscosity measurements (*144*) which have indicated that human transferrin assumes a more spherical shape with iron-saturation, the axial ratio decreasing from 2.5 (apo) to 2.0 (ferric). This latter investigation also indicates a slight expansion (15.4 → 16.9) of the hydrated volume concomitant with the binding of iron. Another parameter of interest is the 'hydration value' (h) which appears to be relatively high for this protein but, while *Bezkorovainy* (*143*) found h to decrease with iron-binding, *Rosseneu-Motreff et. al.* (*144*) reported that it increases from 0.6 to 0.7 grams of water per grams of protein. In this regard it is interesting to note that water proton magnetic relaxation dispersion studies of *Koenig* and *Schillinger* (*145, 146*) have indicated an increase from 13 to about 15 irrotationally bound water molecules when the apo protein binds two Co^{3+} ions. Thus these two last sets of data yield about 16% hydration increase concomitant with metal-binding.

Near its isoelectric point, binding of iron to transferrin results in the introduction of a negative charge due to uptake of one bicarbonate ion by the molecule. *Aisen et al.* (*86*) first showed this effect when an increase in the anodic electrophoretic mobility of the protein was correlated with its extent of iron saturation. The observation has been corroborated by isoelectric fractionation experiments (*87*). It is curious that although DEAE-cellulose chromatography of partially iron-saturated transferrin yields three peaks, as would be expected from the three species present in the mixture, the 1 Fe-transferrin eluted first, followed successively by the iron-saturated molecule and the apoprotein (*88*). Thus the order of elution does not agree with the expectations based on the net electric charge of each species. As the interaction between the ion-exchange bed and the protein is determined mainly by the electrostatic interactions between the two surfaces rather than by their net charges, *Lane* (*88*) has suggested that iron-binding causes significant alterations in the distribution and extent of exposure of the protein charged groups. Indeed, dielectric dispersion measurements have detected a decrease of about 60 Debyes in the dipole moment of apotransferrin as a result of iron-binding, without reorientation of the dipole relative to the equivalent molecular ellipsoid axis (*144*). Furthermore, and in good agreement with the reticulocyte transferrin uptake behavior discussed above, it is binding of the first iron atom that causes the most dramatic effect as the chromatographic difference between 1 Fe-transferrin and apotransferrin is greater than between the latter and 2 Fe-transferrin (*88*). It is thus likely that there are three conformational states according to the extent of iron chelation. Interestingly, rabbit apo- and iron-transferrin, which show almost equal extents of association with reticulocytes, also exhibit similar affinities for binding to DEAE-cellulose (*147*).

Upon iron-binding, changes in the $\tau \sim 7$ and $\tau \sim 9$ regions of the PMR spectrum have been observed which were interpreted as arising from a conformational change (*148*).

The optical activities of free and complexed conalbumin have been examined in the visible and ultraviolet regions. In the range 300—675 nm apotransferrin and conalbumin exhibit a plain negative rotary dispersion which upon iron-binding becomes anomalous due to a negative Cotton effect centered at about the 470 nm absorption band of the complex (*149, 150*). The CD spectrum of iron-conalbumin exhibits a negative band at 460 nm in agreement with the ORD observations (*151*). In the range between 195 to 600 nm the ORD spectrum of conalbumin does not differ significantly from that of other globular proteins such as ovalbumin or lysozyme; a peak occurs at 199 nm and a trough at about 232 nm (*152*).

ORD and CD criteria have indicated 17—18% α-helix in human serum transferrin, the helicity being unaffected by metal-binding (*153*). Similarly, the helix content of conalbumin has been estimated from ORD data to range between 5 and 31% at *p*H 5.8. β-structures or hydrophobically bonded areas other than random coil and helices in the globular protein may be responsible for the wide variability in the helical content as calculated from the A coefficients derived from *Schlechter* and *Blout* plots (*152*). A conformation containing 28% helix, 32% β-structure and 40% random coil has been estimated for the native apoprotein by the method of *Greenfield* and *Fasman* (*151*). The modified two-term Drude equation is satisfied at wavelengths above 300 nm, the rather anomalous dispersion below this wavelength arising probably from optically active, oriented, aromatic amino acids (*152*). In a similar fashion, Moffit plots are not linear below this wavelength. Taking the Moffit parameter b_0 as an indicator of helix content, a loss of helicity was found in an 8 *M* urea solution, or on lowering the *p*H from 8.7 to 2.8 (a process which is concomitant with denaturation of the protein), or on increasing chloroethanol concentration from 0 to 20% (*154*). When the latter treatment was carried up to 100% chloroethanol, a significant increase in helicity was observed which suggests this chaotropic agent[2]) acts at low concentration by breaking the native structure and at high concentrations, because of its lower dielectric constant, by protecting intramolecular H bonds and hence favoring helix formation. Interestingly, all these denaturing treatments accompanied a decrease and disappearance of the 250—300 nm perturbations in the ORD spectrum, suggesting a reorientation of the side chain chromophores exposing them to the solvent with a gain in conformational freedom. Indeed, Koshland's reagent, specific for tryptophan, modified 1.8 of conalbumin's 13 tryptophanyls in aqueous solution, 5.3 in 20% chloroethanol, and 7.4 in 100% chloroethanol (*154*).

Upon iron-binding, the ORD curve below 300 nm was not significantly perturbed (*154, 156*). The CD spectrum shows, however, noticable effects. Native conalbumin exhibited a negative dichroic band at 296 nm, a broad negative band centered at 274 nm, and a positive band at 252 nm. Iron coordination shifted the most positive regions of the spectrum to higher values so that the two spectra are essentially identical except for the magnitude of the ellipticities. In particular, the bands at 293—296 nm may be associated with the tryptophanyl residues and the observed changes could arise from slight conformational alterations between certain residues of the native protein and the iron complex (*154*). Analysis of the 250—310 nm region of these spectra has led Tan (*151*) to suggest reorientation of tryptophanyl and tyrosyl residues and of some disulfide bonds in or near the metal-combining site.

It is interesting to note that upon dilution of the apoprotein in 0.05 *M* sodium dodecyl sulfate at *p*H 7.3, a loss of the rotatory perturbations in the 250—300 nm range was detected while the iron-conalbumin complex dissociated, as detected by a disappearance of the absorption band at 270 nm, suggesting the detergent affects the protein secondary structure in a way such that the resulting conformation is unfavorable for metal-binding (*154*). Recent low temperature EPR studies have

---

[2]) Generalizing the concept of *Hatefi* and *Hanstein* (*155*) by *chaotropic* agents we mean solutes such that by affecting the structure and lipophilicity of water favor the aqueous solubility of membrane-bound enzymes and, in general, of proteins structural at the supramolecular level.

indicated that perchlorate, a well known chaotropic anion, affects the ligand fields at the ferric ion sites (*157*). This exemplifies again the tight coupling that exists between the protein and the ferric moieties. In small peptides, like the ferrichromes, the metal is a major conformational determinant while in large proteins, like the transferrins, the interactions of the residues in the primary sequence outweigh the conformational effects of metal-binding.

Urea titration showed that while changes in the UV absorption spectrum are evident in the apoprotein at 4 *M* urea, no effect is manifested by the iron complex even up to 5 *M* urea (*111*). Upon dilution in 8 *M* urea the UV spectrum of conalbumin changes immediately while that of the ferric complex does so slowly. Difference spectra for conalbumin and Fe-saturated conalbumin in 8 *M* urea referred to identical, but urea-free, solutions of the same species each show peaks at 292 and 286 nm, the former band atributed to exposure of buried tryptophanyl residues. The curve for conalbumin shows much higher absorbancy at 240 nm.

Conalbumin and Fe^{3+}-conalbumin exhibit similar fluorescence spectra showing activation and emission maxima at 282 and 350 nm respectively, the ferric chromophore quenching some of the fluorescence in the complexed protein. Within the range in which the complex exists, *p*H and temperature have a greater effect on the 350 nm fluorescence (F_{350}) of the apoprotein than on the ferric complex. The F_{350}'s of Fe^{3+}-conalbumin and of N^{α}-acetyl-L-tryptophanamide (a convenient reference fluorescent substance) exhibit a reversible decrease as the temperature is raised from 0 to 75 °C. Apoconalbumin exhibits a similar trend up to 60 °C but above this temperature its F_{350} increases until at 64 °C it coincides with that of the reference. Above 64 °C it decreases while the protein precipitates. Furthermore, the curve for the apoprotein exhibits hysteresis: a temperature decrease at any point of the curve yields F_{350} values that are lower than on the preceding ascending-temperature curve. This is only a kinetic effect as the original F_{350} was recovered upon overnight standing at 0 °C. These experiments indicate that both the fluorescent standard and the ferric protein are stable towards heat within the studied temperature range. The increase of the conalbumin fluorescence previous to its precipitation at 64 °C suggests a gradual unfolding prior to denaturation, which exposes the buried tryptophanyl residues thus freeing them from the quenching influence of neighbor groups (*111*). The results are consistent with the experiments of *Azari* and *Feeney* (*124*) which showed stability for the Fe^{3+}-conalbumin at 63 °C but denaturation of the apoprotein at this temperature.

The fluorescence decrease in the ferric peptide on going from *p*H 5 to *p*H 7 is concomitant with metal chelation and hence reflects fluorescence quenching. For the apoprotein the fluorescence remains practically constant in the *p*H range from 5 to 7.5 and then it gradually decreases to 50% of its initial value at *p*H 11.8. It was suggested by the authors (*111*) that this quenching could arise from an energy transfer from excited tryptophanyl residues to ionized tyrosyl residues. The F_{350} increase on raising the *p*H above11.8 in conalbumin and above 12 in the ferric protein might, again, be a reflection of a freeing of tryptophanyl residues during denaturation. Studies on the conalbumin CD spectra in the aromatic (250—310 nm) region as a function of *p*H confirm these conclusions (*151*).

According to *Tan* and *Woodworth* (*111*) the color appearance and the concomitant fluorescence quenching that are simultaneous with metal-binding are suggestive of a charge transfer complex involving tryptophan. Furthermore, the difference of $\sim 5 \times 10^3$ in $\Delta\varepsilon_{292}$ between the colored metal-conalbumins and conalbumin are indicative that two tryptophanyl residues are bound to, or protected by, metal ions in the complexed but not in the metal-free protein. Fron this evidence the authors conclude that binding of Fe^{3+} "*confers on conalbumin a stabilization similar to that conferred on many enzymes by their substrates and on globin by heme*" (*111*). Moreover, the complexing of iron does not alter the CD spectrum of conalbumin in the 190—240 nm wavelength region at *p*H levels ranging between 7.2 and 8.0, suggesting that the secondary structure of the protein is unaffected by iron binding (*156*). A detailed CD study of transferrin, conalbumin, and their copper complexes has recently been published (*158*).

Hydrogen exchange studies have been performed on both conalbumin and transferrin (*156, 159*). Apoconalbumin, preequilibrated in tritiated water, was separated from the radioactive solvent by gel filtration and the exchange-out of the labeled hydrogen atom was monitored in time by allowing the protein to equilibrate with an aqueous solution, buffered at *p*H 8, followed by separation of the solvent by gel filtration. According to *Ulmer* (*159*), about 225 labeled hydrogen atoms per mole are retained (zero exchange-out time) at 4 °C and, after 24 hours, approximately 45 of these are still unexchanged. Upon addition of iron to the tritiated apoprotein the rate of exchange-out was reduced, so that ferric conalbumin retained approximately 50 additional tritium atoms per mole. Similar results were obtained by *Emery* (*156*) at 25 °C and the same *p*H. If the retarded hydrogen exchange is attributed to any of the 650 peptide groups present in the protein, these results indicate that upon binding of iron 7—8% of the peptide hydrogens become unavailable for interaction with the solvent. Furthermore, addition of iron to tritiated conalbumin during any stage of the exchange-out reaction retards a fraction of the unexchanged hydrogens in the apoprotein indicating that metal-binding affects hydrogens exchanging both in the rapid and slow classes. This has been confirmed by a careful analysis of the *p*H- and temperature-dependence of the kinetics (*160*). Iron titration of this effect gave a value of two metal atoms/mole to saturate the exchange retardation, in good agreement with a variety of data concerning the iron-binding capacity of conalbumin (*156*).

The data reported by *Emery* (*156*) and by *Ulmer* (*159*) for the exchange-in kinetics are contradictory: while the latter finds that the apoprotein takes up more radioactivity than ferric conalbumin, *Emery* finds no significant difference between the two species in the rates of tritium labeling. Thus, *Ulmer* suggests that the apo- and metallo-protein differ in their conformation, the latter structure being more compact, while *Emery* believes that metal-binding is a complex process such that protection of local hydrogens is enhanced in certain regions of the protein, while those in other regions may become more exposed for exchange. It may well be that these contradictions are more apparent than real as the two authors have done their experiments at different temperatures. As has been observed in the ferrichromes (*58, 59*), the rates of hydrogen exchange at a fixed temperature do not mean much in terms of the dynamic mechanism involved; in a given conformation the relative kinetics for the different amides differ according to the temperature of the reaction. Only temperature-dependence studies give mechanistic insights since the weight of $\Delta H^{\dagger}$ and of $\Delta S^{\dagger}$ in determining the exchange rate is a function of temperature.

B. Discussion

As may be judged from the 190—250 nm spectropolarimetric data, iron-binding appears not to perturb significantly the polypeptide secondary and tertiary structure. It seems, however, to affect the conformational state of the protein through a local relative reorientation of a small number of amino acid side chains which become locked in a defined spatial configuration by complexation to the metal. As the size of this molecule is relatively large, the overall conformational effect is highly buffered by the interactions of the residues among themselves and with the solvent. That the metal-coordination effect is more dynamic than structural is suggested by the *p*H dependence of the hydrogen-tritium exchange (*156*). As the *p*H is lowered from 8 to 6.5 the exchange rate difference between the ferric complex and the apoprotein disappears. This occurs notwithstanding the protein being mostly complexed at both *p*H values. It may be proposed that, as in the case of ferrichrome, lowering the *p*H has not driven the protein to a new static conformation (as could be judged by complex formation) but rather has increased the conformational fluctuation contribution to the exchange mechanism, so that the "protective" effect of iron chelation is no longer important. At low *p*H it is likely that the hydrogen exchange proceeds through the metal-loose state with which the native protein is in dynamic equilibrium. *Aisen* and *Leibman* (*161*) have indeed found that transferrin does not exchange iron at *p*H 7.3 unless citrate is added to the reaction mixture. At *p*H 6, however, the exchange occurs spontaneously, *i.e.* without mediation by an external chelator, thus suggesting that, as in the case of the ferrichromes, acid-catalyzed metal exchange is a way of activating the molecular conformational fluctuations. Such a dynamic view of the change in the conformational state of conalbumin has been implied by *Faust* and *Tengerdy* (*134, 140*) as an explanation for the differences in antigenicity between conalbumin and its ferric complex. Thus, although a minor change in the shape of the molecule from a less to a more compact structure accompanies iron-binding, this does not affect its antigenicity as judged by the primary antigen-antibody reaction rates. However, free native conalbumin is a flexible molecule — 'high motility protein', in the sense of *Linderstrøm-Lang* (*162*) — so that it can adapt with relative ease to assume that conformation which is thermodynamically more stable for a given chemical environment. The structural stabilization conferred by iron-binding could prevent a conformational adaptation upon changing chemical environments, such as when the protein interacts with the antibody. The antigenic differences between conalbumin and its iron complex, which are manifest in the secondary phase of the immunochemical reaction, might thus arise by iron restricting some kind of conformational stretching of antigen and antibody in the aggregation reaction that forms the precipitin lattice.

Similarly, the lower accessibility of tyrosyl residues to base titration in ferric lactoferrin compared to ferric transferrin has been suggested to be linked to their relative stabilities as metallocomplexes (*67, 163*). Ferric conalbumin also appears to be more stable than ferric transferrin (*157*). These stability differences may be of biological significance given that a role of the serum protein is to distribute iron in the body rather than to sequester the metal.

IV. Hemerythrin

Hemerythrin is an oxygen carrier iron protein found in certain species of marine invertebrates within the Annelida, Brachiopoda, Priapulida and Sipunculoidea. This respiratory pigment can be considered a rather remote non-haem evolutionary relative of haemoglobin. It binds one mole of oxygen per two atoms of iron in a way that, in case of the sipunculids, is non-cooperative (*164—166*).

Under normal conditions, the sipunculid hemerythrin is an aggregate of eight essentially identical subunits with a net molecular weight of about 100,000 to 108,000 dalton (*165—167*). The octamer is readily dissociated into monomers by treatment with mercaptan-blocking reagents (*168*), detergents or exposure to extreme *p*H conditions (*167*) suggesting that the subunits are held together by non-covalent bonds. The monomer consists of a single polypeptide chain of 113 amino acid residues, containing two iron atoms yielding a molecular weight of 13,500 dalton per subunit (*165, 167, 169*). Hydrodynamic data suggest the native octamer is compact and symmetrical. Hence it is likely that the monomeric units are arranged in a closed, *e.g.* cubic or square antiprism, configuration rather than in a linear array (*167*). The aggregation is cooperative, involving an average free energy (8 monomers → octamer) of ~6 kcal per mole of monomer (*170*). The enthalpy of association is very small, the process being mainly entropic with $\Delta S_{assoc} \sim$ 15—23 eu per mole of monomer (*171, 172*).

Iron is readily released from the protein in dilute acid (*169*) and decoloration of oxyhemerythrin by Hg^{2+}, Cu^{2+}, Pb^{2+} and Zn^{2+} has been interpreted as displacement of iron by these ions (*173*). Methemerythrin binds a variety of anions such as HS^-, N_3^-, SCN^-, CNO^-, Cl^-, F^- and HO^- at the Fe^{2+} site, which results in a modified reactivity of the protein towards, *e.g.*, mercaptan reagents (*174*). It also binds strongly ClO_4^- and other anions, which markedly affect the O_2 binding properties of the protein (*175*). This has been interpreted as implying that arginyl 48 and 49 and cysteinyl 50 (see below) might be located close to the iron locus (*176*). The reactivity of lysyl residues also appears to be affected by binding of perchlorate (*177*).

The electronic configuration of the active site has been extensively studied by magnetic susceptibility (*178*) and optical (*174, 179, 180*), EPR (*178*) and Mössbauer (*178, 181, 182*) spectroscopies. On the basis of these physical studies, *Klotz* and collaborators (*180*) have proposed a model in which the iron atoms in both oxy- and met-hemerythrin are bridged by an O^{2-} ion (Fig. 9). According to these authors the two molecules of water which act as external ligands on each iron atom of deoxyhemerythrin are sequentially released upon oxygenation. In the process Fe(II) turns into Fe(III), the metal atoms becoming bridged by O_2^{2-}. Parallels between the sulfur bridging ligand in iron-sulfur proteins and the oxygen ligand in hemerythrin, have recently been proposed (*3, 183*). In case of methemerythrin, the external ligand can be CN^-, Br^-, etc. The Mössbauer evidence further suggests that while the antiferromagnetically coupled high spin Fe(III) atoms are in similar environments in the methemerythrin complexes, such is not the case at the μ-oxo-μ-peroxo Fe(III) dimeric active site (*181, 182*). Oxidation of the two iron atoms is highly cooperative, intermediate oxidation states being notoriously unstable relative to the fully oxidized or fully reduced forms (*184*). The enthalpy of oxygen-binding to sipunculid hemerythrins has been measured to be about − 10 kcal/mole under normal conditions (*166, 185*). For more general aspects of the hemerythrins the reader is referred to several comprehensive reviews (*186—189*), while the specific problem of the electronic state of the iron center has already been surveyed in this series (*190, 191*).

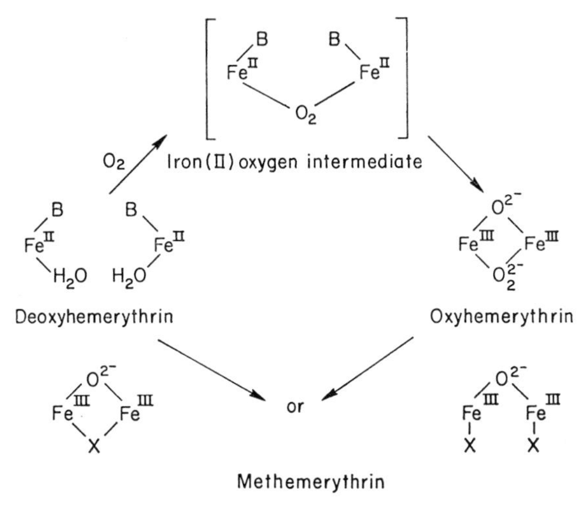

Fig. 9. Models for the central core complexes of native octameric hemerythrin (*182*). Besides being complexed to the protein through amino acid side chains, each iron atom is coordinated to an external, variable ligand X which can be Cl^-, Br^-, F^-, CN^-, NCS^-, N_3^-, etc., when bridging the two Fe(III) centers or H_2O, OH^-, F^-, etc., when two ligands bind. B stands for either OH^- or an amino acid side chain derived from the protein

A. The Conformational State

The amino acid sequence of hemerythrins from the sipunculids *Golfingia gouldii* (*169, 192—194*) and *Dendrostomum pyroides* (*195—196*) have been completely and partially elucidated, respectively. Fig. 10 shows the proposed primary structure for these proteins, which seem to differ in four amino acid sequences. The latter might be of relevance in the apparent higher aggregation affinity exhibited by the *G. gouldii* (*170, 197*) relative to the *D. pyroides* (*165*) hemerythrin. *Langerman* and *Klotz*

Fig. 10. The primary structure of *Golfingia gouldii* hemerythrin (*189*). The amino acid sequence of the *Dendrostomum pyroides* pigment differs in having Gly, Ala, Glu, and Arg at positions 9, 76, 79 and 82, respectively, and possibly an extra Lys between positions 107 and 111 (*196*). It should be noted that monomeric *Golfingia* hemerythrin is heterogeneous, sequence differences with that shown above having been established at positions 79 and 96 where Thr and Ala substitute for Gly and Ser, respectively (*193, 194*). Heavier circles distinguish potential iron ligand residues; as described in the text, however, some of these are unlikely candidates

(*170*) have indicated that single residue replacements in the amino acid sequence can result in significant perturbations of the oligomeric state of the protein. However, partial identity is revealed when antisera to *D. pyroides* hemerythrin is reacted with the *G. gouldii* protein (*165*). The conservancy of single cysteinyl and methionyl residues at positions 50 and 62 might have structural implications as blockage of the –SH group leads to dissociation of the octameric protein and the methionyl thioether (by analogy to cytochrome c) is a potential ligand to iron (*189*).

Succinylation of hemerythrin dissociates it into monomers, presumably by the introduction of negative charges, but it does not affect the spectral properties of methemerythrin thus indicating that it does not perburb the iron chromophore (*167*). The modification does, however, interfere with the O_2 binding ability of the pigment even after reduction of the metal to the ferrous state. Apparently ten lysine residues are succinylated, suggesting they are located on the surface of the subunit. This is consistent with experiments which show that, besides the amino terminal residue, eleven free amino groups are exposed both in hemerythrin and in apohemerythrin as detected by reaction with trinitrobenzenesulfonic acid (*198*) or by amidination (*199*). The latter reactions did not affect the 330 nm absorption of hemerythrin either, indicating that the iron complex was unperturbed.

All 18 carboxyl groups in the *G. gouldii* protein could be modified by reaction with glycine methyl ester in the presence of 1-ethyl-3-dimethylaminopropylcarbodiimide (*200*). Although the reaction appeared not to affect the iron site, blockage of as few as 8 carboxyl groups resulted in complete dissociation into the monomeric subunits. Thus, while carboxyl ligands are not present at the active center, on the basis of kinetic data the author concluded that one such group is involved in the subunit interaction. Similarly, exposure of all four tryptophanyl residues in each subunit was suggested by reaction of apo- and iron-proteins with N-bromosuccinimide (*201*).

Histidine reacts with 5-diazo-1-H-tetrazole (DHT) to form di- or mono-DHT derivatives according to the number of free ring nitrogens on the residue side chain. Reaction of DHT with the native, aggregated oxyhemerythrin and with hemerythrin subunits, showed three histidines to be exposed. The modified protein retained the iron. However, when the iron-free protein was acted upon by DHT all seven histidines present in the polypeptide reacted to form the di-DHT derivative thus implying that four histidines, the four that yielded the mono-DHT derivative in the iron-protein, are somehow protected by iron-binding (*198*). Furthermore, this study showed a red shift in the 360 nm absorption band of mono-DHT-hemerythrin relative to actameric oxyhemerythrin or to the monomer in 8 *M* urea. This was attributed to changes in the quaternary or tertiary structure of the polypeptide which affect the ionic or hydrophobic nature of a hypothetical histidine iron-binding site. Since the histidyl at position 82 in the *G. gouldii* hemerythrin is replaced by arginine in the *D. pyroides* protein (see Fig. 10), this particular residue cannot be involved in iron complexation (*196*).

Controlled reaction of *G. gouldii* hemerythrin with tetranitromethane has shown that tyrosyl residues 8 and 109 and, to a lesser extent, tyrosyl 67 are protected from nitration in the iron-protein (*201—202*). The more readily nitrated residues, namely tyrosines 18 and 70 are, hence, likely to lie on the surface of the monomer. This partial modification did not affect either iron-binding, as judged by its visible absorption spectrum, nor the polypeptide conformation, as suggested by the lack of changes in the far ultraviolet region of the CD spectrum (*202*). However, when the nitration reaction was allowed to proceed so that only tyrosyls 18, 67 and 70 were modified, the 330 nm peak absorption decreased, indicating that the antiferromagnetic coupling (*203*) between the two iron atoms had been broken (*201*). Under similar controlled conditions, all five tyrosines in the polypeptide were nitrated in the heat-denatured apohemerythrin (*201*). Similarly, only three tyrosyl residues in the native protein can be O-acetylated without affecting the spectral characteristics of the iron chromophore (*199*). The modification, however, lead to dissociation of the octamer into its subunits, suggesting some of the three exposed tyrosines may be involved in structuring the oligomer. This should be contrasted with the amidination experiments which were, in this regard, of no effect, implying that neither lysyl residues nor the terminal amine group are directly involved in the subunit binding site (*199*). As the authors suggest, it would be of interest to attempt the chemical modification on the undenatured iron-free protein so that direct metal protection can be separated from purely conformational hindrance of certain hydroxyls towards the external reagents. Unfortunately, '*procedures which unfold protein and render these hypothetical* (metal-binding) *groups available also cause the loss of iron and the release of the potential ligands*' [*Fan* and *York* (*148*)].

Ulmer and *Valée* (*150, 179*) have found significant differences in the 320—670 nm ORD spectra of *G. gouldii* methemerythrin as compared to its oxygenated form. Similar observations have been made on the protein from *S. nudus* in a study that also showed no detectable optically active transition for deoxyhemerythrin in this region (*204*). Ligand-binding to the iron center results in observable changes in the near ultraviolet ($\sim$290 nm) region of the optical rotatory spectra. Since the CD around this wavelength is sensitive to nitration of phenolic groups, tyrosyls have been suggested to be involved in the optical activity in this region (*185*). The $n \rightarrow \pi^*$ $\pi \rightarrow \pi^*$ peptide transition region exhibits negative ellipticity bands at about 222 and 209 nm and a positive band at about 197 nm, the latter band being about 35% higher in the *S. nudus* relative to the *G. gouldii* hemerythrin (*185, 204*). Furthermore, the CD in this region is unperturbed either by ligand-binding at the active site or by dissociation of the octamer into its subunits (*185, 204*). This indicates that neither of these two events, that appear to drive local conformational changes at the active site, seem to affect the structure of the polypeptide backbone. This is significant because the oxygen-binding kinetics of *S. nudus* hemerythrin is apparently more complex than a simple bimolecular reaction as two relaxation times are detected in temperature jump experiments (*166*). The implication is that one of the processes might involve local aromatic side chain reorientation at the O_2-binding locus rather than an overall protein conformational change (*204*). From the far ultraviolet CD spectra helix contents of about 70% have been estimated for the *G. gouldii* (*185*) and *D. pyroides* (*205*) hemerythrins, while for the *S. nudus* protein (*205*) similar data suggests 20% random coil, 30% β-structure and 50% helix; *i.e.*, the sipunculids' hemerythrins appear to be mainly helical.

Although mercaptans seem to be involved in the aggregation of the subunits, Ag^+ itself was found to be ineffective in producing dissociation (*168*). Titration of hemerythrin with $AgNO_3$ has indeed shown a lack of reactivity of the silver ion unless the protein is dissolved in 8 *M* urea (*178*). Under such conditions iron is released as the titration proceeds. Since the single cysteinyl residue is not coordinated to iron (*168*), this experiment suggests a conformational drift in urea that exposes the mercaptan which, upon binding Ag^+, in turn relaxes the conformation while releasing the iron atoms.

After a 75-hour HCl hydrolysis of the native, iron-containing, *G. gouldii* protein only 5% of the threonine was destroyed while the same treatment on the iron-free species resulted in 37% degradation of this residue (*169*). A related finding is that nitrated, heat-denatured apo-hemerythrin is hydrolyzed by pepsin faster than acid denatured nitro-hemerythrin (*201*). Furthermore, carboxypeptidase A, which reacted immediately on the iron-free protein, did not release any amino acid from native methemerythrin even after incubation for 8 hours, suggesting the conformation of the undenatured protein prevents access of the peptidase to the carboxyl terminus residue of the polypeptide chain (*169*). Of similar implications, but related to the amino terminus, is the earlier observation that pork kidney leucine amino peptidase does not release any amino acid from hemerythrin (*206*).

A monomer of aquohemerythrin, obtained by blocking the free cysteine-SH with a mercurial or tetranitromethane, is relatively unstable; on standing at room temperature it tends to precipitate while releasing iron (*207*). In the process no qualitative change in the visible absorption and CD spectra of the remaining iron-protein is observed. This implies no perturbation of the iron locus and thus suggests an intimate coupling between iron-binding ability, conformational stabilization and quaternary structure. Furthermore, extensive nitration of native hemerythrin, which modifies all five tyrosyl residues, is concomitant with the release of only one iron atom per subunit as if the two iron atoms were not similar (*207*). This would appear to contradict the Mössbauer data, indicative of identical chemical environments around each iron atom when in this oxidation state (*181*). It may well be, however, that nitration could cause minor conformational changes which propagate differently to each iron-binding site so as to generate subtle differences between them which would, in turn, be responsible for the release of only one of the two metal ions. The event, quoting *Rill* and *Klotz* (*207*), "*could remove some of the restrictions on the protein conformation and allow some unfolding which would expose additional tyrosine residues that were previously inaccessible. The far ultraviolet circular dichroic spectra of the samples nitrated to various extents show that distinct changes in the protein spectrum occur upon nitration and the subsequent release of iron. These changes in the far ultraviolet (as in the visible) also parallel the extent of iron loss and the production of nitrotyrosine with time and are in a direction corresponding to conversion of part of the protein from helix to random coil.*"

B. Discussion

The circumstantial evidence presented above clearly indicates that iron is intimately involved in determining both the conformational stability of the hemerythrin subunit and its resistance towards hydrolytic degradation. Similarly, the primary structure would appear to affect the state of the iron locus and of the polypeptide chain as is hinted by the subtle differences observed in the visible and ultraviolet CD spectra among hemerythrins extracted from various tissues of *D. pyroides* and which happen to differ in their amino acid composition (*205*). Relative to other iron proteins, research on the hemerythrins is still in its infancy. It should be evident from our survey that additional investigations of its soluble free and chelated forms are mandatory if further understanding of the metal-polypeptide interaction in this protein is to be achieved. The ease of availability of the crystalline octamer and the low molecular weight of the monomeric unit should make hemerythrin especially suited for X-ray crystallographic studies.

V. The Rubredoxins

The rubredoxins (meaning "red-redox agents") form a group of non-haem iron proteins of relatively low molecular weight, apparently involved in electron transfer reactions. Early observed in *Clostridium thermosaccharolyticum* (*208*), the protein from *Clostridium pasteurianum* (*209*) was the first to be isolated, purified and crystallized. Rubredoxin has since been found in other anaerobic (*210*), nitrogen-fixing (*211*) and photosynthetic green sulfur bacteria (*212*). A related molecule is present in *Pseudomonas oleovorans*, an aerobic bacterium (*213, 214*). Except for the

Ps. oleovorans rubredoxin, which may bind one or two, they all complex a single iron atom per mole of protein. In all cases the iron atoms are ligated each only by four cysteinyl mercaptides (*217—219*): this is the characteristic defining this class of proteins.

Most reviews dealing with iron-sulfur proteins in their broadest sense, include the rubredoxins [see *e.g.*, (*215, 216* a, b)].

A. The 1-Fe Rubredoxins

The 1-Fe rubredoxins are single polypeptide chain proteins of about 55 amino acid residues and 6,000 dalton molecular weight. The primary structure of the *Micrococcus aerogenes* (*220, 221*) and *Peptostreptoccus elsdenii* (*210*) proteins have been determined. The two amino acid sequences (Fig. 11) reveal a relatively high mutation frequency, with a noticeable conservancy around the four cysteinyl residues which are critically involved in binding the iron and hence in the proposed electron transfer role of the protein (*210*). Another curious feature of the anaerobic rubredoxins is the presence of N-formyl methionine as N-terminus amino acid (*222*). Synthesis of the polypeptide is in progress (*223*).

The biological function of the 1-Fe rubredoxins is unknown. They undergo reversible one-electron oxidation and reduction with a redox potential of about -5.7 mV (*209, 219*). Both the oxidized and reduced forms of the protein are stable. Magnetic susceptibility measurements and Mössbauer spectroscopic data have shown that the reduced and oxidized forms of the proteins are high spin Fe(II), S=2, and high spin Fe(III), S=5/2, respectively (*224, 225*). Although the reduced rubredoxins do not show significant EPR, the oxidized form exhibits absorptions at $g = 4.3$ and $g = 9.4$ (*211, 217, 218, 225*). These resonance are characteristic of Fe(III) in highly asymmetric environments (*226, 227*) and, as already described, have also been observed in ferrichrome A, enterobactin and the transferrins. X-ray (*228, 229*) and laser-Raman spectroscopy (*230*) studies on the crystalline oxidized material shows the iron center environment to be close to tetrahedral. The symmetry appears to be maintained in solution whether the protein is in the oxidized or reduced state, according to laser-Raman (*231*) and near infrared CD (*232*) spectroscopic evidence, respectively. *Eaton* and *Lovenberg* (*232*) have mentioned unpublished X-ray studies by *L. H. Jensen* indicating persistence of the tetrahedral geometry around the iron center upon one-electron reduction of the oxidized crystals.

The Conformational State

A remarkable stability of the clostridial rubredoxin to acid has been noticed by *Lovenberg* and *Sobel* (*209*). Similarly, *Bachmayer et al.* (*221*)

```
            1                  5                    10                          15                20                  25
P.E.     f-Met-Asp-Lys-Tyr-Glu-Cys-Ser-Ile-Cys-Gly-Tyr-Ile-Tyr-Asp-     -Glu-Ala-Glu-Gly-Asp-Asp-Gly-Asn-Val-Ala-Ala-
M.A.     f-Met-Gln-Lys-Phe-Glu-Cys-Thr-Leu-Cys-Gly-Tyr-Ile-Tyr-Asp-     -Pro-Ala-Leu-Val-Gly-Pro-Asp-Thr-Pro-Asp-Gln-
               (1)                                                                                                (24)
P.O.a          Ala-Ser-Tyr-Lys-Cys-Pro-Asp-Cys-Asn-Tyr-Val-Tyr-Asp-     -Glu-Ser-Ala-Gly-Asn-Val-His-Glu-Gly-Phe-Ser-
              (119)                                                                                               (143)
P.O.c          Leu-Lys-Trp-Ile-Cys-Ile-Thr-Cys-Gly-His-Ile-Tyr-Asp-Trp-Glu-Ala-Leu-Gly-Asp-Glu-Ala-Glu-Gly-?he-Thr-

                            30                    35                 40                   45
P.E.     Gly-Thr-Lys-   -Phe-Ala-Asp-Leu-   -Pro-Ala-Asp-Trp-Val-Cys-Pro-Thr-Cys-Gly-Ala-Asp-Lys-Asp-Ala-Phe-    -
M.A.     Asp-Gly-Ala-   -Phe-Glu-Asp-Val-   -Ser-Glu-Asn-Trp-Val-Cys-Pro-Leu-Cys-Gly-Ala-Gly-Lys-Glu-Asp-Phe-Glu-
         (25)                                                                                                   (48)
P.O.a    Pro-Gly-Thr-   -Pro-Trp-His-Leu-Ile-Pro-Glu-Asp-Trp-Asp-Cys-Pro-Cys-Cys-    -Ala-Val-Arg-Asp-Lys-Leu-Asp-
        (144)                                                                                                  (167)
P.O.c    Pro-Gly-Thr-Arg-Phe-Glu-Asp-Ile-   -Pro-   -Asp-Trp-Asp-Cys-Cys-Trp-Cys-Asx,Pro-Gly-Ala-Thr-Lys-Glu-Asn-

          50
P.E.     Val-Lys-Met-Asp-COOH
M.A.     Val-Tyr-Glu-Ala-COOH
         (49)
P.O.a    Phe-Met-Leu-Ile-
        (168)
P.O.c    Tyr-Val-Leu-Tyr-Glu-Glu-Lys-COOH
```

Fig. 11. Primary structures of the *P. elsdenii* (P.E.) and *M. aerogenes* (M.A.) 1-Fe rubredoxins (*210, 220, 221*). For comparison, the amino (P.O.$_a$) and carboxy (P.O.$_c$) terminii fragments of the 2-Fe *P. oleovorans* homologue (*239*) are included. The sequence of the Asx, Pro dipeptide at positions 160—161 in P.O.$_c$ is uncertain. Gaps (insertions or deletions) have been introduced as to achieve maximum number of constant (shown underlined) residues. Residues suggesting evolutionary invariance in the 1-Fe type protein are blocked. The numbering of residues at the top refers to the 1-Fe rubredoxins while the residues along the *P. oleovorans* protein sequence are numbered in parenthesis. The intermediate fragment of the 2-Fe protein, missing above, is:

```
52                                   60
Ile]—Glu—Ser—Gly—Val—Gly—Glu—Lys—Gly—Val—Thr—Ser—Thr—His—

                          70
Thr—Ser—Pro—Asn—Leu—Ser—Glu—Val—Ser—Gly—Thr—Ser—Leu—Thr—

80                                        90
Ala—Glu—Ala—Val—Val—Ala—Pro—Thr—Ser—Leu—Glu—Lys—Leu—Pro—

                               100
Ser—Ala—Asp—Val—Lys—Gly—Gln—Asp—Leu—Tyr—Lys—Thr—Glu—Pro—

          110                                            119
Pro—Arg—Ser—Asp—Ala—Glu—Gly—Gly—Lys—Aly—Tyr—[Leu
```

have found the *M. aerogenes* rubredoxin to be stable to 8 *M* urea and 50% ethanol and only slowly denatured by 5 *M* guanidine. Since these authors also noticed a relative resistance of the native protein to the action of proteases, a compact, stable conformation was suggested for the chelate structure.

From the EPR spectrum, the iron:cysteine content, and some preliminary characterization of tryptic and chymotryptic peptides of the *M. aerogenes* rubredoxin, *Bachmayer et al.* (*218*) proposed a simple model in which the iron atom plays a fundamental structural role by holding the molecule together (Fig. 12). The four cysteine residues are grouped in

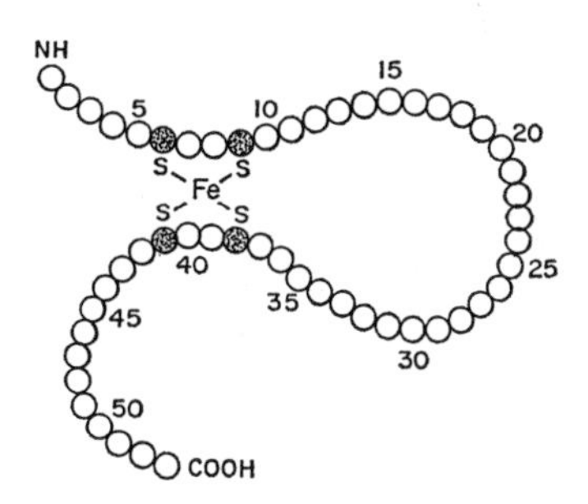

Fig. 12. Model for the 1-Fe rubredoxins proposed by Bachmayer, Piette, Yasunobu and Whiteley (*218*). A structural feature to be noted is the role of iron in bringing together relatively distant parts of the polypeptide chain. Such a loop has been confirmed by X-ray crystallography (see Fig. 14)

pairs of which one occupies sites 6 and 9, near the amino-terminal end, and the other, near the carboxyl-terminal end, occupies sites 38 and 41; *i.e.* two residues separate the cysteinyls of each pair. In order to satisfy tetrahedral coordination of the iron atom, hinted at by the EPR data, these authors suggested a folding of the polypeptide chain that generates a loop of 28 residues. It was explicitly proposed that the iron atom is not only serving as an electron carrier but that it also stabilizes the structure of the protein. Experiments with model peptide fragments have indicated that the relative separation between the four cysteinyl residues in the primary structure is important in order to achieve stable $Fe-(S_{cys})_4$ complexes (*233*).

The ORD and CD spectra of 1-Fe rubredoxins from various sources have been published (*211*, *217*, *219*, *225*, *234*, *235*). The changes in the ORD spectra of the *C. pasteurianum* aporubredoxin and of the iron protein (Fig. 13) in the oxidized and reduced forms where the aromatic amino acids absorb, namely from 260 to 300 nm, are suggestive that the environments around these side chain chromophores might change upon binding and reduction of the Fe^{3+} ion (*235*). In case of the *M. aerogenes* rubredoxin, it has been noticed (*221*) that the 280 nm extinction coefficient of the native protein is about twice what could be accounted for in terms of its tryptophanyl and tyrosyl contents, which more closely

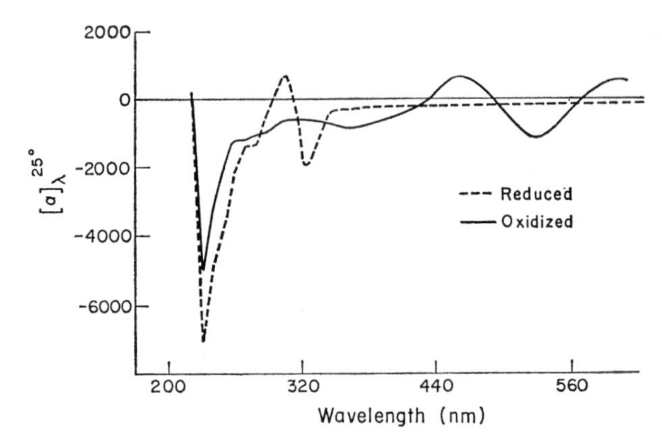

Fig. 13. The ORD spectra of reduced (---) and oxidized (—) *C. pasteurianum* rubredoxin at *p*H 7.3 (*219*)

account for the absorbancy of the aporubredoxin. This obviously complicates the interpretation of the observed ORD changes since the chelate contribution to the ultraviolet absorption could add up to that due to any conformational change. The trough at about 233 nm is indicative that some secondary structure might be present in the apoprotein. A significant increase in laevo rotation can be observed at this frequency upon reduction of the native protein (Fig. 13), suggesting the possibility of some conformational change during reduction (*219*). The magnitude of such trough would indicate a relatively high helical content for the protein as calculated from Moffit-Yang plots; the meaning of the data are, again, uncertain as the contribution of the iron chromophore to the observed rotations could not be substracted (*210, 219*). Furthermore, both α-helical and β-sheet structures show troughs in the 230—233 nm range of the dispersion mode so that it would be difficult to distinguish from such data which of the two structures is dominant. The CD spectrum, however, allows a better differentiation between these two possibilities (*144, 236*). Indeed, the shape of the negative band, peaked at 224 nm (*235*), is suggestive of a β-pleated sheet predominance. Such a contention is further supported by the X-ray data for the oxidized protein (*228*), to be discussed below. In view of the different location of proline residues in the *M. aerogenes* and the *P. elsdenii* proteins, *Bachmayer et al.* (*210*) had indicated that a helix content of any significance was unlikely in rubredoxin.

Reversible chemical reduction of the red oxidized crystals yields the bleached reduced species without damage of the crystalline particles, thus indicating that the conformational change which accompanies the redox transition is not a major one (*221*). The 220 MHz PMR spectra show changes upon reduction of the oxidized protein (*224*). Although the difference could, in part, be due to some conformational change, the authors did not speculate on this, perhaps because of the strong possibility of resonance displacements arising from paramagnetic interactions with the iron center concomitant with the electron transfer.

The order of reactivity of native rubredoxin with mercurials, namely

mercuric acetate > *p*-mercuri-sulfonate
> *p*-mercuri-benzoate > sodium mersalyl,

suggested that accessibility of the sulfides is limited by the bulkiness of the reagent (*219*). Furthermore, the cysteines were readily alkylated with iodoacetate in the metal-free polypeptide while no significant reaction occurred with the ferrous protein even in 8 *M* urea or in 8 *M* urea with 0.5 *M* 2-mercaptoethanol, indicating that the cysteinyl sulfhydryl reactivity depends on the removal of iron despite the presence of reductive agents (*219*). In the same article, *Lovenberg* and *Williams* (*219*) report iron exchange kinetics measurements in *C. pasteurianum* rubredoxin. For exchange to occur it was found that a sulfhydryl reagent was required, presumably because of involvement of the cysteine ligands. The reaction took about one hour for completion at the optimum *p*H (~7). According to the interpretation given to the metal exchange in the ferrichromes, these experiments may show a "breathing" of the protein so that its conformation can fluctuate and result in a time dependent exposure of internal groups. Unfortunately, the hydrogen exchange of the rubredoxins has not yet been studied.

Using mercuri-iodide and uranyl derivatives, *Jensen* and collaborators have been able to study by X-ray techniques the crystalline, oxidized *C. pasteurianum* protein at 3 Å resolution (*228*) and refined it, in the crystallographic sense, to 1.5 Å resolution (*229*). The proposed model is shown in Fig. 14. The electron density map shows the polypeptide chain to be rather irregularly folded to generate a globular structure of about 20 Å diameter. The higher resolution data has allowed recognition of amino acid sequences within certain peptide fragments (*229*). Apparently no α-helix is present but extensive twisted antiparallel β-pleated sheet is generated between the terminal ends of the chain and also within the intermediate folded loop.

By comparing the X-ray data with the amino acid sequences of the *P. elsdenii* and *M. aerogenes* rubredoxin, the N- and C-chain terminii have been identified as noted in the figure. The tentative identification

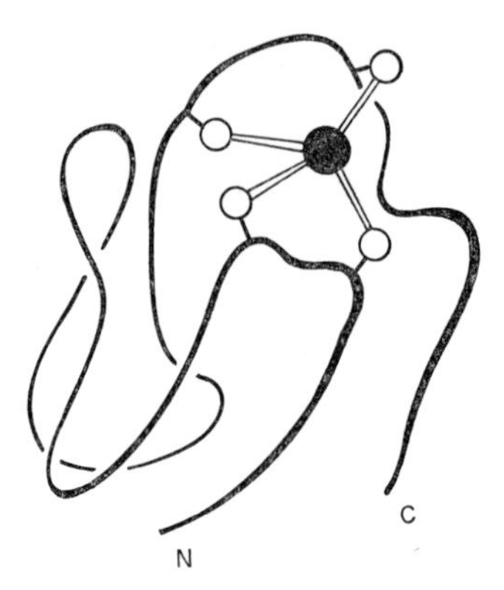

Fig. 14. Crystallographic model for the *C. pasteurianum* rubredoxin at 1.5 Å resolution (*229*). Only the polypeptide backbone and the Fe—S_{cys} complex are shown. N and C designate the N and C terminus, respectively. The sulfur and iron atoms are denoted by empty and filled circles, respectively

of the residues at 2.5 Å resolution (*228*) has suggested that the oil drop model might apply to this protein as the polar, hydrophilic side chains project from the surface of the molecule and all but one of the seven bulky side groups (there are seven aromatic residues in the protein) appear to be internal. The metal center is located at one side of the molecule removed from the middle peptide section (residues 11—36). The analogy between the X-ray model (Fig. 14) and the loop scheme (Fig. 12) proposed by *Bachmayer et al.* (*218*) is striking. It is evident that although the protein lacks disulfide bonds, the iron bridge between the amino and carboxy terminal fragments of the polypeptide chain would serve a similar structural role (*221*).

B. The 2-Fe Rubredoxin

A rubredoxin-type protein of about 19,000 dalton, which binds up to two iron atoms per molecule (*237, 238*), has been found as a component of the ω-hydroxylation system of the aerobic bacterium *Pseudomonas oleovorans* (*213, 214*). Its primary structure suggests two sequences homologous with that of rubredoxin from anaerobic bacteria (Fig. 11) (*239*). On this basis the authors have speculated on the possibility of an evolutionary divergence from a common ancestor for all these organisms so that the heavier type rubredoxin may have resulted by gene duplication. In the process, the specificity of the enzyme was so affected that 1-Fe rubredoxins cannot substitute for the 2-Fe protein in the ω-hydroxylation reaction.

Lode and *Coon* (*238*) have found no significant optical or EPR spectroscopic difference between the iron-saturated (2-Fe) and partially iron-saturated (1-Fe) *Ps. oleovorans* rubredoxin suggesting that both iron atoms are located in similar environments. The two species of rubredoxin exhibit the characteristic g = 9.4 and g = 4.3 spin resonances observed in the anaerobic bacteria rubredoxins and these have been attributed to transitions within the lowest and middle *Kramer's* doublets of the high spin ferric ion, respectively (*227*). The optical absorption, ORD and CD spectra of the reduced protein show bands and Cotton effects, respectively, which are similar to those of the ~6,000 dalton rubredoxins.

The *Ps. oleovorans* rubredoxin contains a single methionyl and ten cysteinyl residues. The amino acid sequence (Fig. 11) shows that the cysteinyl residues are located in two groups of five each, one located close to the amino-terminal end (*239*). Reaction of the aporubredoxin with cyanogen bromide cleaves the molecule at the methionyl site. Each of the resulting two peptides contains five cysteinyl residues and binds one atom of iron. Furthermore, reaction of the 1-Fe (*i.e.* the partially iron-saturated) species with idoacetate bromide and subsequent cleavage with cyanogen bromide yielded two fragments which indicated that the exposed cysteines are those close to the amino-terminal end. The structural implications of these experiments are depicted in Fig. 15, which shows two possible models for the diferric protein, model C being favored by

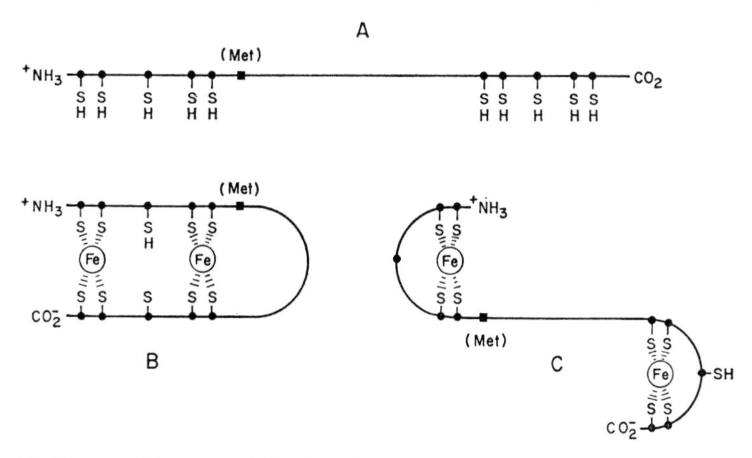

Fig. 15. Structural features of the *Ps. oleovorans* rubredoxin (*238*). A is a schematic representation of the polypeptide chain showing the relative positions of the cysteinyl and methionyl residues on the primary structure. B and C depict two possible ways of coordinating the iron atoms. Model C appears to be favored by the data

the authors over model B (*238*). Apparently, only the iron atom located closer to the carboxy end of the molecule, which is more tightly bound, is involved in the redox action (*238*).

C. Discussion

The crystallographic bond lengths and angles of the *C. pasteurianum* rubredoxin iron-cysteine complex, reveal one of the Fe—S bonds to be much shorter (1.97 Å) than the other three (ranging between 2.31 and 2.39 Å). Such a distorted tetrahedral symmetry was explained on the basis of "*cumulative effects stemming from features in the protein part*" of the molecule (*229*). The large Dq values (d→d transition tetrahedral splitting parameter) revealed by near infrared CD studies (*232*) have been interpreted to hint at serious structural constraints imposed by the protein that would somehow result in increasing the ligand field strength of the cysteine sulfurs in rubredoxin above unusual values. Similarly the *C. pasteurianum* and *Ch. ethylica* rubredoxins exhibit differences in their EPR and CD spectra which have been attributed to conformational effects arising from their distinct primary structures (*225*). The dual nature of the protein-metal interaction is, hence, again exposed; on the one hand the state of the iron atom is affected by strains generated by the peptide moiety, while on the other hand the metal ion (and its ligands) are among the most important elements in stabilizing the conformation of the molecule (*231*). From this standpoint, it is obvious that further conformational studies on the *Ps. oleovorans* protein ought to be rewarding since features of its secondary and tertiary structure should contain the clue to understanding the differences in catalytic specificity between the 1-Fe and 2-Fe rubredoxins.

VI. The Iron-Sulfur Proteins

Consistent with the definition used by other authors (*240*), by "iron-sulfur proteins" we mean those iron proteins containing "inorganic" sulfur, i.e. sulfur that is released as H_2S in acidic medium, is stoichiometrically related to the iron content in a one-to-one ratio, and which is a structural component of the active site. All iron-sulfur proteins known to-date are of the non-haem type and display a cysteine content that is related to the iron and sulfur complement. This type of protein is ubiquitous in living organisms exhibiting a variety of functions which are always related to electron transfer processes.

In this article a distinction is made between the "high potential iron-sulfur proteins" (HiPISP) and the low potential type, the "ferredoxins", for reasons that will become apparent later but that are primarily based on a manifest lack of structural homology between the two classes. Among the ferredoxins the 8 Fe: 8S "bacterial type" should be distinguished from the 2 Fe: 2S "plant type" (which also occurs in bacteria) and these two should in turn be contrasted with the 4 Fe: 4S HiPISP. A 6 Fe: 6S type, reported in nitrogen-fixing bacteria (*241*, *242*) has recently been disproven (*243*). As will be apparent from the text that follows, the experimental approaches that have been used and the conclusions drawn are of wide applicability and illustrate the general features of the metal-polypeptide interaction in this class of protein. These are a consequence of the peculiar makeup of their active sites and of the redox transitions they can undergo.

A. The High Potential Iron-Sulfur Proteins

The high potential iron-sulfur proteins are single polypeptide chain 4 Fe: 4S proteins of about 10,000 molecular weight having a particularly high redox potential [$\sim +350$ mV (*244*)]. Of unknown biological function, these proteins have been purified from the photosynthetic bacteria *Rhodopseudomonas gelatinosa*, *Thiocapsa pfennigii* and the obligate photoanaerobe *Chromatium vinosum* [see (*245*) and references therein].

Fig. 16. Primary structure of the *Chromatium* HiPISP (*245*). The only ambiguity in the proposed sequence is for the assignment of residues 77 and 78:

1 (Ser) … 10 (Asp)
Ser—Ala—Pro—Ala—Asn—Ala—Val—Ala—Ala—Asp—Asn—Ala—Thr—Ala—

20 (Asn)
Ile—Ala—Leu—Lys—Tyr—Asn—Gln—Asp—Ala—Thr—Lys—Ser—Glu—Arg—

30 (Ala) … 40 (Glu)
Val—Ala—Ala—Ala—Arg—Pro—Gly—Leu—Pro—Pro—Glu—Glu—Gln—His—

50 (Ala)
Cys—Ala—Asp—Cys—Gln—Phe—Met—Gln—Ala—Asx—Ala—Ala—Gly—Ala—

60 (Glu) … 70 (Lys)
Thr—Asp—Glu—Trp—Lys—Gly—Cys—Gln—Leu—Phe—Pro—Gly—Lys—Leu—

80 (Ala)
Ile—Asn—Val—Asn—Gly—Trp(Cys,Ala)Ala—Ser—Trp—Thr—Leu—Lys—Ala—

86
Gly—COOH

The amino acid sequence of the *Chromatium* HiPISP has recently been reported, except for some ambiguity at residues 77 and 78 (*245*). It possesses four cysteinyl residues, which are located at sites 43, 46, 63 and 77 (Fig. 16).

Besides their high redox potential the *Chromatium* HiPISP also has the distinguishing characteristic of being diamagnetic while in the reduced state and paramagnetic when oxidized (*246*), a single electron being transferred during the redox transition (*247*). The paramagnetic state is $S = 1/2$ and shows a rather peculiar $g = 2.115$ and $g = 2.37$ EPR signal (*240*).

The iron and inorganic sulfur atoms have been visualized by X-ray crystallography to integrate a single 4 Fe: 4 S cluster in a cubic array as depicted in Fig. 17 (*248–250*). The iron-sulfur cluster is anchored to the protein by coordination of each of the iron atoms, located at alternate corners of the cube, by the cysteinyl side chain sulfides. Average bond lengths, determined at 2.25 Å resolution, are given in the legend to Fig. 17, the distances suggesting some bonding interaction between all four

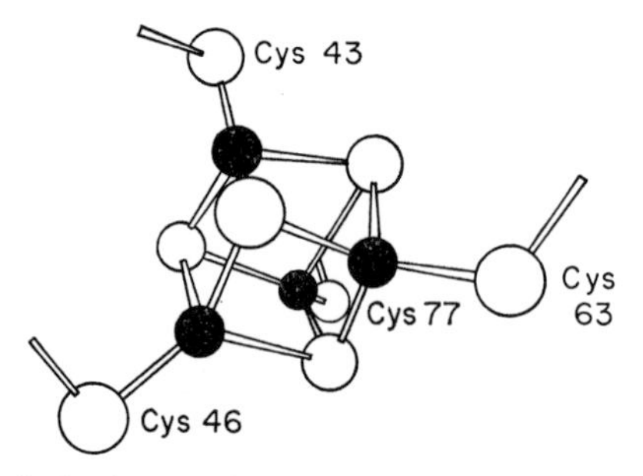

Fig. 17. Structure of the iron-sulfur cluster of the *Chromatium* HiPISP (*250*). The iron atoms, represented by smaller filled spheres, are coordinated each to one cysteinyl and three inorganic sulfides (larger empty spheres) in a tetrahedral fashion. The average bond lengths revealed by the crystallographic studies are:

Bond type	Distance (Å)
Fe—Fe	3.09 ± 0.14
$Fe—S_{inorg}$	2.35 ± 0.24
$Fe—S_{cys}$	2.01 ± 0.26

iron atoms. Furthermore, within standard deviations and the crystallographic limits of resolution, a tetrahedral environment was indicated at each iron atom. ^{57}Fe Mössbauer (*251*) and PMR (*252*) studies have, however, indicated the possibility of an imperfect tetrahedral symmetry, perhaps due to geometric effects that were unobserved by the X-ray studies.

The Conformational State

Dus et al. (*244*) have found that the native iron protein from *Chromatium* and from *Rps. gelatinosa* is only slightly susceptible to tryptic digestion. That the anti-lytic protection is conferred by metal-binding was suggested by the fact that after exhaustive reduction with mercaptoethanol and alkylation of the cysteinyl residues, the resulting S-β-aminoethylated derivative was readily digested by trypsin. The authors also observed a higher stability of the *Chromatium* relative to the *Rps. gelatinosa* protein as indicated by the relative reactivity of their inorganic sulfides. It should be noted that the HiPISP from these two species differ, at least, in the composition of their N-terminal decapeptide and exhibit pI′ (0°) at *p*H 3.68 and 9.50, respectively, when in their reduced states. HiPISP also appeared to be much more stable than the bacterial ferredoxins with regard to the reactivity of both their iron and inorganic sulfide moities.

Unlike the native protein, urea treatment resulted in a measurable reactivity of the iron with *o*-phenanthroline, the ferrous-*o*-phenanthroline complex remaining attached to the protein (*244*). The metal is so tightly bound that attempts to remove it without complete denaturation of the protein failed. Release of H_2S was promoted in 8 *M* urea even at *p*H 6.0, a process which, in absence of the denaturant, requires a *p*H below 1.0. This should also be contrasted with the bacterial ferredoxins, which spontaneously release H_2S at about *p*H 4.

The CD spectrum shows lack of any ellipticity band around 222 nm, a position where the characteristic n-π^* amide transition in right-handed α-helices occurs (*247*). Furthermore, the overall CD spectrum of the HiPISP differs qualitatively, as well as quantitatively, from those of other non-heam iron-sulfur proteins. This is consistent with the Mössbauer (*253*) and EPR (*240*) data that have indicated a different nature of the iron center in HiPISP relative to the ferredoxins and justifies, in part, the independent treatment we are giving these two classes of proteins.

A preliminary description of the geometry of the *Chromatium* molecule has been provided by *Kraut* and collaborators on the basis of X-ray studies (*250*). They state: "*The 2.25 Å structure of HiPIP confirms the suggestion (248) that the four iron atoms form a tetrahedral cluster near the center of an approximately spherical protein molecule. The molecule is contained in a prolate ellipsoid whose axes are approximately 35 Å and*

20 Å. Cysteine residues at positions 43, 46, 63 and 77 bind the four iron atoms to the cluster. Residues 1—42 lie in one end of the circumscribing ellipsoid and residues 47—86 lie in the other. A turn of 3_{10} helix formed by residues Cys 43-Cys 46 can be considered as a hinge connecting the two halves of the molecule. Each half separately is held together by a network of hydrogen bonds. All but two of the apolar side chains in the molecule line the interphase where the two halves come together enclosing the cluster and provide a highly hydrophobic environment for the redox center. Solvent accessibility to the cluster through this hydrophobic interface is evidently prevented in the model by close van der Waals contacts between side chains. It is possible, however, that in the actual molecule these contacts are sufficiently relaxed to provide access to the cluster."

These X-ray studies have indicated a negligible effect on the structure of the 4 Fe: 4S cluster upon oxidation of the protein. There is, however, chemical and spectroscopic evidence that suggests some change in the conformational state concomitant with the redox transition. Dinitrophenylation of HiPISP from *Chromatium* and *Rps. gelatinosa* completely derivatized the single histidyl residue in the oxidized proteins but not in the reduced species (*244*). Furthermore, reduction shifts the pI' (0°) from pH 3.88 to pH 3.68 (*244*). A 0.2 pH unit difference corresponds to *ca.* one unit charge difference between the two states of the protein. As the authors noted, at $3.5 < p\mathrm{H} < 3.9$ the oxidized protein should exhibit an excess of two charges over the reduced protein: one because of the reductive electron loss and the other from a single proton gain. The anomaly in the observed charge might be explained by assuming that reduction results in masking of an undetected negative counter ion (*244*). Consistently, the CD in the 430—650 nm wavelength region shows significant qualitative differences between the oxidized and reduced forms of the *Chromatium* HiPISP (*247*). In conjuntion with the X-ray crystallographic data this would indicate that the observable alterations in the symmetry of the chromophore have to arise from changes in the spatial arrangement of the polypeptide chain.

B. The Ferredoxins

The name 'ferredoxin' was first proposed for a non haem iron-redox-protein (hence its name) isolated from *Clostridium pasteurianum* and presumably involved in the hydrogen gas evolution from pyruvate by this bacterium (*254*). The smallest of the known iron-sulfur proteins, ~6,000 dalton, the clostridial ferredoxins are, however, the most complex in terms of the iron and inorganic sulfur content: 8 Fe: 8S. They are single chain polypeptides of about 55 residues of which eight are cysteines (Fig. 18).

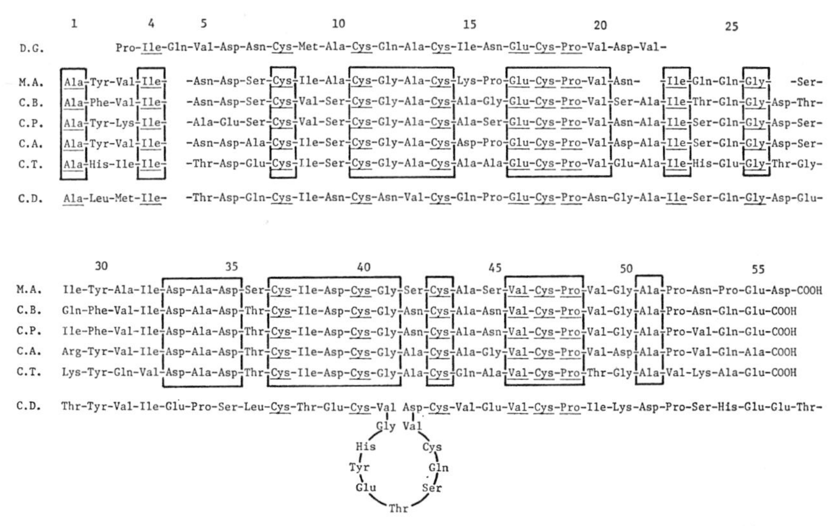

Fig. 18. Primary structures of the bacterial ferredoxins (*255*). M. A.: *Micrococcus aerogenes*, C. B.: *Clostridium butyricum*, C. P.: *Clostridium pasteurianum*, C. A.: *Clostridium acidi-urici*, and C. T.: *Clostridium tartarivorum*. Constant residues are blocked; the homology between the two halves of the molecule should be noted. Also included are N-terminal fragments of the *Desulfovibrio gigas* (D. G.) and the *Chromatium* strain D (C. D.) ferredoxins at the top and the bottom of the figure, respectively (*256, 257*). The latter exhibit some homology with the *M. aerogenes* and the clostridial ferredoxins to which the residue numbering (starting from the N-terminus) applies. Residues found invariant *in all* the ferredoxins are underlined. Deletions have been inserted as to achieve maximum coincidence of residues among the different polypeptide chains. The oligopeptides that complete the sequence in the *D. gigas* and *Chromatium* proteins are:

D. G. —Val]—Phe—Gln—Met—Asp—Glu—Gln—Gly—Asp—Lys—Ala—Val—Asn—Ile—Pro—Asn—Ser—Asn—Leu—Asp—Asp—Gln—Cys—Val—Glu—Ala—Ile—Gln—Ser—Cys—Pro—Ala—Ala—Ile—Arg—Ser—COOH.

C. D. —Thr]—Glu—Asp—Glu—Leu—Arg—Ala—Lys—Tyr—Glu—Arg—Ile—Thr—Gly—Glu—Gly—COOH.

At *p*H 7 their redox potential is about −400 mV. Anaerobic titration of the *C. pasteurianum* ferredoxin (*258*) distinctly shows that two electrons are singly transferred to two different locii as two different ESR signals arise during the reduction. NMR and magnetic susceptibility studies on the clostridial ferredoxins support this view, while indicating extensive antiferromagnetic coupling between the iron atoms (*259, 260*). X-ray studies at 2.5 Å resolution by *Sieker et al.* (*261, 262*) on the related *Micrococcus aerogenes* ferredoxin have revealed the presence of two roughly cubic aggregates each similar to that described for HiPISP, and exhibiting a distance of about 12 Å between them. The interatomic distances are such that the iron atoms appear displaced into, and the inorganic sulfides out from, the cubic array. The structural resemblance between the ferredoxin and the HiPISP active sites (Fig. 17) suggests that the reason for the major redox potential difference between the two proteins may have to be sought in the polypeptide part of the molecule. Indeed, recent evidence indicates close similarity in the geometric and electronic structure of the 4 Fe: 4S clusters of these proteins and that of synthetic analogues (*263*).

The 2 Fe: 2S 'plant type' ferredoxins, MW ~12,000 dalton, $E_{m7} = -430$ mV, were first isolated from chloroplast and photosynthetic bacteria. Similar proteins have been purified from the bacteria *E. coli* (*264*) and *Pseudomonas putida* ["putidaredoxin", $E_{m7} = -235$ mV, (*215*)] and from mammalian adrenal cortex mitochondria ["adrenodoxin" $E_{m7} = -367$ mV, ~13,100 dalton (*165*)] among other sources.

Differences in the visible region ORD spectra of spinach and clostridial ferredoxins led to the proposal that the bonding of iron at these two chromophores should differ (*150, 178, 234*). The electronic nature of the 2 Fe: 2S ferredoxins has been studied by PMR (*266*), and EPR, magnetic susceptibility, near infrared, electron-nuclear double resonance and *Mössbauer* spectroscopies [see (*267*) and references therein]. A critical comparative evaluation of these extensive data has indicated the two iron atoms to be non equivalent (*268*), each in high spin form. Both ions are ferric in the oxidized state, one of them becoming ferrous upon reduction, the two centers remaining always antiferromagnetically coupled (*267*). While indicating similar iron sites in both the oxidized and reduced states, these studies have excluded octahedral symmetry at the ferrous site of the reduced protein; an axial, rhombically distorted, tetrahedral environment has been suggested (*266, 267*). A plausible model, showing a binuclear tetrahedral configuration, is depicted in Fig. 19. Two excellent comprehensive reviews have recently appeared that deal specifically with the electronic and structural aspects of the iron-sulfur chromophores (*191, 215*).

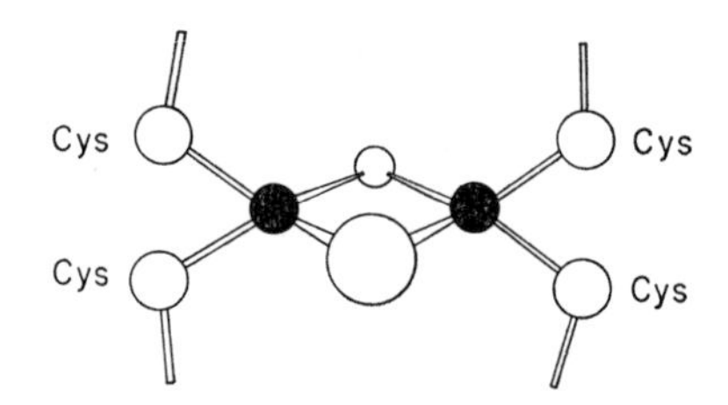

Fig. 19. Proposed model for the iron sulfur center in plant ferredoxin (*266, 267, 269, 270*)

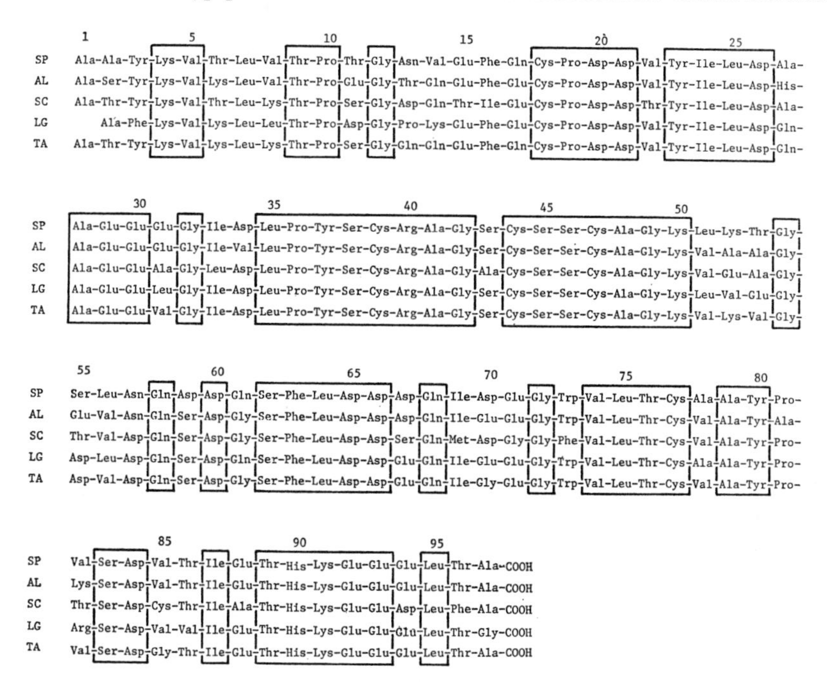

Fig. 20. Primary structures of plant ferredoxins. SP: spinach, AL: alfalfa, SC: *Scenedesmus*, LG: *Leucena glauca* ("koa"), and TA: taro. The deletion at position 1 in the *L. glauca* ferredoxin has been assumed so as to exhibit sequence homology with the other ferredoxins. (See (*271*) for original references)

All the known ferredoxins consist of a single polypeptide chain. The amino acid sequences of a number of ferredoxins are shown in Figs. 18, 20 and 21. The occurrence of these proteins throughout the living world in a variety of situations demanding specialized electron transfer roles and the simplicity of its inorganic chromophore have motivated wide speculation regarding their likely evolution from a simple polypeptide prototype (*273*, *274*) and have provided a convenient system to trace the origin and evolution of life itself (*275*). As soon as the first of the bacterial ferredoxins was sequenced by *Yasunobu* and collaborators (*276*) a striking symmetry of the polypeptide chain about its central residue was noticed. This has been confirmed in four other bacterial ferredoxins (Fig. 18) and has reinforced the hypothesis of an early doubling of a primordial ferredoxin gene coding for a 28 residue polypeptide chain (*274*). Indeed, *Chromatium* ferredoxin [not to be confused with the *Chromatium* HiPISP with which it shows little homology (*245*)] exhibits *Mössbauer* and optical absorption spectra which suggests close similarity with the clostridial proteins; it is, however, 26 residues longer (*274*).

Fig. 21. The primary structure of bovine adrenodoxin (*272*). Potential iron-binding cysteinyls are found in positions 46, 52, 55, 92, 95:

1 10
Ser—Ser—Ser—Gln—Asp—Lys—Ile—Thr—Val—His—Phe—Ile—Asn—Arg—

20
Asp—Gly—Glu—Thr—Leu—Thr—Thr—Lys—Gly—Lys—Ile—Gly—Asp—Ser—

30 40
Leu—Leu—Asp—Val—Val—Val—Gln—Asn—Asn—Leu—Asp—Ile—Asp—Gly—

50
Phe—Gly—Ala—Cys—Glu—Gly—Thr—Leu—Ala—Cys—Ser—Thr—Cys—His—

60 70
Leu—Ile—Phe—Glu—Gln—His—Ile—Phe—Glu—Lys—Leu—Glu—Ala—Ile—

80
Thr—Asn—Glu—Glu—Asn—Asn—Met—Leu—Asp—Leu—Ala—Tyr—Gly—Leu—

90
Thr—Asp—Arg—Ser—Arg—Leu—Gly—Cys—Gln—Ile—Cys—Leu—Thr—Lys—

100 110
Ala—Met—Asp—Asn—Met—Thr—Val—Arg—Val—Pro—Asp—Ala—Val—Ser—

114
Asp—Ala—COOH

Its amino acid sequence has been determined by *Matsubara et al.* (*257*), who have shown that by inserting a gap between residues 4 and 5 and a loop between residues 52 and 50, all cysteinyl residues could be arranged symmetrically in two half chains in a manner that shows homologies between the *Chromatium* and the bacterial ferredoxins (Fig. 18). The two distinguishing features of this ferredoxin are, then, the extra segments at the loop region and at the carboxy terminal region, the latter being longer and more acidic. These features might account for the fact that the *Chromatium* ferredoxin is the most reductive of all the ferredoxins, with $E' = -490$ mV.

By comparing the primary structure of alfalfa and spinach ferredoxins, *Keresztes-Nagy et al.* (*277*) observed a significant statistical presence of Ala-Ala segments every 26 residues along their polypeptide chains. In view of the fact that all clostridial and several plant ferredoxins start with alanine (see Figs. 18 and 20) and since repeating sequences have been detected in the plant ferredoxins [notice, *e.g.*, the similarity between segments 1—9 and 78—86 in spinach ferredoxin (*278*), Fig. 20], and between the plant and clostridial ferredoxins (*279*), an evolutionary connection between the bacterial and plant types appears as a strong possibility (*274*). The rigorous conservancy of cysteine-containing segments within both the bacterial and plant ferredoxins would be understandable in view of the participation of this residue in the active site.

It should also be pointed out here that the amino acid sequence of ferredoxin from the sulfate-reducing bacterium *Desulfovibrio gigas*, a protein which contains four atoms each of iron and inorganic sulfur but six cysteinyl residues, shows a strong degree of homology between the first half of the molecule (residues 1—29) and other bacterial ferredoxins (Fig. 18). The second half, which contains two cysteinyl residues, is, however, curiously homologous with plant type ferredoxins suggesting it may be derived from a prototype ferredoxin intermediate between green plants and bacteria (*256*).

It is not our purpose to survey here the intriguing aspects of the molecular evolution of these proteins but rather to stress, at this stage, the underlying unity of their chemical design. This justifies discussion of the conformational properties of the bacterial and plant ferredoxins under a common approach in spite of their distinctive molecular weights and iron-sulfur contents. The evolutionary relationship should not, however, be overemphasized; indeed, the compositional differences provide valuable information regarding the structural factors that determine the conformational state of iron proteins in general.

A number of reviews have been published recently concerning biochemical and functional characteristics of the ferredoxins (*216* a, b, *265*, *271*, *280—282*). In what follows we will center our attention on the conformational aspects only.

The Conformational State

a) *The Roles of Iron, Sulfur and the Polypeptide Chain.* At *p*H 6.3 and ambient temperature the stability of *C. pasteurianum* ferredoxin should be relatively high as no significant exchange of its own iron with external iron was observed under conditions in which ionically bound Fe^{3+} in ferrichrome does exchange (*52*). Exchange was achieved upon denaturation of the protein with the sulfhydryl reagent sodium mersalyl, followed by regeneration in excess 2-mercaptoethanol. Similar experiments showed that ^{35}S-sulfide does not exchange with either native or mersalyl-treated ferredoxins. At alkaline *p*H's, however, clostridial ^{59}Fe-ferredoxins can exchange their iron for external ferrous iron (*283*). The reaction is base-catalyzed, the exchange rate constant increasing about 20 fold with every unit increase in *p*H. At a given *p*H, iron exchanges 130 fold faster in the presence of 6 *M* urea. A similar base-catalyzed, urea-accelerated, exchange of sulfide for ^{35}S-sulfide-ferredoxin was also observed. The effects of urea indicate that the dynamic accessibility of the chromophoric constituents is significantly coupled to the conformational state of the polypeptide portion of the molecule. And conversely: reactivity of the polypeptide backbone groups is tightly dependent on the structural integrity of the metal-binding site. Thus, native *C. acidi-urici* ferrodoxin is negligibly acylated with acetic anhydride and is completely stable towards digestion by carboxypeptidase A under conditions in which apoferredoxin can be acetylated or digested by the enzyme causing it to release carboxy terminal Ala^{55} and Gln^{54} (*284*).

If iron is such an important structural determinant of ferredoxin, it might be expected that apoferredoxin and the iron-sulfide protein could exhibit different conformations. Such differences are not obvious from the ORD spectra in the far ultraviolet peptide absorption region (Fig. 22); native clostridial ferredoxin has a secondary structure which upon

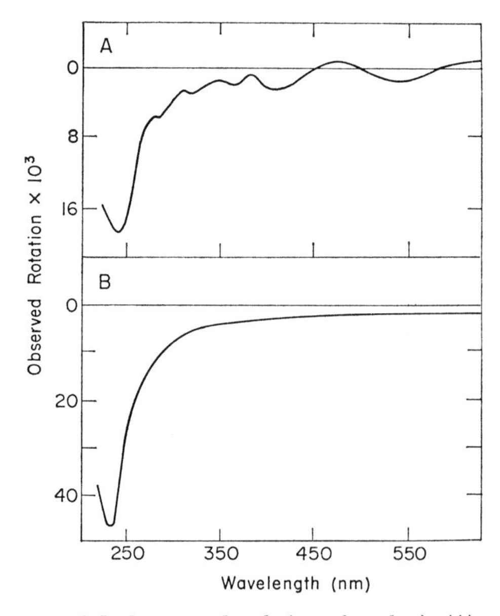

Fig. 22. ORD spectra of *C. thermosaccharolyticum* ferredoxin (A) and apoferredoxin (B) (*285*). The protein was at concentrations of 0.15 mg/ml (A) and 0.12 mg/ml (B) in aqueous solution, *p*H 7.6

metal-release yields a random coil apoprotein of similar optical rotatory properties as the iron protein (*285*). The PMR spectrum of clostridial apoferredoxin (*286*) differs, however, from that of the oxidized iron protein (and the latter, in turn, from that of the reduced ferredoxin) (*287* a, b). The apoferredoxin spectrum may be closely accounted for by its amino acid content. The fit is not exact, suggesting enough intramolecular interactions exist to generate some non-random structure; these forces could be of relevance in determining the stability of native ferredoxin. As in the case of rubredoxin, a plain conformational interpretation of the PMR of the iron protein is not granted because of the presence of significant magnetic interactions with the active center which makes dubious a direct spectral peak assignment.

Early experiments attempting to detect cross-reactivity between anti-*C. pasteurianum* ferredoxin serum and the iron-sulfide-free protein were equivocal in that they showed higher homology of the antiserum with apoferredoxin (or even its alkylated form) than with the native iron-containing antigen (*288*). However, in view of the fact that antiserum against the performic acid-oxidized (cysteine → cysteic acid) derivative showed a higher cross reactivity towards the homologous antigen than towards native ferredoxin and since some cross reactivity was always observed between any of the two antisera and the performic acid-oxidized and alkylated derivatives as well as the native protein, the authors were led to conclude that cysteinyl residues are not antigenic determinants. Furthermore, the distinct cross reactivities suggested conformational differences among the various species. *Hong* and *Rabinowitz* (*289*) have later used the immunological tests with purer preparations of the antigenic protein. Rabbit antisera against the *C. acidi-urici* protein gave a single precipitin curve when reacted with *C. acidi-urici* ^{59}Fe-labelled ferredoxin, while no reaction was observed with apoferredoxin. Similarly, microcomplement fixation tests with ferredoxin and apoferredoxin showed that while the *C. acidi-urici* ferredoxin reacted at 1:800 dilution of antiserum, the oxidized apoferredoxin did not fix complement at this concentration. A similar response was obtained by reacting anti-*C. pasteurianum* ferredoxin serum with the *C. pasteurianum* ferredoxin and apoferredoxin. *C. pasteurianum* antiserum did not react, however, with *C. acidi-urici* ferredoxin. Since these two ferredoxins differ in 14 of the 55 amino acids, it is likely that this lack of cross reactivity reflects differences in amino acid side chains as antigenic determinants rather than conformational differences between the two proteins. Indeed, a strong cross reactivity has been found between anti-*C. pasteurianum* ferredoxin serum and the *C. butyricum* ferredoxin, the latter differing from the homologous protein at only nine positions (*290*). The reservations (*289*) with regard to a purely conformational explanation of the antibody recognition are thus partly removed, especially so in view of the studies of *Mitchell et al.* (*290*) which indicate a partial reactivity of the C-terminal octapeptide against both anti-*C. pasteurianum* ferredoxin and anti-oxidized ferredoxin sera. [The haptenic nature of C- and N-terminal peptides has been further investigated by the same group (*291, 292*).] In any event, *C. acidi-urici* ferredoxin, reconstituted from apoferredoxin by reacting it with iron and sulfide, restored the microcomplement fixation to the level exhibited by the native protein (*289*). This is of interest since the reconstituted protein shows identical spectral properties and enzymatic activity as the native protein.

Rates of tritium-hydrogen exchange have been monitored for the metal-free and native *C. acidi-urici* ferredoxins by the Erlander gel filtration technique (*289*). The experiments were performed at *p*H 8.0, 4 °C, by preequilibrating the protein with tritiated water for 41 hours. A major change in the isotope exchange lability seems to be correlated with the removal of the iron-sulfur chromophore. It is interesting to note that when oxidized ferredoxin was reconstituted from fully tritiated apoferredoxin, the exchange-out process exhibited 35 slowly exchanging hydrogens compared to the 27 exhibited by the oxidized native ferredoxin (preequilibrated with tritiated water for the same period of time). This indicates that a substantial fraction of the exchangeable hydrogens are, in the ferric protein, either "buried" or participating in stable, intramolecular hydrogen bonds. From the standpoint of the conformational dynamics of the bacterial ferredoxins, it should be mentioned here that these authors have found evidence that the reconstitution process and the iron-sulfur release (by reacting native ferredoxin with α,α'-bipyridyl and mercurials) appears to occur in an all-or-none fashion (*293*). The experimental results as well as theoretical calculations indicate the existence of interactions between the iron-sulfide binding sites such that binding (release) of one mole of iron and sulfide facilitates the binding (release) of the seven other iron and sulfur atoms to (from) the molecule.

Thermal denaturation of proteins is another way of probing their conformational stability. Ferredoxins isolated from thermophilic and mesophilic clostridia, which differ by a relatively small number of amino acid substitutions, have clearly illustrated the importance of the primary structure in determining the lability of proteins toward heat and have, in particular, indicated the relatively tight coupling that exists between integrity of the metallic chromophore and the protein conformation as a whole (*255, 285*). The optical absorption spectra, molecular weight, iron, sulfur and half cystine content of the two thermophilic *C. tartarivorum* and *C. thermosaccharolyticum* ferredoxins are typical of those derived from other clostridia. However, upon heating at 70 °C, *p*H 7.4, the activity of the proteins, as assayed by the phosphoroclastic reaction, decreased faster for the mesophilic- than for the thermophilic-derived species. In particular, that from *C. thermosaccharolyticum* was the most stable. The far ultraviolet CD spectra suggested that the enhanced stability should not be related to any significant conformational differences between the proteins. During incubation at elevated temperature the absorbancy at 390 nm decreased, the rate of decrease being slower for the thermophile ferredoxin, indicating that the inactivation may be attributed to iron and sulfur release from the protein. Indeed, a satisfactory recovery of activity could be achieved by recombining the apoferredoxin resulting from the heat-treatment with iron and sulfur.

The very sharp crystalline formation of the *C. acidi-urici* ferredoxin relative to that from *C. pasteurianum* has been correlated with their different amino acid compositions, in particular with the extra proline at site 16 (*294*). It is hence interesting that the kinetics of heat-inactivation at 70 °C indicates a higher stability for the former. Similarly, the aerobic reaction of the *C. acidi-urici* and *C. tartarivorum* ferredoxins with *o*-phenanthroline is faster for the most thermolabile of the two. This was reflected clearly in the temperature dependence of the reaction (*285*). Along these lines, *Gillard et al.* (*295*) have reported that iron is removed from native *Peptostreptococcus elsdenii* and from *C. pasteurianum* ferredoxins by *o*-phenanthroline while this ferrous chelator was found inactive on the *C. acidi-urici* protein. The magnitude of these differences needs further substantiation since other authors have claimed that *o*-phenanthroline can sequester iron from the *C. acidi-urici* protein as well (*296*).

The thermophile ferredoxins are the only ferredoxins known to contain histidine. In addition to histidine they contain four other basic amino acids; this should be compared with zero or one for the mesophiles. Furthermore, the ratio of free acidic to basic amino is seven for *C. pasteurianum* and 1.4 and 1.2 for *C. tartarivorum* and *C. thermosaccharolyticum* ferredoxins, respectively. The amino acid sequences of the clostridial ferredoxins reported to-date reveal extensive conservancy with only a few point mutations (Fig. 18). The thermostability of the *C. tartarivorum* ferredoxin is thus not being reflected in gross changes in the primary structure. The basic amino acids are not located in the vicinity of the cysteinyl residues within the primary sequence. Furthermore, since the X-ray data shows that only inorganic and cysteine sulfides are involved in complexation of the iron atom, the extra stabilization of the thermophile protein cannot be accounted for, *e.g.*, by assuming that the extra histidine residues participate in coordinating the metal. The suggestion (*285*) that the heat stability might arise from a more stable chelate structure than that found in mesophilic ferredoxins, thus seems to us untenable. As our studies on the ferrichromes have shown, more stable chelate structures might be a consequence rather than a cause of an overall enhanced conformational stability, the latter being a reflection of only slight modifications in the amino acid composition.

The extent to which clostridial ferredoxin stability as well as enzymatic activity are dependent on the polypeptide chain composition has been further demonstrated by *Hong* and *Rabinowitz* (*284*). Des-(Gln^{54}-Ala^{55}) ferredoxin was made by reconstitution of *C. acidi-urici* apoferredoxin whose C-terminal dipeptide had been removed by carboxypeptidase A. This derivative was about 78% as active as the native protein and its activity decayed with a half life of about 31 hr, concomitant with denaturation. A number of aminoacyl ferredoxins were also prepared with a similar loss in both activity and stability, the decay being proportional to the size of the substituent side chain group. Furthermore, since glutamyl ferredoxin was less stable than lysyl ferredoxin, the sign of the electrostatic charge change appears to be of importance. Similarly N-acetyl, N-acetimido and N-*t*-butyloxycarbonyl derivatives of ferredoxin were all less active and less stable than the unmodified protein. Succinylation and iodination of apoferredoxin did not allow reconstitution with iron and sulfur probably, in iodoapoferredoxin, because of steric effects. *Lode* (*297*) has recently reported a significant dependence of the stability and activity of the *C. acidi-urici* ferredoxin towards its N-terminal dipeptide. In particular, the presence of an aromatic residue at position 2 seems to be critical; thus, while substitution of Tyr^2 by phenylalanine or tryptophane yields 100% and 70% of the original protein activity, its substitution by glycine results in failure to reconstitute a ferredoxin derivative.

The structural role of iron in these proteins has finally been unveiled by the 2.8 Å (refined to 2 Å) resolution X-ray crystallographic study of *Adman et al.* (*262, 298*) already mentioned. The data reported for the *Peptococcus aerogenes* (*M. aerogenes*) ferredoxin reveals a prolate ellipsoid shape of $\sim$27 Å $\times$ 22 Å. Concurring with the spectropolarimetric studies (Fig. 22) and with predictions based on the amino acid sequence of *C. pasteurianum* ferredoxin (*276*), no α-helix is present although there is room for some β-structure. Thus, the amino and carboxy termini fragments are rather parallel and linear, exhibiting turns extending from residues 7 and 46 to residues 17 and 43, respectively (Fig. 23). Cysteines 8, 11, 14 and 45 coordinate one of the 4Fe:4S clusters while the polypeptide chain shows a number of convolutions between residues 17 and 43; the second 4Fe:4S cluster is held by cysteinyls 18, 35, 38 and 41. This means that the iron-sulfide clusters are not anchored by sets of consecutive cysteines but rather that the polypeptide chain secures each cube with residues contributed by both halves of the molecule. This array obviously confers further structural compactness while posing serious questions to the hypothesis of a protein evolution by gene duplication. Interestingly, the two tyrosines present in the molecule are each located at rather equivalent positions relative to the

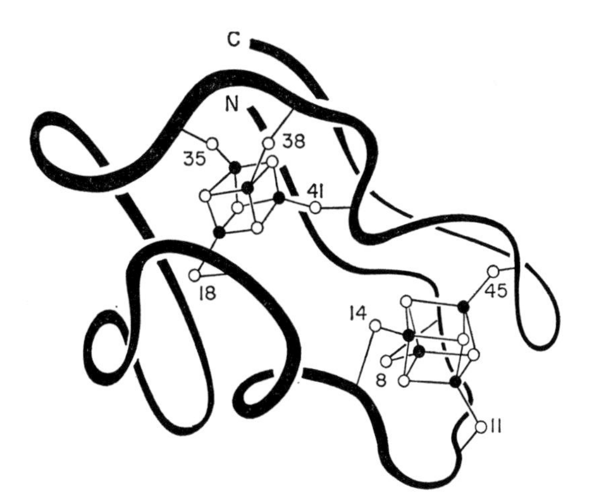

Fig. 23. The X-ray crystallographic model for the *P. aerogenes* ferredoxin (*298*). Individual atoms are not shown except for the two 4 Fe:4 S:4 S(Cys) clusters, the continuous line representing the polypeptide backbone

4Fe:4S clusters, namely parallel, at about 4 Å, to one of the faces of the cubes, consistent with ^{13}C-NMR observations on the *C. acidi-urici* protein in solution (*260*).

With regard to the conformation of a 2Fe:2S-type ferredoxin, *Kimura* and collaborators (*299*) have found no difference in the peptide absorption region of the CD spectrum of adrenodoxin that might suggest a change in the conformation of this protein concomitant to the release of the iron and sulfur moieties. However, a slight difference in the single tyrosyl fluorescence excitation maximum of the iron protein relative to its apo-derivative could be due to some conformational shift (*300, 301*).

b) *The Effects of Denaturants.* A marked stimulation of the reactivity of iron in native clostridial ferredoxin with the chelating agent *o*-phenanthroline by the presence of urea or guanidine · HCl has been observed (*296*). Similarly, the iron-chelating agent Tiron did not react unless the protein had been dissolved in 4*M* guanidine · HCl. Under such conditions the inorganic sulfide was also labilized towards 5,5′-dithio-bis(2-nitrobenzoic acid); the cysteinyl sulfur, however, reacted only when the metal was simultaneously chelated with EDTA. Interestingly, the reactions with *o*-phenanthroline or with Tiron only achieved completion under aerobic conditions. In contrast with its relative stability under unaerobic conditions, denaturants also caused degradation of the protein in the presence of air. It is obvious that urea or guanidine · HCl favor degradation of the protein by oxygen, metal sequestration or sulfur blockage. These observations also indicate that the protein conformational state is such that the iron atoms are mostly unexposed (*296*).

The effect of chaotropic agents on the reactivity of active site components has also been studied in the plant-type proteins. Alkylation of spinach ferredoxin with iodoacetamide is considerably faster in the presence of 5*M* urea (*302*). By spectrophotometric monitoring of the initial phase of the reaction, little perturbation was found during the formation of the first mole of carboxymethylcysteine, suggesting one of the five cysteines present in the molecule is not involved in the active site. Chaotropic agents also stimulated the reaction of adrenodoxin iron with *o*-phenanthroline and Tiron under aerobic conditions (*303*). In the absence of air, while urea facilitated the reaction with Tiron it inhibited chelation to *o*-phenanthroline. In any event, the denaturants affect the reactivity of the atoms at the active site. On the basis of kinetic measurements and comparisons with model system, *Kimura* and *Nakamura* (*303*) have suggested the following two mechanisms for the reaction of oxidized adrenodoxin with the iron chelators:

$$\text{adrenodoxin-(2 Fe}^{3+}) \xrightarrow[k\,=\,10^{-6}\,\text{M min}^{-1}]{o\text{-phenanthroline}} \text{adrenodoxin-(2 Fe}^{3+})\text{-}o\text{-phenanthroline}$$

$$\xrightarrow[k\,=\,10^{-3}\,\text{M min}^{-1}]{\text{intramolecular reductants}} \text{adrenodoxin-(2 Fe}^{2+}\text{-}o\text{-phenanthroline)} \qquad (1)$$

$$\text{adrenodoxin-(2 Fe}^{3+}) \xrightarrow[k\,=\,10^{-7}\,\text{M min}^{-1}]{\text{Tiron}} \text{adrenodoxin-(2 Fe}^{3+}\text{-Tiron)} \qquad (2)$$

The authors proposed that chaotropic agents affect the first step, namely, a rate limiting migration of *o*-phenanthroline molecules in the protein matrix.

The denaturant effect of oxygen in the presence of chaotropic agents has been noticed in adrenodoxin (*265*), putidaredoxin and plant ferredoxin (*304*). In fact, oxygen deteriorates alfalfa ferredoxin even in the absence of denaturants (*305*). The proposal of *Malkin* and *Rabinowitz* (*296*) that the sulfide is the aerobically sensitive moiety in ferredoxin has recently been proven correct by *Petering et al.* (*304*) who have demonstrated the generation of zero-valence state sulfur by the action of molecular oxygen.

To what extent is the effect of denaturants on the reactivity of the active site a result of changes in the secondary and tertiary structure of the protein? To answer this question it would be desirable to monitor changes in the ultraviolet optical activity of the protein. Such experiments are handicapped by the high absorbance of urea in the ultraviolet, which makes CD measurements extremely difficult; furthermore, interpretation of ORD changes in the peptide absorption region are confounded by the iron-sulfur chromophore itself (*302*). However, to the extent that the optical activity in the visible region reflects symmetry conditions at the active site arising from the polypeptide foldings around it (*178*), such measurements should be of certain value. 8 *M* urea was found to affect neither the absorption nor the ORD spectrum in the visible region of oxidized *C. acidi-urici* ferredoxin (*234*). This suggested that the changes in the chromophore accessibility do not affect the symmetry in the environment of the iron atoms. A most interesting finding reported in these early studies was that the urea-labilized reactivity of iron towards α,α-bipyridyl released ——*tris* bipyridyl iron(II). This implies a highly asymmetric site in the molecule that forces the reaction between the external chelator and the metal to occur in a stereoselective way.

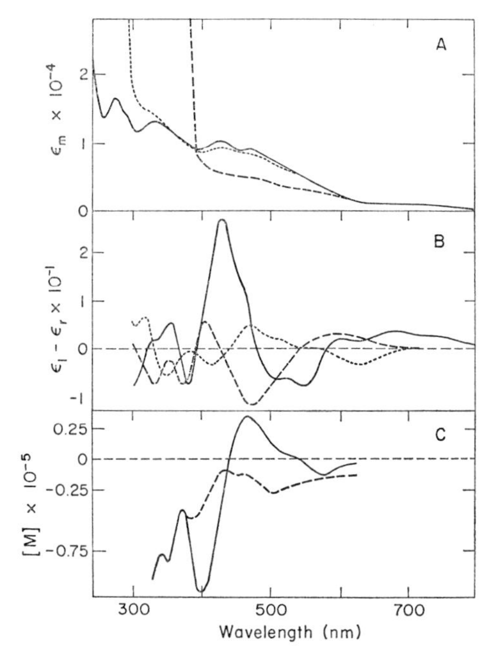

Fig. 24. Effects of urea and of the redox transition on the optical properties of spinach ferredoxin (*306*). In A, B and C the absorption, CD and ORD spectra of the reduced (---) and oxidized (—; in 8 *M* urea: · · ·) proteins are shown, respectively

In contrast with its effect on clostridial ferredoxin, 8 *M* urea clearly induces changes (Fig. 24) in the absorption, ORD and CD spectra of oxidized spinach ferredoxin (*306*). An overall gain in symmetry around the iron chromophore is suggested. A similar pattern is observed when the *p*H of the solution is raised to 11, which led the authors to suggest breakage of secondary structure in the neighborhood of the iron chromophore leaving a residual Cotton effect intrinsic to the iron-sulfur complex. Indeed, this reduced Cotton effect was of about the same magnitude as that found in the clostridial ferredoxin, hinting that conformational differences between the two types of ferredoxins may partly account for their different optical activities in the native state. Consistently, antibodies to Swiss chard ferredoxin do not cross react with the clostridial protein (*307*).

Effects of urea and guanidine · HCl on adrenodoxin have also been observed (*308*). The kinetics of denaturation is faster for adrenodoxin than for spinach ferredoxin, which was indicative of a more stable structure for the plant protein (*309*). The spectropolarimetric data was interpreted to suggest either that the strong visible region Cotton effects are intrinsic to the iron-sulfur site or that the optical activity arises by a coupling of the active site chromophore to aromatic side chain transitions in the ultraviolet rather than to a backbone α-helix (*308, 309*).

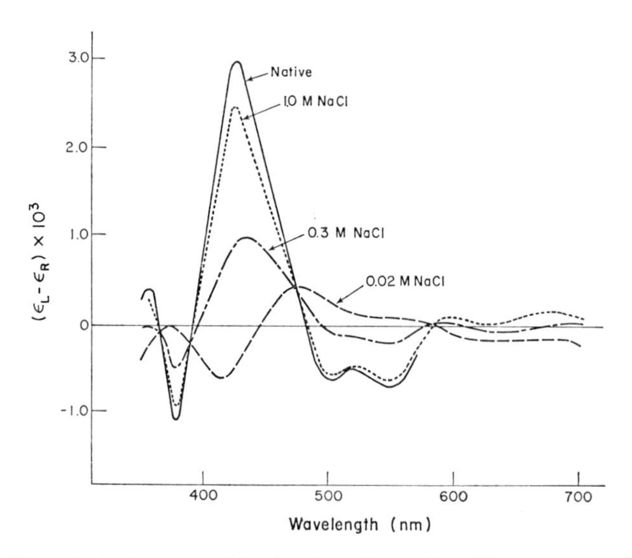

Fig. 25. The effect of urea and ionic strength on the CD spectrum of spinach ferredoxin (*302*). The data was recorded anaerobically for protein solutions at *p*H 7.3 and 23 °C. Spectra labelled with varying salt concentrations are for the protein dissolved in 5 *M* urea

Chaotropic agents other than urea and guanidine · HCl have been found to affect oxidized spinach ferredoxin similarly, as judged by the effects on the protein ORD spectrum (*310*). *Petering* and *Palmer* (*302*) have pointed out that the effect of denaturants on the optical rotatory properties of iron-sulfur proteins might not be so simple. Under anaerobic conditions spinach ferredoxin dissolved in 5 *M* urea, 1 *N* NaCl, exhibits optical absorption and CD spectra that are very similar to that of the native protein. However, as the salt concentration is reduced to 0.02 *M* the CD pattern changes (and, to a lesser extent, so does the absorption spectrum), the most remarkable effect being the replacement of the intense Cotton effect at 428 nm in high salt by a weak negative band at 420 nm (Fig. 25). The effect is reversible, as the native CD spectrum was recovered upon addition of salt to 1 *N* NaCl, and is not ion specific as KCl or $(Na)_2SO_4$ produced the same effect. Thus, it was suggested that urea and high ionic strength are each affecting the polypeptide conformation independently resulting in an alteration of the structural symmetry of the iron-sulfur chromophore in such a way that the site itself is left intact (*302*).

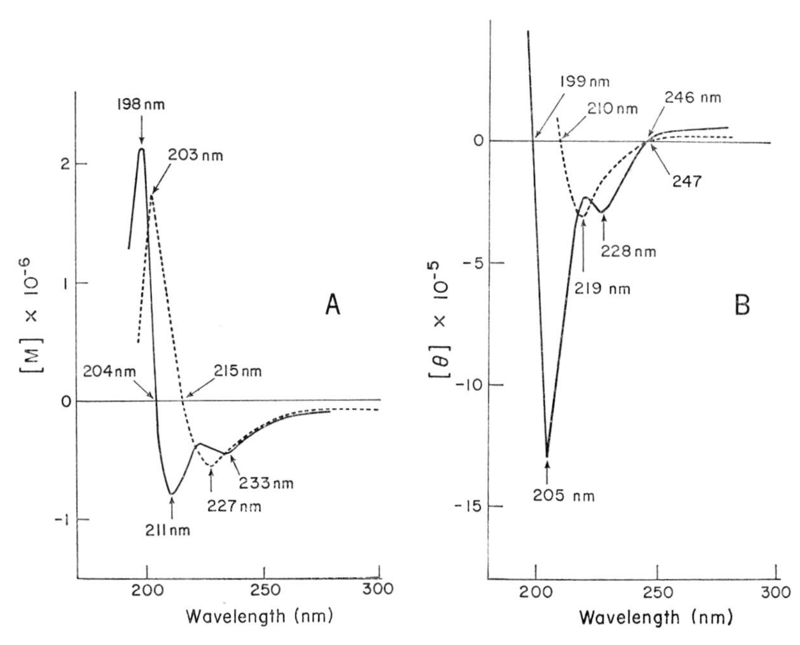

Fig. 26. The polypeptide backbone absorption region optical activity of two 2Fe:2S proteins (*299*). The ultraviolet ORD (A) and CD (B) spectra of bovine adrenodoxin (—) and spinach ferredoxin (---), taken in aqueous solutions, *p*H 7.4

It has been suggested by *Keresztes-Nagy et al.* (*277*) that the occurrence in alfalfa ferredoxin of overwhelmingly clustered distribution of strongly hydrophilic and of hydrophobic residues in its primary structure is likely to generate a low helix content. The rationale was that in order to have a high helix content the non polar residues should be located along the sequence at intervals such that their side chains can always point towards the same side, namely, the hydrophobic "inside" of the molecule. In agreement with these speculations, the $n \rightarrow \pi^*$, $\pi \rightarrow \pi^*$ peptide absorption region CD of alfalfa (*185*) and spinach (*299*) ferredoxins as well as the ORD and infrared absorption (*299*) of the latter, have indicated negligible α-helical content for the plant ferredoxins and have suggested the presence of some β-structure (Fig. 26). For adrenodoxin, the same data indicated about 20% helix content (*299, 309*). The differences in secondary structure between these two classes of 2Fe:2S proteins are of interest as the optical absorption, ORD, CD and EPR spectra and magnetic susceptibility data have indicated close similarity of their iron-sulfur chromophores [see (*265, 299*) and references therein].

A globular shape (a/b $\sim$5) has been estimated for adrenodoxin from intrinsic viscosity measurements (*299*). This axial ratio differs from the value of 14 to 18 obtained for plant ferredoxin in 1961 (*311*) with a protein preparation that probably was aggregated (estimated molecular weight $\sim$17,000 dalton). Adrenodoxin has a larger number of basic amino acid residues than plant ferredoxin and the two proteins also differ in the tyrosyl, tryptophanyl and methionyl content (*299*). The primary structure indicates no homology between either the bacterial or plant ferredoxins and the bovine adrenodoxin (Figs. 18, 20, 21). Furthermore, the amino acid sequence provides no indication of the segment duplication patterns observed among the plant and bacterial counterparts. This might then account for the differences in stability and in secondary structure already discussed and might also be related to the fact that neither can spinach (or euglena) ferredoxin substitute for the adrenal 2Fe:2S protein in the steroid 11β-hydroxylation reaction, nor can adrenodoxin replace plant ferredoxin in the photosynthetic pyridine nucleotide reduction (*265*).

c) *The Effect of Reduction and the Strains at the Active Site.* Reduction of clostridial ferredoxin generates drastic changes in the ORD (*234*) and CD (*235*) spectra of the iron-sulfur chromophore, an overall gain in symmetry being indicated. A labilization of the iron for bipyridyl substitution, even in the absence of denaturants, was also detected in these studies. Since the resulting tris bipyridyl iron(II) complex lacked optical activity, this further enforced the interpretation of a lower asymmetry in the reduced, relative to the oxidized protein (*234*). Consistent with these results, *Malkin* and *Rabinowitz* (*296*) have found a higher initial reactivity of the reduced relative to the oxidized forms of the *C. acidi-urici* ferredoxin towards the ferrous chelating agent *o*-phenanthroline. However, after the first mole of iron had been chelated, the remaining iron reacted more slowly. Similar results were observed with *C. pasteurianum* ferredoxin (*295*).

The problem has also been approached by measuring the rates of tritium-hydrogen exchange of *C. acidi-urici* ferredoxin in both redox states (*289*). The experiments were performed as already described in comparing apo- and native ferredoxins. While oxidized ferredoxin exhibits about 27 slowly exchanging hydrogens ($t_{1/2} > 20$ min), there are 17 hydrogens of this class in the reduced protein. The authors thus suggested that "*the reduced state has a more compact structure than the oxidized state of the protein since it has fewer sites for equilibration with tritium*". In view of our results with the ferrichromes (*58*, *59*) such interpretation is not granted unless backed by temperature dependence studies of the exchange kinetics. However, as far as the hydrogen-tritium exchange experiments show differences between the two redox states, these data are consistent with the other experiments already mentioned in that they indicate a change in the protein conformational state in the broader sense assumed in this review. Thus, the ^{13}C-NMR study of a clostridial ferredoxin by *Packer*, *Sternlicht* and *Rabinowitz* (*260*) suggests that the two tyrosines which are resolved in the spectra, may not change their relative orientation with regard to the Fe:S cluster upon the redox transition.

Changes in the optical properties of the iron-sulfur chromophore were also induced by reduction of the plant ferredoxins, as shown in Fig. 24. *Garbett* and *Stangroom* (*306*) mention unpublished evidence which indicates differences between the oxidized and reduced states of the protein with regards to its reactivity towards *o*-phenanthroline or dipyridyl. *Kimura* and *Nakamura* (*303*) found that reduced and oxidized adrenodoxin react with *o*-phenanthroline at comparable rates, further suggesting conformational differences with the plant ferredoxins. The latter, however, in contrast with the trend exhibited by the bacterial protein, is more reactive in the oxidized than in the reduced state. Consistent with this trend, *Cammack et al.* (*310*) have found that while high concentrations ($> 1\,M$) of methanol and chaotropic agents degenerate the ORD spectrum of spinach ferredoxin towards that of the apoprotein, independent of its initial redox state, at low concentration the denaturants failed to modify the ORD of the reduced protein while drastically affecting that of the oxidized species.

Reduction of plant ferredoxin is a one electron transfer process that generates a characteristic $g = 1.94$ EPR band as the only signal in the 0—5 kG range. Oxidized spinach ferredoxin does not exhibit any signal in this range, being insensitive, in this regard, to treatment with urea up to 7.5 *M* final concentration of the denaturant in 1 *N* NaCl (*302*). The reduced species, however, shows 58% of the initial amplitude at 7.5 *M* urea and 1 *M* NaCl, but the signal is practically unaffected up to 5 *M* urea and high salt. When the salt concentration is reduced to 0.015 *M*, most of the signal disappears which shows, again, that high ionic strength opposes the effects caused by the denaturant as detected at the active site (*302*). In these studies, *Petering* and *Palmer* made the interesting observation that as the urea concentration was increased from 0 to 7.5 *M*, g_z and g_y moved upfield by 2 and 4 G, respectively, while g_x moved by some 6 G to lower magnetic fields. The studies by *Cammack et al.* (*310*) have shown that the chaotropic agents perchlorate, trichloroacetate, thiocyanate, iodine, urea and guanidine · HCl at concentrations as low as 10—50 m*M* cause changes in linewidths and shifts in the apparent g-values of the EPR spectrum of spinach ferredoxin. Thus, when dissolved in a trichloroacetate solution, the apparent g_y and g_z move upfield by 13 and 18 G respectively, while g_x moves downfield 9 G, making the spectrum to look more like that of adrenodoxin (Fig. 27A). Raising the concentration of guanidine · HCl (Fig. 27B, c and d) resulted in the introduction of a transient new signal at $g = 1.95$. The process was reversible as the original EPR could be restored by reoxidation of the sample and removal of the denaturant (Fig. 27B, e). Other chaotropic agents generated the same effects but at relatively higher concentrations. Thus, the spectrum shown in Fig. 27C, similar to that in Fig. 27B, c, required 6.5 *M* urea. Higher concentration of the chaotropes (or 70% v/v methanol) resulted in rapid, irreversible denaturation of the reduced protein, probably through drastic effects on its conformation. Since low concentrations of the denaturants affected the EPR signal while leaving the optical rotatory properties unperturbed, it is suggested that the electron spin resonance is sensitive to changes that do not manifest themselves in the static architecture of the molecule. Similar studies by *Coffman* and *Stavens* (*312*) have shown that in the concentration range 0—20% v/v, methanol causes no changes in the CD spectrum of reduced spinach ferredoxin while the EPR g values shifted, as already found for other denaturants, towards a less rhombic (*i.e.* more axial) type of symmetry. The interesting observation was that by plotting the g values of a number of 2Fe:2S proteins versus an adequate rhombic magnetic asymmetry parameter (defined as $x = g_2 - g_3$), the linear plots could be adequately matched by the plant ferredoxin parameters measured from spectra taken at different methanol concen-

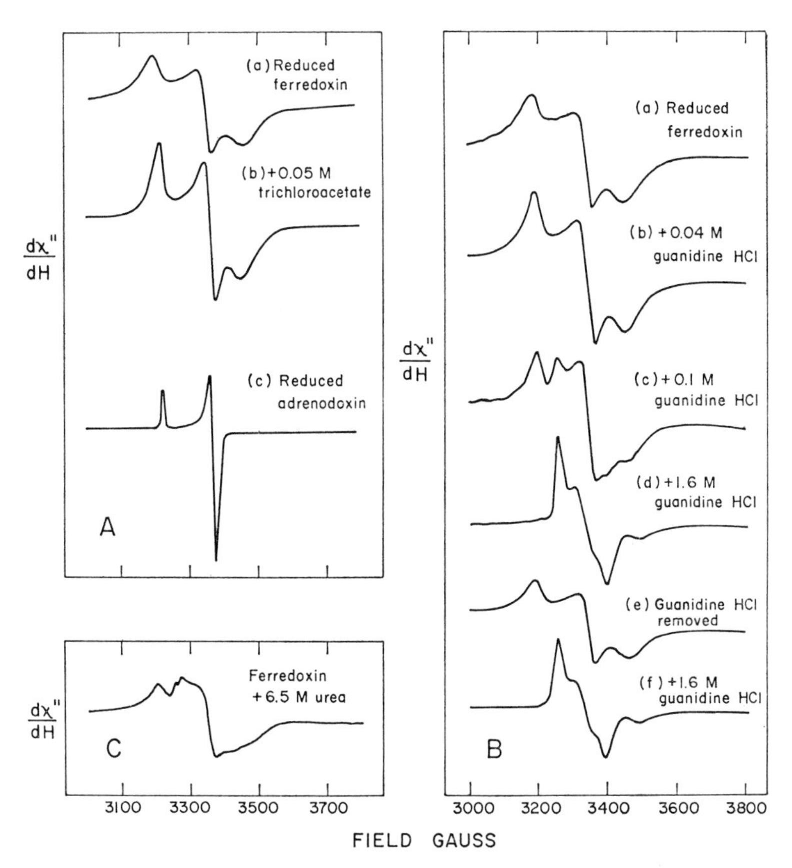

Fig. 27. Effect of several denaturants on the EPR of reduced spinach ferredoxin (*310*). In A, 0.005 *M* trichloroacetate is shown to modify the spinach protein signal approximating it to that of adrenodoxin. In B, the presence of varying concentrations of guanidine · HCl is seen to affect reversibly the appearance of the EPR, the effect of the denaturant being similar to that of 6.5 *M* urea (shown in C)

trations (Fig. 28). In other words: the effects on the electron configuration of the active site arising from the different primary structures of these proteins could be mimicked by gently affecting the state of spinach ferredoxin with a denaturant without perturbing the conformation of the protein itself to any observable extent. The authors conclude that the way the magnetic anisotropy of a 2Fe:2S protein changes *"suggests an effect due to a conformation-related strain variable in the paramagnetic group of atoms"* (*312*).

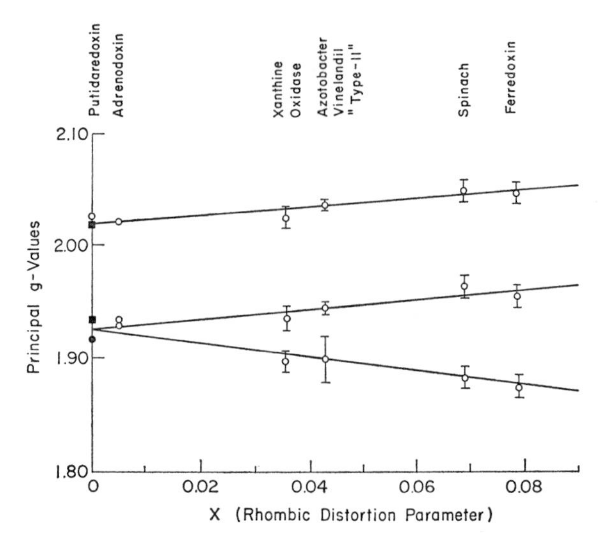

Fig. 28. Reported principal g values for various 2 Fe—2 S proteins versus x ($=g_2-g_3$), a rhombic distortion parameter. The straight lines are least squares fits of the data points as explained in the original reference (*312*)

C. Discussion

An intimate coupling exists between the polypeptide chain and the active site complex in the iron-sulfur proteins. It has recently been suggested that the notorious differences in redox potential exhibited by HiPISP and the clostridial ferredoxins may be a consequence of such interaction (*313*). A diamagnetic 'paired-spin' state is assumed for the active center which can become paramagnetic either by taking up (ferredoxin) or giving out (HiPISP) a single electron. At the basis of the hypothesis is the assumption that although both proteins can be in the intermediate state only one of the two paramagnetic levels is compatible with each polypeptide moiety. The way in which this occurs is not clear. Thus, *e.g.* "...*although both polypeptides surround the Fe_4S_4 cluster with non-polar amino-acid side chains, the two ferredoxin clusters are more accessible to solvent*" (*312*). However, as we have shown, solvent accessibility is as much a dynamical as a static structural phenomenon. Similarly, there is abundant evidence indicating that a change in the conformational state of these proteins accompanies the redox transition. This change was not detected (HiPISP) nor is it expected to be detected (ferredoxin) by the crystallographic studies (*313*). Indeed, a change in the molecular plasticity could be directly related to its biological mode of action in allowing the protein either to fit into the reductase (electron donor substrate) or the oxidase (electron acceptor substrate) active sites (*14*). It is such subtle aspects of the interaction between polypeptide configuration and the iron chromophore that our survey of these proteins has attempted to illustrate.

VII. Summary and Conclusions

A potential energy surface in the protein coordinate space exhibits a number of energy minima that permit the protein to assume metastable conformations that do not correspond to the absolute energy minimum. By allowing a higher number of vibrational modes, a local non-absolute energy minimum might still result in an absolute minimal conformational free energy due to some entropy gain (*314*). In such a view, ligand-binding is considered to stabilize one of these metastable states thus resulting in the possibility of significant conformational differences between the ligand-free and ligand-bound protein (*315*). The iron-proteins reviewed in this paper have exhibited this effect to different degrees.

The extent to which the metal can control the protein conformation is reflected in the existence of non-planar peptide bonds in complexed enterobactin (*66*), ferrichrome A (*316*) and rubredoxin (*294*). Obviously, a situation of tension prevails that is relaxed upon metal release. It may thus be asserted that in those cases where the molecular weight is not too high ($\lesssim$ 15,000 dalton) the energetics of the native structure is mainly determined by the complexation event. In the extreme low molecular weight case of the ferrichromes, with an iron-binding constant of the order of 10^{30}, the resulting free energy for metal complexation ($\sim$41 kcal) appears to be large enough to account for the conformational energy differences that have been estimated for cyclic hexapeptides by different criteria (*317*). Indeed, if constancy of a minimum number of amino acid residues is insured, such as glycyls located at peculiar chain turning points (in general, residues that are evolutionary invariant), the overall conformation of the iron-polypeptide globule is practically unaffected by substitutions of amino acids of different charge or polarity. In the ferrichromes, this kind of substitution results in conformational shifts in the iron-free oligopeptide but not in the chelates.

As clearly shown by the ferrichromes and the bacterial ferredoxins, the stability of the coordinated globule immediately reflects the single, non-essential amino acid substitutions. These further introduce positive and negative strains that are not strong enough to overcome the tight conformational control imposed by the iron complexation but which modify the free energy trough where the molecule lies. However, the forces arising from the protein part of the molecule should become increasingly important as the polypeptide chain is lengthened. This might explain why the ~14,000 dalton, 8 Fe:8 S *A. vinelandii* ferredoxin exhibits a significantly higher stability in 4 *M* guanidine · HCl and towards reaction with *p*-mercuribenzoate than the ~6,000 dalton *C. pasteurianum* protein (*243*).

In high molecular weight proteins, the energetics of the residue-residue non-bonding interactions appear to outweigh the conformational influence of metal-binding so that the effect of the ligand is mainly to stabilize the molecule in a conformation that is very similar to that assumed by the apoprotein. However, although the geometry of the molecule is only slightly altered, the state of the protein as a whole is affected. On the basis of the hydrogen exchange kinetics, a minimal conformational stabilization free energy of about 1.5 kcal/mole, has been estimated by *Ulmer* for the iron-binding to transferrin (*160, 318*). Nevertheless, in spite of their size, even the transferrins appear to exhibit the kind of stability effects discussed above and which arise from the glycoprotein composition: it has been shown (*163*) that human lactoferrin binds Fe(III) about 8 times more avidly than the bovine milk protein and the latter complex is about 35 times more stable than the human serum species, the relative stabilities being in good agreement with previous data on the H^+ *vs* Fe^{3+} binding equilibria of these homologous proteins (*319*).

By limiting our survey to the non-haem type iron proteins, the role of the iron-porphyrin complex as a conformational factor has been deliberately excluded. There is, however, at least one instance where the separate contributions of the iron and porphyrin moieties of haem have been clearly resolved, namely in the case of horse heart cytochrome c (~12,000 dalton). In this protein the tetrapyrrole ring is held through two thioether linkages to cysteines 14 and 17, the central iron atom being axially coordinated by sulfur and nitrogen atoms belonging to the residues methionyl 80 and histidyl 18, respectively (*7, 51* a). While the protein lacks disulfide bridges, the crystallographic structure reveals the haem group itself as a main folding influence on the molecule which, upon reduction, assumes a relatively more compact structure [see (*14*) and references therein]. The polypeptide backbone is 104 residues long and provides the haem with a hydrophobic environment. By treating the native protein with anhydrous HF, the iron atom can be displaced and a porphyrin-cytochrome c species obtained. On the basis of viscosity, CD spectroscopic and fluorescence emission studies, *Fisher et al.* (*320*) have shown that while ferricytochrome c and the demetallo derivative have a similar, compact, globular shape in solution, under the same conditions apo-cytochrome c (*i.e.*, the haem-free protein) exhibits a more disordered structure. Consistently, the porphyrin-cytochrome c and the native protein show approximately the same elution patterns upon filtration in Sephadex G-50, being more retained than the apo-protein. Judged by their relative lability towards heat and guanidine · HCl, however, the porphyrin-containing derivative possesses a less stable structure than the haem-containing molecule. Since upon removal of the denaturant both the demetallo- and the haem-proteins spontaneously refold, it is obvious that the main factor responsible for the native conformation is the porphyrin moiety. Thus, the ferric ion plays, in this regard, mainly a structure-stabilizing role.

It is hoped that the examples reviewed and the discussion given above have clarified why the title of this article refers to the 'conformational state' rather than to the 'conformation' of iron proteins. At times, the stability of the molecules was discussed in terms of certain manifestations at the iron locus. This lead us, rather surreptitiously, into probing the 'entatic' nature of the active site. Indeed, the stability of the iron protein can be drastically affected by the action of denaturants or by solvent perturbation, and this appears to be more clearly reflected in the electronic state of the metal center than in any significant (static) conformational effect.

Acknowledgements. Although the author bears sole responsibility for the contents of this article, many of the ideas here presented arose from a fruitful reasearch collaboration with Dr. *M. P. Klein* and Professor *J. B Neilands*. The author is grateful to Professor *L. H. Jensen* for the availability of information prior to publication and to Professor *K. Raymond* for helpful discussions. Thanks are also due to Drs. *E. T. Lode, E. L. Packer, L. E. Vickery* and to Professor *J. B. Neilands* and other members of his laboratory for reading and checking the manuscript.

This work was supported by the U.S. Atomic Energy Commission, and by grants N° AI-04156 and GB 5276 from the U.S. Public Health and the National Science Foundation, respectively.

References

1. *Neilands, J. B.:* Struct. Bonding *11*, 145 (1972).
2. *Vallee, B. L., Williams, R. J. P.:* Proc. Natl. Acad. Sci. U.S. *59*, 498 (1968).
3. *Williams, R. J. P.:* Cold Spring Harbor Symp. Quant. Biol. *36*, 53 (1972).
4. *Malmström, B. G.:* Pure Appl. Chem. *24*, 393 (1970).
5. *Busch, D. H., Farmery, K., Goedken, V., Katovic, V., Melnyk, A. C., Sperati, C. R., Tokel, N.:* Advan. Chem. Ser. *100*, 44 (1971).
6. *Hoard, J. L.:* Science *174*, 1295 (1971).

7. *Wüthrich, K.:* Struct. Bonding *8*, 53 (1970).
8. *Spiro, T. G., Saltman, P.:* Struct. Bonding *6*, 116 (1969).
9. *Midvan, A. S., Cohn, M.:* Advan. Enzymol. *33*, 1 (1970).
10. *Feder, J., Garrett, L. R., Wildi, B. S.:* Biochemistry *10*, 4552 (1971).
11. *Suelter, C. H.:* Science *168*, 789 (1970).
12. *Neilands, J. B.:* Bact. Rev., *21*, 101 (1957).
13. — In: Inorganic biochemistry, Vol. I p. 167. Ed. *Eichhorn, G.* New York: Elsevier 1973.
14. *Takano, T., Swanson, R., Kallai, O. B., Dickerson, R. E.:* Cold Spring Harbor Symp. Quant. Biol. *36*, 397 (1972).
15. *Keller-Schierlein, W., Prelog, V., Zähner, H.:* Fortschr. Chem. Org. Naturstoffe *22*, 279 (1964).
16. *Neilands, J. B.:* Struct. Bonding *1*, 59 (1966).
17. *Nüesch, J. Knüsel, F.:* Antibiotics. Vol. I, p. 499. Ed. *Gottlieb, D.* and *Shaw, P. D.* Berlin-Heidelberg-New York: Springer 1967.
18. *Emery, T.:* Advan. Enzymol. *35*, 135 (1971).
19. *Maehr, H.:* Pure Appl. Chem. *28*, 603 (1971).
20. *Hutner, S. H.:* Ann. Rev. Microbiol. *26*, 313 (1972).
21a. Different chapters in: Microbial iron metabolism. Ed. *Neilands, J. B.* New York: Academic Press, in press.
21b. *Lankford, C. E.:* CRC Crit. Rev. Microbiol. *2*, 273 (1973).
22. *Ehrenberg, A.:* Nature *178*, 379 (1956).
23. *Wickman, H. H., Klein, M. P., Shirley, D. A.:* Phys. Rev. *152*, 345 (1966).
24. *Klein, M. P.:* In: Magnetic resonance in biological systems, p. 407. Ed. *Ehrenberg, A., Malström, B. G., Vänngård, T.:* London: Pergamon Press 1967.
25. *Bracket, G., Richards, P.:* J. Chem. Phys. *54*, 43 (1971).
26. *Wickman, H. H., Klein, M. P., Shirley, D. A.:* J. Chem. Phys. *42*, 2113 (1965).
27. *Zalkin, A., Forrester, J. D., Templeton, D. H.:* Science *146*, 261 (1964).
28. — — — J. Am. Chem. Soc. *88*, 1810 (1966).
29. *Llinás, M., Klein, M. P., Neilands, J. B.:* J. Mol. Biol. *52*, 399 (1970).
30. — — — J. Mol. Biol. *68*, 265 (1972).
31. *Schwyzer, R.:* In: Amino acids and peptides with antimetabolic activity, p. 171. Ed. *Wolstenholme, G. E. W.* London: Churchill 1958.
32. *Karle, I. L., Karle, J.:* Acta Cryst. *16*, 969 (1963).
33. — *Gibson, J. W., Karle, J.:* J. Am. Chem. Soc. *92*, 3755 (1970).
34. *Bovey, F. A., Brewster, A. I., Patel, D. J., Tonelli, A. E., Torchia, D. A.:* Acc. Chem. Res. *5*, 193 (1972).
35. *Kopple, K. D., Go, A., Logan, R. H., Šavrda, J.:* J. Am. Chem. Soc. *94*, 973 (1972).
36. *Warren, R. A. J., Neilands, J. B.:* J. Biol. Chem. *240*, 2055 (1965).
37. *Emery, T.:* Biochemistry *10*, 1483 (1971).
38. *Neilands, J. B.:* Science *156*, 1 (1967).
39. — Exper. Suppl. *IX*, 222 (1964).
40. *Llinás, M., Klein, M. P., Neilands, J. B.:* Intern. J. Peptide Protein Res. *4*, 157 (1972).
41. *Urry, D. W., Ohnishi, M.:* In: Spectroscopic approaches to biomolecular conformation, p. 263. Ed. *Urry, D. W.* Chicago: A.M.A. 1970.
42. *Ulmer, D. D.:* Biochemistry *4*, 902 (1965).
43. *Llinás, M.:* Dissertation. Berkeley, University of California 1971.
44. *Gratzer, W. B., Cowburn, D. A.:* Nature *222*, 426 (1969).
45. *Emery, T.:* Biochemistry *6*, 3858 (1967).

46. *Bystrov, B. F., Portnova, S. L., Tsetlin, V. I., Ivanov, V. T., Ovchinnikov, Y. A.*: Tetrahedron *25*, 493 (1969).
47. IUPAC-IUB Commission on Biochemical Nomenclature: J. Mol. Biol. *52*, 1 (1970).
48. *Ramachandran, G. N., Chandrasekaran, R., Kopple, K. D.*: Biopolymers *10*, 2113 (1971).
49. *Llinás, M., Neilands, J. B.*: to be published.
50. *Venkatachalam, C. M.*: Biopolymers *6*, 1425 (1968).
51a. *Dickerson, R. E., Takano, T., Eisenberg, D., Kallai, O. B., Samson, L., Cooper, A., Margoliash, E.*: J. Biol. Chem. *246*, 1511 (1971).
51b. *Crawford, J. L., Lipscomb, W. N., Schellman, C. G.*: Proc. Natl. Acad. Sci. U.S. *70*, 538 (1973).
52. *Maehr, H., Pitcher, R. G.*: J. Antibiotics *24*, 830 (1971).
53. *Berger, A., Linderstrøm-Lang, K.*: Arch. Biochem. Biophys. *69*, 106 (1957).
54. *Hvidt, A., Nielsen, S. O.*: Advan. Protein Chem. *21*, 287 (1966).
55. *Klotz, I. M.*: J. Colloid Interface Sci. *27*, 804 (1968).
56. *Lovenberg, W., Buchanan, B. B., Rabinowitz, J. C.*: J. Biol. Chem. *238*, 3899 (1963).
57. *Turková, J., Mikeš, O., Schraml, J., Knessl, O., Šorm, F.*: Antibiotiki *9*, 506 (1964).
58. *Llinás, M., Klein, M. P., Neilands, J. B.*: J. Biol. Chem. *248*, 915 (1973).
59. — — — J. Biol. Chem. *248*, 924 (1973).
60. *Anderegg, G., L'Eplattenier, F., Schwarzenbach, G.*: Helv. Chim. Acta *46*, 1409 (1963).
61. *Bickel, H., Hall, G. E., Keller-Schierlein, W., Prelog, V., Vischer, E., Wettstein, A.*: Helv. Chim. Acta *43*, 2129 (1960).
62. *Bock, J. L., Lang, G.*: Biochim. Biophys. Acta *264*, 245 (1972).
63. *O'Brien, I. G., Cox, G. B., Gibson, F.*: Biochim. Biophys. Acta *237*, 537 (1971).
64. *Bryce, G. F., Brot, N.*: Biochemistry *11*, 1708—1715 (1972).
65. *Langman, L., Young, I. G., Frost, G. E., Rosenberg, H., Gibson, F.*: J. Bact. *112*, 1142—1149 (1972).
66. *Llinás, M., Wilson, D. M., Neilands, J. B.*: Biochemistry *12*, 3836 (1973).
67. *Teuwissen, B., Masson, P. L., Osinski, P., Heremans, J. F.*: Europ. J. Biochem. *31*, 239 (1972).
68. *Neilands, J. B.*: In Structure and function of oxidation reduction enzymes, p. 541. Ed. *Åkeson, Å* and *Ehrenberg, Å*. Oxford: Pergamon Press 1972.
69. *Shemyakin, M. M., Ovchinnikov, Y. A., Ivanov, V. T., Antonov, V. K., Vinogradova, E. I., Shkrob, A. M., Malenkov, G. G., Evstratov, A. V., Laine, I. A., Melnik, E. I., Ryabova, I. D.*: J. Membrane, Biol. *1*, 402 (1969).
70. *Patel, D. J., Tonelli, A. E.*: Biochemistry *12*, 486 (1973).
71. *Steinrauf, L. K., Czerwinski, E. W., Pinkerton, M.*: Biochem. Biophys. Res. Commun. *45*, 1279 (1971).
72. *Czerwinski, E. W., Steinrauf, L. K.*: Biochem. Biophys. Res. Commun. *45*, 1284 (1971).
73. *Duax, W. L., Hauptman, H., Weeks, C. M., Norton, D. A.*: Science *176*, 911 (1972).
74. *Palmour, R. M., Sutton, H. E.*: Biochemistry *10*, 4026 (1971).
75. *Fletcher, J., Huehns, E. R.*: Nature *218*, 1211 (1969).
76. *Williams, R. J. P.*: R. I. C. Rev. *1*, 13 (1968).
77. *Garibaldi, J. A.*: Food Res. *25*, 337 (1960).
78. *Bullen, J. J., Rogers, H. J., Leigh, L.*: Brit. Med. J. *1*, 69 (1972).

79. *Perkins, D. J.*: In: Protides of the biological fluids, Vol. 14, p. 83. Ed. *Peeters, H.* Amsterdam: Elsevier 1966.
80. *Ford-Hutchinson, A. W., Perkins, D. J.*: Europ. J. Biochem. *21*, 55 (1961).
81. *Tan, A. T., Woodworth, R. C.*: Biochemistry, *8*, 3711 (1969).
82. *Gaber, B. P., Aisen, P.*: Biochim. Biophys Acta *221*, 228 (1970).
83. *Aasa, R., Malmström, B. G., Saltman, P., Vänngård, T.*: Biochim. Biophys. Acta *75*, 203 (1963).
84. *Warner, R. C., Weber, I.*: J. Am. Chem. Soc. *75*, 5094 (1953).
85. *Davis, B., Saltman, P., Benson, S.*: Biochem. Biophys. Res. Commun. *8*, 56 (1962).
86. *Aisen, P., Leibman, A., Reich, H. A.*: J. Biol. Chem. *241*, 1666 (1966).
87. *Wenn, R. V., Williams, J.*: Biochem. J. *108*, 69 (1968).
88. *Lane, R. S.*: Biochim. Biophys. Acta *243*, 193 (1971).
89. *Jeppson, J. O.*: Acta Chem. Scand. *21*, 1686 (1967).
90. *Baker, E., Shaw, D. C., Morgan, E. H.*: Biochemistry *7*, 1371 (1968).
91. *Efremov, G. D., Smith, L. L., Barton, B. P., Huisman, T. H. J.*: Anim. Blood Groops. Biochem. Genet. *2*, 159 (1971).
92. *Bezkorovainy, A., Grohlich, D.*: Biochim. Biophys. Acta *147*, 497 (1967).
93. — — *Gerbeck, C. M.*: Biochem. J. *110*, 765 (1968).
94. — *Zschocke, R., Grohlich, D.*: Biochim. Biophys. Acta *181*, 295 (1969).
95. — *Grohlich, D.*: Biochim. Biophys Acta *263*, 645 (1972).
96a. *Elleman, T. C., Williams, J.*: Biochem. J. *116*, 515 (1970).
96b. *Mann, K. G., Wayne, W. F., Cox, A. C., Tanford, C.*: Biochem. *9*, 1348 (1970).
97. *Aisen, P., Leibman, A., Sia, C. L.*: Biochemistry *11*, 3461 (1972).
98. *Aasa, R.*, Biochem. Biophys. Res. Commun. *49*, 806 (1972).
99. *Aisen, P., Aasa, R., Redfield, A. G.*: J. Biol. Chem. *244*, 4628 (1969).
100. *Luk, C. K.*: Biochemistry *10*, 2838 (1971).
101. *Woodworth, R. C.*: In: Protides of the biological fluids, Vol. 14, p. 37. Ed. *Peeters, H.* Amsterdam: Elsevier Press 1966.
102. — *Tan, A. T., Virkaitis, L. R.*: Nature *223*, 833 (1969).
103. *Stratil, A.*: Comp. Biochem. Physiol. *22*, 227 (1967).
104. *Williams, J., Phelps, C. F., Lowe, J. M.*: Nature *226*, 858 (1970).
105a. *Aisen, P., Koenig, S. H., Schillinger, W. E., Scheinberg, I. H., Mann, K. G., Fish, W.*: Nature *226*, 859 (1970).
105b. — *Lang, G., Woodworth, R. C.*: J. Biol. Chem. *248*, 649 (1973).
106. *Wishnia, A., Weber, I., Warner, R. C.*: J. Am. Chem. Soc. *83*, 2071 (1961).
107. *Komatsu, S. K., Feeney, R. E.*: Biochemistry *6*, 1136 (1967).
108. *Line, F. W., Grohlich, D., Bezkorovainy, A.*: Biochemistry *6*, 3393 (1967).
109. *Windle, J. J., Wiersema, A. K., Clark, J. R., Feeney, R. E.*: Biochemistry *2*, 1341 (1963).
110. *Bezkorovainy, A., Grohlich, D.*: Biochem. J. *123*, 125 (1971).
111a. *Tan, A. T., Woodworth, R. C.*: J. Polymer Sci., Part C *30*, 599 (1970).
111b. *Lehrer, S. S.*: J. Biol. Chem. *244*, 3613 (1969).
112. *Phillips, J. L., Azari, P.*: Arch. Biochem. Biophys. *151*, 445 (1972).
113. *Aasa, R., Aisen, P.*: J. Biol. Chem. *243*, 2399 (1968).
114. *Pinkowitz, R. A., Aisen, P.*: J. Biol. Chem. *247*, 7830 (1972).
115a. *Aisen, P., Aasa, R., Malmström, B. G., Vänngård, T.*: J. Biol. Chem. *242*, 2484 (1967).
115b. *Price, E. M., Gibson, J. F.*: Biochem. Biophys. Res. Commun. *46*, 646 (1972).
116a. *Jamieson, G. A., Jett, M., de Bernardo, S. L.*: J. Biol. Chem. *246*, 3686 (1971).
116b. *Hudson, B. G., Ohno, M., Brockway, W. J., Castellino, F. J.*: Biochemistry *12*, 1047 (1973).

117. *Feeney, R. E., Komatsu, S. K.:* Struct. Bonding *1,* 149 (1966).
118. — *Allison, R. G.:* In: Evolutionary biochemistry of proteins: Homologous and analogous proteins from avian egg whites, blood sera, milk and other substances, Chap. 6, p. 144. New York: Wiley-Interscience 1969.
119. *Groves, M. L.:* In: Milk proteins, chemistry and molecular biology, Vol. II, p. 367. Ed. *McKenzie, H. A.* New York: Academic Press 1971.
120. *Aisen, P.:* In: Inorganic biochemistry Vol. I. Ed. *Eichhorn, G.* Amsterdam: Elsevier 1973.
121. *Fraenkel-Conrat, H., Feeney, R. E.:* Arch. Biochem. *29,* 101 (1950).
122. *Koechlin, B. A.:* J. Am. Chem. Soc. *74,* 2649 (1952).
123. *Warner, R. C.:* Trans, N. Y. Acad. Sci. *16,* 182 (1954).
124. *Azari, P. R., Feeney, R. E.:* J. Biol. Chem. *232,* 293 (1958).
125. — — Arch. Biochem. Biophys. *92,* 44 (1961).
126. *Glazer, A. N., McKenzie, H. A.:* Biochim. Biophys. Acta *71,* 109 (1963).
127. *Zschocke, R. H., Chiao, M. T., Bezkorovainy, A.:* Europ. J. Biochem. *27,* 145 (1972).
128. *Kourilsky, F. M., Burtin, P.:* In: Protides of the biological fluids, Vol. 14, p. 103. Ed. *Peeters, H.* Amsterdam: Elsevier 1966.
129. — — Nature *218,* 375 (1968).
130. *Tengerdy, C., Azari, P., Tengerdy, R. P.:* Nature *211,* 203 (1966).
131. *Zschocke, R. H., Bezkorovainy, A.:* Biochim. Biophys. Acta *200,* 241 (1970).
132. *Tengerdy, R. P.:* Immunochemistry *3,* 463 (1966).
133. — J. Immunol. *99,* 126 (1967).
134. — *Faust, C. H.:* Nature *219,* 195 (1968).
135. *Jandl, J. H., Katz, J. H.:* J. Clin. Invest. *42,* 314 (1963).
136. *Fletcher, J., Huehns, E. R.:* Nature *215,* 585 (1967).
137. *Kornfeld, S.:* Biochim. Biophys. Acta *194,* 25 (1969).
138. — Biochemistry *7,* 945 (1968).
139. *Charlwood, P. A.:* Biochem. J. *125,* 1019 (1971).
140. *Faust, C. H., Tengerdy, R. P.:* Immunochemistry *8,* 211 (1971).
141. *Fuller, R. A., Briggs, D. R.:* J. Am. Chem. Soc. *78,* 5253 (1956).
142. *Bezkorovainy, A., Rafelson, M. E.:* Arch. Biochem. Biophys. *107,* 302 (1964).
143. — Biochim. Biophys. Acta *127,* 535 (1966).
144. *Rosseneu-Motreff, M. Y., Soetewey, F., Lamote, R., Peeters, H.:* Biopolymers *10,* 1039 (1971).
145. *Koenig, S. H., Schillinger, W. E.:* J. Biol. Chem. *244,* 3283 (1969).
146. — — J. Biol. Chem. *244,* 6520 (1969).
147. *Lane, R. S.:* Brit. J. Haematol. *22,* 309 (1972).
148. *Woodworth, R. C., Morallee, K. G., Williams, R. J. P.:* Biochemistry *9,* 839 (1970).
149. *Vallee, B. L., Ulmer, D. D.:* Biochem. Biophys. Res. Commun. *8,* 331 (1962).
150. *Ulmer, D. D., Vallee, B. L.:* Biochemistry *2,* 1335 (1963).
151. *Tan, A. T.:* Can. J. Biochem. *49,* 1071 (1971).
152. *Tomimatsu, Y., Gaffield, W.:* Biopolymers *3,* 509 (1965).
153. *Nagy, B., Lehrer, S. S.:* Arch. Biochem. Biophys. *148,* 27 (1972).
154. *Gaffield, W., Vitello, L., Tomimatsu, Y.:* Biochem. Biophys. Res. Commun. *25,* 35 (1966).
155. *Hatefi, Y., Hanstein, W. G.:* Proc. Natl. Acad. Sci. U.S. *62,* 1129 (1969).
156. *Emery, T. F.:* Biochemistry *8,* 877 (1969).
157. *Price, E. M., Gibson, J. F.:* J. Biol. Chem. *247,* 8031 (1972).
158. *Tomimatsu, Y., Vickery, L. E.:* Biochim. Biophys. Acta *285,* 72 (1972).
159. *Ulmer, D. D.:* Biochim. Biophys. Acta *181,* 305 (1969).

160. — *Vallee, B. L.:* Advan. Chem. Ser. *100,* 365 (1971).
161. *Aisen, P., Leibman, A.:* Biochem. Biophys. Res. Commun. *32,* 220 (1968).
162. *Linderstrøm-Lang, A.:* Symp. Peptide Chemistry, p. 1. Spec. Publ. No. 2, Chem. Soc., London (1955).
163. *Aisen, P., Leibman, A.:* Biochim. Biophys. Acta *257,* 314 (1972).
164. *Boeri, E., Ghiretti-Magaldi, A.:* Biochim. Biophys. Acta *23,* 489 (1957).
165. *Ferrell, R. E., Kitto, G. B.:* Biochemistry *9,* 3053 (1970).
166. *Bates, G., Brunori, M., Amiconi, G., Antonioni, E., Wyman, J.:* Biochemistry *7,* 3016 (1968).
167. *Klotz, I. M., Keresztes-Nagy, S.:* Biochemistry *2,* 445 (1963).
168. *Keresztes-Nagy, S., Klotz, I. M.:* Biochemistry *2,* 923 (1963).
169. *Groskopf, W. R., Holleman, J. W., Margoliash, E., Klotz, I. M.:* Biochemistry *5,* 3779 (1966).
170. *Langerman, N. R., Klotz, I. M.:* Biochemistry *8,* 4746 (1969).
171. *Langerman, N., Sturtevant, J. M.:* Biochemistry *10,* 2809 (1971).
172. *Rao, A. L., Keresztes-Nagy, S.:* Arch. Biochem. Biophys. *142,* 514 (1972).
173. *Klotz, I. M., Klotz, J. A., Fiess, H. A.:* Arch. Biochem. Biophys. *68,* 284 (1957).
174. *Keresztes-Nagy, S., Klotz, I. M.:* Biochemistry *4,* 919 (1965).
175. *de Phillips, H. A.:* Arch. Biochem. Biophys. *144,* 122 (1971).
176. *Darnall, D. W., Garbett, K., Klotz, I. M.:* Biochem. Biophys. Res. Commun. *32,* 264 (1968).
177. *Garbett, K., Darnall, D. W., Klotz, I. M.:* Arch. Biochem. Biophys. *142,* 455 (1971).
178. *Okamura, M. Y., Klotz, I. M., Johnson, C. E., Winter, M. R. C., Williams, R. J. P.:* Biochemistry *8,* 1951 (1969).
179. *Vallee, B. L., Ulmer, D. D.:* In: Non-heme iron proteins, p. 43. Ed. *San Pietro, A.* Yellow Springs: Antioch Press 1965.
180. *Garbett, K., Darnall, D. W., Klotz, I. M., Williams, R. J. P.:* Arch. Biochem. Biophys. *135,* 419 (1969).
181. *York, J. L., Bearden, A. J.:* Biochemistry *9,* 4549 (1970).
182. *Garbett, K., Johnson, C. E., Klotz, I. M., Okamura, M. Y., Williams, R. J. P.:* Arch. Biochem. Biophys. *142,* 574 (1971).
183. *Bayer, E., Krauss, P., Röder, A., Schretzmann, P.:* Bioinorg. Chem. *1,* 215 (1972).
184. *Brunori, M., Rotilio, G. C., Rotilio, G., Antonini, E.:* FEBS (Federation Europ. Biochem. Soc.) Letters *16,* 89 (1971).
185. *Darnall, D. W., Garbett, K., Klotz, I. M., Aktipis, S., Keresztes-Nagy, S.:* Arch. Biochem. Biophys. *133,* 103 (1969).
186. *Ghiretti, F.:* In: Oxygenases, Chap. 12. Ed. *Hayaishi, O.* New York: Academic Press 1962.
187. *Manwell, C.:* In: Oxygen in the animal organism, p. 49. Ed. *Dickens, F.* and *Neil, E.* London: Pergamon Press 1964.
188. *Florkin, M.:* In: Chemical zoology, Vol. 4, Chap. 4, p. 111. Ed. *Florkin, M.* New York: Academic Press 1969.
189. *Klotz, I. M.:* In: Biological macromolecules, Vol. 5, p. 55. Ed. *Timasheff, S. N.,* and *Fasman, G. D.* New York: M. Dekker, Inc. 1971.
190. *Bayer, E., Schretzmann, P.:* Struct. Bonding *2,* 181 (1967).
191. *Bearden, A. J., Dunham, W. R.:* Struct. Bonding *8,* 1 (1970).
192. *Groskopf, W. R., Holleman, J. W., Margoliash, E., Klotz, I. M.:* Biochemistry *5,* 3783 (1966).
193. *Subramanian, A. R., Holleman, J. W., Klotz, I. M.:* Biochemistry *7,* 3859 (1968).

194. *Klippenstein, G. L., Holleman, J. W., Klotz, I. M.:* Biochemistry *7,* 3868 (1968).
195. *Ferrell, R. E., Kitto, G. B.:* FEBS (Federation Europ. Biochem. Soc.) Letters *12,* 322 (1971).
196. — — Biochemistry *10,* 2923 (1971).
197. *Klapper, M. H., Barlow, G. H., Klotz, I. M.:* Biochem. Biophys. Res. Commun. *25,* 116 (1966).
198. *Fan, C. C., York, J. L.:* Biochem. Biophys. Res. Commun. *36,* 365 (1969).
199. — — Biochem. Biophys. Res. Commun. *47,* 472 (1972).
200. *Klippenstein, G. L.:* Biochem. Biophys. Res. Commun. *49,* 1474 (1972).
201. *York, J. L., Fan, C. C.:* Biochemistry *10,* 1659 (1971).
202. *Rill, R. L., Klotz, I. M.:* Arch. Biochem. Biophys. *147,* 226 (1971).
203. *Gray, H. B.:* Advan. Chem. Ser. *100,* 365 (1970).
204. *Bossa, F., Brunori, M., Bates, G. W., Antonini, E., Fasella, P.:* Biochim. Biophys. Acta *207,* 41 (1970).
205. *Klippenstein, G. L., Van Riper, D. A., Oosterom, E. A.:* J. Biol. Chem. *247:* 5959 (1972).
206. *Holleman, J. W., Biserte, G.:* Bull. Soc. Chim. Biol. *40,* 1417 (1958).
207. *Rill, R. L., Klotz, I. M.:* Arch. Biochem. Biophys. *136,* 507 (1970).
208. *Wilder, M., Valentine, R. C., Akagi, J.:* J. Bacteriol. *86,* 861 (1963).
209. *Lovenberg, W., Sobel, B. E.:* Proc. Natl. Acad. Sci. U.S. *54,* 193 (1965).
210. *Bachmayer, H., Yasunobu, K. T., Peel, J. L., Mayhew, S.:* J. Biol. Chem. *243,* 1022 (1968).
211. *Newman, D. J., Postgate, J. R.:* European J. Biochem. *7,* 45 (1968).
212. *Meyer, T. E., Sharp, J. J., Batsch, R. G.:* Biochim. Biophys. Acta *234,* 266 (1971).
213. *Peterson, J. A., Basu, D., Coon, M. J.:* J. Biol. Chem. *241,* 5162 (1966).
214. *— Kusunose, M., Kusunose, E., Coon, M. J.:* J. Biol. Chem. *242,* 4334 (1967).
215. *Tsibris, J. C. M., Woody, R. W.:* Coord. Chem. Rev. *5,* 417 (1970).
216a. *Rabinowitz, J. C.:* Advan. Chem. Ser. *100,* 322 (1971).
216b. *Mason, R., Zubieta, J. A.:* Angew. Chem. Internat. Edit. *12,* 390 (1973).
217. *Lovenberg, W.:* In: Protides of the biological fluids, Vol. 14, p. 165. Ed. *Peeters, H.* Amsterdam: Elsevier 1967.
218. *Bachmayer, H., Piette, L. H., Yasunobu, K. T., Whiteley, H. R.:* Proc. Natl. Acad. Sci. U.S. *57,* 122 (1967).
219. *Lovenberg, W., Williams, W. M.:* Biochemistry *8,* 141 (1969).
220. *Bachmayer, H., Yasunobu, K. T., Whiteley, H. R.:* Biochem. Biophys. Res. Commun. *26,* 435 (1967).
221. *— Benson, A. M., Yasunobu, K. T., Garrard, W. T., Whiteley, H. R.:* Biochemistry 7, 986 (1968).
222. *McCarthy, K., Lovenberg, W.:* Biochem. Biophys. Res. Commun. *40,* 1053 (1970).
223. *Stevenson, D., Cook, R. M., Weinstein, B.:* Intern. J. Peptide Protein Res. *4,* 101 (1972).
224. *Phillips, W. D., Poe, M., Weiker, J. F., McDonald, C. C., Lovenberg, W.:* Nature *227,* 574 (1970).
225. *Rao, K. K., Evans, M. C. W., Cammack, R., Hall, D. O., Thompson, C. L., Jackson, P. J., Johnson, C. E.:* Biochem. J. *129,* 1063 (1972).
226. *Blumberg, W. E.:* In: Magnetic resonance in biological systems, p. 119. Ed. *Ehrenberg, A., Malström, B. G.,* and *Vänngård, T.* London: Pergamon Press 1967.
227. *Peisach, J., Blumberg, W. E., Lode, E. T., Coon, M. J.:* J. Biol. Chem. *246,* 5877 (1971).

228. *Herriot, J. R., Sieker, L. C., Jensen, L. H., Lovenberg, W.:* J. Mol. Biol. *50*, 391 (1970).
229. *Watenpaugh, K. D., Sieker, L. C., Herriot, J. R., Jensen, L. H.:* Cold Spring Harbor Symp. Quant. Biol. *36*, 359 (1972).
230. *Long, T. V., Loehr, T. M.:* J. Am. Chem. Soc. *92*, 6384 (1970).
231. — — *Allkins, J. R., Lovenberg, W.:* J. Am. Chem. Soc. *93*, 1809 (1971).
232. *Eaton, W. A., Lovenberg, W.:* J. Am. Chem. Soc. *92*, 7195 (1970).
233. *Ali, A., Fahrenholz, F., Garing, J. C., Weinstein, B.:* J. Am. Chem. Soc. *94*, 2556 (1972).
234. *Gillard, R. D., McKenzie, E. D., Mason, R., Mayhew, S. G., Peel, J. L., Stangroom, J. E.:* Nature *208*, 769 (1965).
235. *Atherton, N. M., Garbett, K., Gillard, R. D., Mason, R., Mayhew, S. J., Peel, J. L., Stangroom, J. E.:* Nature *212*, 590 (1966).
236. *Fasman, G. D.:* In: Protides of biological fluids, Vol. 14, p. 453. Ed. *Peeters, H.* Amsterdam: Elsevier 1967.
237. *Peterson, J. A., Coon, M. J.:* J. Biol. Chem. *243*, 329 (1968).
238. *Lode, E. T., Coon, M. J.:* J. Biol. Chem. *246*, 791 (1971).
239. *Benson, A., Tomoda, K., Chang, J., Matsueda, G., Lode, E. T., Coon, M. J., Yasunobu, K. T.:* Biochem. Biophys. Res. Commun. *42*, 640 (1971).
240. *Palmer, G., Brintzinger, H., Estabrook, R. W., Sands, R. H.:* In: Magnetic resonance in biological systems, p. 150. Ed. *Ehrenberg, A., Malström, B. G.,* and *Vänngård, T.* Oxford: Pergamon Press 1967.
241. *Yoch, D. C., Beneman, J. R., Valentine, R. C., Arnon, D. I.:* Proc. Natl. Acad. Sci. U.S. *64*, 1404 (1969).
242. *Shethna, Y. I.:* Biochim. Biophys. Acta *205*, 58 (1970).
243. *Yoch, D. C., Arnon, D. I.:* J. Biol. Chem. *247*, 4514 (1972).
244. *Dus, K., de Klerk, H., Sletten, K., Bartsch, R. G.:* Biochim. Biophys. Acta *140*, 291 (1967).
245. — *Tedro, S., Bartsch, R. G., Kamen, M. D.:* Biochem. Biophys. Res. Commun. *43*, 1239 (1971).
246. *Moss, T. H., Petering, D., Palmer, G.:* J. Biol. Chem. *244*, 9 (1969).
247. *Flatmark, T., Dus, K.:* Biochim. Biophys. Acta *180*, 377 (1969).
248. *Kraut, J., Strahs, G., Freer, S. T.:* In: Structural chemistry and molecular biology, p. *55*. Ed. *Rich, A.,* and *Davidson, N.* New York: Freeman, W. H. 1968.
249. *Strahs, G., Kraut, J.:* J. Mol. Biol. *35*, 503 (1968).
250. *Carter, C. W., Freer, S. T., Xuong, Ng. H., Alden, R. A., Kraut, J.:* Cold Spring Harbor Symp. Quant. Biol. *36*, 381 (1972).
251. *Evans, M. C. W., Hall, D. O., Johnson, C. E.:* Biochem. J. *119*, 289 (1970).
252. *Phillips, W. D., Poe, M., McDonald, C. C., Bartsch, R. G.:* Proc. Natl. Acad. Sci. U.S. *67*, 682 (1970).
253. *Moss, T. H., Bearden, A. J., Bartsch, R. G., Cusanovich, M. A., San Pietro, A.:* Biochemistry 7, 1591 (1968).
254. *Mortenson, L. E., Valentine, R. C., Carnaham, J. E.:* Biochem. Biophys. Res. Commun. 7, 448 (1962).
255. *Tanaka, M., Haniu, M., Matsueda, G., Yasunobu, K. T., Himes, R. H., Akagi, J. M., Barns, E. M., Davanathan, T.:* J. Biol. Chem. *246*, 3953 (1971).
256. *Travis, J., Newman, D. J., le Galle, J., Peck, H. D.:* Biochem. Biophys. Res. Commun. *45*, 452 (1971).
257. *Matsubara, H., Sasaki, R. M., Tsuchiya, D. K., Evans, M. C. W.:* J. Biol. Chem. *245*, 2121 (1970).

258. *Orme-Johnson, W. H., Beinert, H.:* Biochem. Biophys. Res. Commun. *36,* 337 (1969).
259. *Poe, M., Phillips, W. D., McDonald, C. C.:* Biochem. Biophys. Res. Commun. *42,* 705 (1971).
260. *Packer, E. L., Sternlicht, H., Rabinowitz, J. C.:* Proc. Natl. Acad. Sci. U.S. *69,* 3278 (1972).
261. *Sieker, L. C., Adman, E., Jensen, L. H.:* Nature *235,* 40 (1972).
262. *Adman, E., Sieker, L. C., Jensen, L. H.:* J. Biol. Chem. *248,* 3987 (1973).
263. *Herskovitz, T., Averill, B. A., Holm, R. H., Ibers, J. A., Phillips, W. D., Weiher, J. F.:* Proc. Natl. Acad. Sci. U.S. *69,* 2437 (1972).
264. *Vetter, H., Knappe, J.:* Hoppe-Seylers Z. Physiol. Chem. *352,* 433 (1971).
265. *Kimura, T.:* Struct. Bonding *5,* 1 (1968).
266. *Poe, M., Phillips, W. D., Glickson, J. D., McDonald, C. C., San Pietro, A.:* Proc. Natl. Acad. Sci. U.S. *68,* 68 (1971).
267. *Dunham, W. R., Bearden, A. J., Salmeen, I. T., Palmer, G., Sands, R. H., Orme-Johnson, W. H., Beinert, H.:* Biochim. Biophys. Acta *253,* 134 (1971).
268. *Fritz, J., Anderson, R., Fee, J., Palmer, G., Sands, R. H., Tsibris, J. C. M., Gunsalus, I. C., Orme-Johnson, W. H., Beinert, H.:* Biochem. Biophys. Acta *253,* 110 (1971).
269. *Gibson, J. F., Hall, D. O., Thornley, J. H. M., Whatley, F. R.:* Proc. Natl. Acad. Sci. U.S. *56,* 987 (1966).
270. *Thornley, J. H. M., Gibson, J. F., Whatley, F. R., Hall, D. O.:* Biochem. Biophys. Res. Commun. *24,* 877 (1966).
271. *Buchanan, B. B., Arnon, D. I.:* Advan. Enzymol. *33,* 119 (1970).
272. *Tanaka, M., Haniu, M., Yasunobu, K. T., Kimura, T.:* J. Biol. Chem. *248,* 1141 (1973).
273. *Eck, R. V., Dayhoff, M. O.:* Science *152,* 363 (1966).
274. *Matsubara, H., Jukes, T. H., Cantor, C. R.:* Brookhaven Symp. Biol. *21,* 201 (1968).
275. *Hall, D. O., Cammack, R., Rao, K. K.:* Nature *233,* 136 (1971).
276. *Tanaka, M., Nakashima, T., Benson, A., Mower, H., Yasunobu, K. T.:* Biochemistry *5,* 1666 (1966).
277. *Keresztes-Nagy, S., Perini, F., Margoliash, E.:* J. Biol. Chem. *244,* 981 (1969).
278. *Matsubara, H., Sasaki, R. M., Chain, R. K.:* Proc. Natl. Acad. Sci. U.S. *57,* 439 (1967).
279. — — J. Biol. Chem. *243,* 1732 (1968).
280. *San Pietro, A.:* In: Biological oxidations, p. 515. Ed. *Singer, T. P.* New York: John Wiley and Sons 1968.
281. *Hall, D. O., Evans, M. C. W.:* Nature *223,* 1342 (1969).
282. *Yoch, D. C., Valentine, R. C.:* Ann. Rev. Microbiol. *26,* 139 (1972).
283. *Hong, J.-S., Rabinowitz, J. C.:* J. Biol. Chem. *245,* 6582 (1970).
284. — — J. Biol. Chem. *245,* 4988 (1970).
285. *Devanathan, T., Akagi, J. M., Hersh, R. T., Himes, R. H.:* J. Biol. Chim. *244,* 2846 (1969).
286. *McDonald, C. C., Phillips, W. D.:* J. Am. Chem. Soc. *91,* 1513 (1969).
287a. *Poe, M., Phillips, W. D., McDonald, C. C., Lovenberg, W.:* Proc. Natl. Acad. Sci. U.S. *65,* 797 (1970).
287b. *Packer, E. L.:* Dissertation. Berkeley, University of California (1973).
288. *Nitz, R. M., Mitchell, B., Gerwing, J., Christensen, J.:* J. Immunol. *103,* 319 (1969).
289. *Hong, J.-S., Rabinowitz, J. C.:* J. Biol. Chem. *245,* 4995 (1970).
290. *Mitchell, B., Levy, J. G., Nitz, R. M.:* Biochemistry *9,* 1839 (1970).

291. — — Biochemistry *9*, 2762 (1970).
292. *Levy, J. G., Hull, D., Kelly, B., Kilburn, D. G., Teather, R. M.:* Cell. Immunol. *5*, 87 (1972).
293. *Hong, J.-S., Rabinowitz, J. C.:* J. Biol. Chem. *245*, 6574 (1970).
294. *Rall, S. C., Bolinger, R. E., Cole, R. D.:* Biochemistry *8*, 2486 (1969).
295. *Gillard, R. D., Mason, R., Mayhew, S. G., Peel, J. L., Stangroom, J. E.:* In: Protides of the biological fluids, Vol. 14, p. 159. Ed. *Peeters, H.* Amsterdam: Elsevier 1966.
296. *Malkin, R., Robinowitz, J. C.:* Biochemistry *6*, 3880 (1967).
297. *Lode, E. T.:* Federation Proc., Federation Am. Soc. Expt. Biol. *32*, 542 Abs. (1973).
298. *Jensen, L. H., Sieker, L. C., Watenpaugh, K. D., Adman, E. T., Herriot, J. R.:* Biochem. Soc. Trans. *1*, 27 (1973).
299. *Kimura, T., Suzuki, K., Padmanabhan, R., Samejima, T., Tarutani, O., Ui, N.:* Biochemistry *8*, 4027 (1969).
300. — *Ting, J. J.:* Biochem. Biophys. Res. Commun. *45*, 1227 (1971).
301. — — *Huang, J. J.:* J. Biol. Chem. *247*, 4476 (1972).
302. *Petering, D. H., Palmer, G.:* Arch. Biochem. Biophys. *141*, 456 (1970).
303. *Kimura, T., Nakamura, S.:* Biochemistry *10*, 4517 (1971).
304. *Petering, D., Fee, J. A., Palmer, G.:* J. Biol. Chem. *246*, 643 (1971).
305. *Keresztes-Nagy, S., Margoliash, E.:* J. Biol. Chem. *241*, 5955 (1966).
306. *Garbett, K., Gillard, R. D., Knowles, P. F., Stangroom, J. E.:* Nature *215*, 824 (1967).
307. *Tel-Or, E., Fuchs, S., Avron, M.:* FEBS (Federation Europ. Biochem. Soc.) Letters *29*, 156 (1973).
308. *Padmanabhan, R., Kimura, T.:* Biochem. Biophys. Res. Commun. *37*, 2 (1969).
309. — — J. Biol. Chem. *245*, 2469 (1970).
310. *Cammack, R., Rao, K. K., Hall, D. O.:* Biochem. Biophys. Res. Commun. *44*, 8 (1971).
311. *Appella, E., San Pietro, A.:* Biochem. Biophys. Res. Commun. *6*, 349 (1961).
312. *Coffman, R. E., Stavens, B. W.:* Biochem. Biophys. Res. Commun. *41*, 163 (1970).
313. *Carter, C. W., Kraut, J., Freer, S. T., Alden, R. A., Sieker, L. C., Adman, E., Jensen, L. H.:* Proc. Natl. Acad. Sci. U.S. *69*, 3526 (1972).
314. *Scheraga, H. A.:* In: Symmetry and function of biological systems at the molecular level, p. 43. Ed. *Engström, A.* and *Strandberg, B.* Stockholm: Almqvist and Wiksell 1969.
315. *Némethy, G., Laiken, N.:* Farmaco (Pavia) *25*, 999 (1970).
316. *Ramachandran, G. N., Sasisekharan, V.:* Advan. Protein Chem. *23*, 283 (1968).
317. *Ramakrishnan, C., Sarathy, K. P.:* Intern. J. Protein Res. *1*, 103 (1969).
318. *Ulmer, D. D.:* Federation Proc. *29*, 463 (1970).
319. *Querinjean, P., Masson, P. L., Heremans, J. F.:* Europ. J. Biochem. *20*, 420 (1971).
320. *Fisher, W. R., Taniuchi, H., Anfinsen, C. B.:* J. Biol. Chem. *248*, 3188 (1973).

Received April 17, 1973

Calcium-Binding Proteins

Frank L. Siegel

Joseph P. Kennedy, Jr. Laboratory and the Departments of Pediatrics and Physiological Chemistry, The University of Wisconsin Center for Health Sciences, Madison, Wisconsin 53706/USA

Table of Contents

I. Biological Functions of Calcium

Calcium probably fulfills a greater variety of biological functions than any other cation. Calcium homeostasis is likewise the most elaborate fine control system for cation regulation found in higher organisms. The isolation of calcium-binding proteins from many calcium-sensitive biological systems and the tissue specificity of these proteins points to the possibility that many binding proteins may play obligatory roles in calcium function. We might profitably begin any discussion of calcium-binding proteins with a survey of calcium functions in living organisms, as such a survey should indicate potential sources of binding proteins.

Calcium appears to be an integral part of biological membranes and may impart structural and functional integrity to membrane systems (*1*, *2*). Calcium stimulates the activity of membrane systems, as evidenced by its effects on transport of cations and amino acids into cells and mitochondria (*3*, *4*). The regulation of mitosis (*5*), sea urchin egg fertilization (*6*) and the phytohemagglutinin-induced transformation of lymphocytes (*7*) are under calcium regulation. The secretion of many hormones is a calcium-dependent process. This has been demonstrated for the release of insulin (*8*), thyroid stimulating hormone (*9*), lutenizing hormone (*10*), vasopressin (*11*), oxytocin (*12*), growth hormone (*13*) and thyroxine (*14*). Amylase secretion from salivary glands and pancreas is also calcium-dependent (*15*) and may share a common mechanism with hormone release. In nerve cells, calcium has at least three distinct functions, including roles in the stimulus-provoked release of neurotransmitters such as norepinephrine (*16*) and acetylcholine (*17*), the generation of an action potential via a 'calcium current' in some neurons (*18*), and the binding of transmitters to post-synaptic receptors (*19*). Contraction (*20*) and relaxation (*21*) of muscle are both regulated by tissue levels of calcium. In plant tissues, calcium governs slime mold aggregation (*22*), pollen tube growth (*23*) and cell wall extension (*24*). The formation of bone is governed by the incorporation of calcium into the matrix of that tissue (*25*), as is formation of the avian egg shell (*26*).

On a molecular level, calcium has been implicated as a regulator of glycogenolysis and gluconeogenisis (*27, 28*), in the conversion of prothrombin to thrombin (*29*), and in the assembly of microtubules (*30*). There is a complex interplay between cellular calcium and the several components of the cyclic nucleotide system (*31, 32*), including adenylate cyclase, phosphodiesterase and protein kinase, and the proposal has been made that in some instances calcium may act as a 'third messenger' of hormone action, cyclic AMP being the second messenger (*31*). Lastly, a variety of enzymes have been shown to be calcium-activated, including the calcium-magnesium ATPase (*33*), several hydrolytic enzymes — deoxyribonuclease (*34*), phospholipase (*35*), a variety of proteolytic enzymes (*36*), and the α-amylases (*37*).

We have delineated the sites of calcium action in biological systems — these sites represent one potential source of calcium-binding proteins. Implicit in this suggestion is the assumption, largely unproven, that many actions of calcium are mediated through 'calcium receptors' within cells. A rational search for calcium-binding proteins might consider two other aspects of the biology of calcium. The intracellular levels of free calcium are thought to be very low, less than 10^{-5}M in neurons (*38*) and 10^{-6}M in kidney, for example. Calcium appears to be sequestered to mobile receptors in the cytosol (*38*), to be accumulated by mitochondria (*39, 40*) and by microsomes (*41*), and to be bound to plasma membranes (*42*). These subcellular binding sites which sequester calcium represent another potential location for calcium-binding proteins.

The calcium-transport system is another potential source of binding proteins. Transport into mitochondria, across plasma membranes and through the intestinal mucosa and the renal tubule may be the function of distinct transport systems, each of which may be the source of binding proteins. The possibility of isolating bacterial mutants deficient both in calcium uptake and in calcium-binding proteins offers an interesting approach to the role of binding proteins in calcium transport.

In summary, one might expect to find calcium-binding proteins playing six distinct roles in living systems: 1. binding sites on the outer surface of plasma membranes, 2. transport carriers in cell membranes, 3. intracellular storage reservoirs of calcium, 4. intracellular receptors linked to calcium function (i.e., in contractile systems), 5. as part of matrix of mineralized tissues and, 6. as a co-factor in calcium-activated enzymes.

It is well known that calcium binds to non-protein anionic constituents of cells, such as phospholipids (*42*), gangliosides (*43*), nucleic acids (*44*), and ATP (*45*) and it may be assumed that non-protein sites are also important in calcium binding in some biological systems.

All proteins might be considered to be calcium-binding proteins by virtue of the affinity of carboxylate groups for cations, non-specific binding is low affinity however, with binding constants of about 1 mM. Likewise, many metalloenzymes may be non-specifically stimulated by calcium and a variety of other divalent metal ions. I will therefore restrict the scope of this review to include only those proteins which have binding constants of at least 10^{-4}M, and for which a functional relationship exists between calcium and the physiological function of the protein.

II. Assay of Calcium-Binding Activity

Metal-binding studies require equilibration of the metal ion and protein, a means of separating protein-bound and free metal, and a sensitive method for their quantitative determination. A variety of methods have been used for studying ligand binding to proteins, including optical (*46*), calorimetric (*47*), fluorescence (*48*), titrimetric (*49*), and radiometric procedures. Radiometric methods are almost exclusively used for measurement of calcium binding, as the ready availability of calcium-45 provides a rapid, inexpensive and sensitive means for the analysis of this element. Calcium-45 is a β-emitting isotope which may conveniently be counted by liquid scintillation spectrometry in Bray's scintillator (*50*) or other counting solutions which accept aqueous samples. To minimize extraneous calcium contamination in binding assays the following precautions have been suggested (*51*): 1. all glassware should be washed in 1M HCl and deionized water, and, whenever possible, plastic apparatus should be used. 2. All solvents are passed through columns of either Dowex-50 or Chelex-100; the latter is used when Mg^{++} is also present. 3. Before binding assay, protein solutions are dialyzed against buffer containing 1 mM ethylene glycol (bis-β-aminoethyl ether)-N,N'-tetraacetic acid (EGTA). 4. Acid-washed Whatman No. 41 filter paper is used to wipe pipettes. These precautions reduced calcium levels in solutions to 10^{-6}M, whereas without them the contaminant calcium was increased 5—10 times. Several calcium-binding proteins appear to be unstable in the absence of calcium, however, and this may mitigate against treatment of these proteins with EGTA before binding assay. Dialysis tubing should be exhaustively washed with dilute acid or EGTA before use (*52*).

Several methods for separating protein-bound calcium from free calcium have been utilized as binding assays. These include 1. the use of Chelex-100 to bind free calcium, leaving protein-bond calcium in solution for radioassay, 2. dialysis, either equilibrium or rapid-flow, 3. gel filtration, and 4. ultrafiltration. In the Chelex-100 assay of calcium-binding (*53*) a mixture of protein, varying amounts of $^{45}CaCl_2$ and Chelex-100 resin are allowed to equilibrate in a series of conical centrifuge tubes. The resin is sedimented by centrifugation and an aliquot of each supernatant is removed for liquid scintillation counting. Subtraction of these radioactivity values from those obtained from tubes containing no protein gives the amount of protein-bound calcium. The data can be expressed as a Scatchard plot (*54*) in which the ratio of protein-bound calcium to free calcium is plotted against the amount of protein-bound calcium. The slope of the linear portion of the curve obtained is numerically equal to the binding constant and extrapolation of the linear portion of the curve to its intercept with the x axis gives the number of binding sites. The Chelex method enables one to assay large numbers of samples simultaneously, but has several disadvantages, including the need to use relatively large amounts of protein, and variation of binding activity with the *p*H and ionic strength of the medium. These effects make it more difficult to determine *p*H optima of binding and the competition effects of other cations.

Equilibrium dialysis (*55*) is the most frequently employed method for determining ligand binding to proteins. Even when the method is scaled down, equilibrium dialysis requires at least 0.1 mg per sample and better results are obtained with larger amounts of protein (i.e., larger volumes). Dialysis is usually permitted to take place for a minimum of six hours, at which time determinations of free and protein-bound radioactivity are made. Even though chemical equilibrium for the binding reaction may be reached within a fraction of a second, a period of hours is needed for the attainment of diffusion equilibrium. If the calcium binding protein is labile under conditions, or if a calcium-binding prosthetic group is removed from the protein by dialysis, this procedure may be inappropriate. The method is also limited to proteins of molecular greater than 12,000—15,000, which will not diffuse through dialysis membranes. High concentrations of protein and low ionic strength contribute to anomalous binding due to the Donnan effect and should be avoided. One should also be certain that there is no significant binding of protein to the dialysis membrane.

The need for prolonged dialysis inherent in equilibrium dialysis methods has been circumvented by the development of a method in which a flow-dialysis, or rate of dialysis, procedure is employed (*56*). This method employs a dialysis cell with an upper chamber, containing the protein and a radioactive ligand, separated by a membrane from a lower chamber, through which buffer is pumped at a constant rate, and from which the effluent is sampled for measurement of radioactivity. In flow dialysis, radioactive calcium is introduced into the protein solution in the upper chamber at a molarity considerably below that of the binding protein. Calcium equilibrates immediately with the binding protein, and the rate at which calcium enters the lower chamber is proportional to the concentration of unbound calcium in the upper chamber. This rate is determined by measuring the radioactivity in the effluent from the lower chamber. After steady state conditions have been achieved, a small volume (5—10 μl) of non-radioactive calcium is added to the upper chamber to establish a new equilibrium in which an increased fraction of the total calcium becomes free and diffusable. The concentration of free ^{45}Ca in the upper chamber increases immediately and the concentration in the effluent approaches a new steady state. Additions of non-radioactive calcium are made repeatedly and a series of values are obtained for the fraction of free calcium at each concentration.

Flow dialysis thus uses a single sample of protein to obtain a complete set of binding data at different calcium concentrations. A minimum of 200 μg of protein is required, however, and there is but slight over-all saving of protein expended compared to micro-scale equilibrium dialysis methods. The speed at which the assay can be run permits the collection of reliable data from unstable proteins; however, one disadvantage of the method is the practical inability to measure dissociation constants below 10^{-6} to 10^{-7}M. This arises from the fact that the rate of dialysis is so slow that only about 0.1% of free ligand passes from the upper chamber per minute and the effluent contains less than 0.01% of the concentration of free ligand in the upper chamber; this may be too low to conveniently measure. The restrictions on molecular weight and ionic strength which were discussed for equilibrium dialysis also apply in flow dialysis methods.

An alternative to dialysis methods is the gel filtration technique of *Hummel* and *Dreyer* (*57*), as modified by *Price* (*58*). A column of Sephadex G-25 is equilibrated with a buffer containing a desired concentration of calcium and $^{45}CaCl_2$, and is used for gel filtration of the binding protein. As the process of gel filtration proceeds, the protein migrates in the excluded volume of the column, removing calcium ions from the column buffer until equilibrium is reached. The protein peak, in the void volume of the column, contains above-base line amounts of calcium. The amount of calcium bound to the protein can be found by dividing the molar concentration of calcium above the base line value by the molar protein concentration. A separate gel filtration run is used to determine each point on the binding plot. This is at once time-consuming and rather wasteful in terms of protein required. Gel filtration would seem to be the method of choice only in the case of dialyzable proteins which can not be studied by other methods.

The ultrafiltration method, first introduced by *Flexner* (*59*) and recently modified by *Blatt et al.* (*60*) and by *Paulus* (*61*) appears to have several advantages over the other methods of binding assay. In the Paulus method, solutions of binding protein and $^{45}CaCl_2$ solutions of known calcium concentration are filtered, under pressure, through UM-10 Diaflo membranes (Amicon Corp., Lexington, Mass.). The protein-bound calcium is retained by the membrane and may be assayed by liquid scintillation counting. A series of blanks containing no protein is run to correct for calcium binding to the membrane. In the author's laboratory PM-10 membranes, which do not have ionic groups exposed, have been found to bind much less calcium than UM-10 membranes and thus have lower blanks, and in addition, have much faster flow rates than UM-10 membranes (*62*). The recent report (*63*) of variable results with different lots of Diaflo membranes runs counter to our experience. Ultrafiltration is run in an eight-chambered cell (Metaloglass Corp., Boston, Mass.) and requires only about 20 μg of protein for each binding point. The method is thus both rapid and sensitive. The protein must, of course, be retained by the membrane; PM-2 membranes can be used for proteins of molecular weight below 10,000. The uncertainty of running a binding assay under conditions of constantly changing protein concentration has been discussed (*64*), but since the results from ultrafiltration experiments appear to agree with those obtained by equilibrium dialysis this objection would not appear to represent a serious difficulty.

One practical alternative to the radiometric methods is provided by murexide, which forms a complex with free calcium (*46*). This complex can be determined spectrophotometrically and used for the calculation of unbound calcium. Bound calcium calculated by difference, enables one to construct binding plots. This method has two advantages in special cases; it is very rapid and therefore can be used for labile binding proteins, and since no dialysis or ultrafiltration is needed it can be used with low molecular weight peptides. About 1 mg of protein is required for the murexide method.

III. Survey of Calcium-Binding Proteins

A compilation of those calcium-binding proteins with known affinity for calcium is given in Table 1. In addition to the proteins listed in this table we will, in this review, consider proteins which bind calcium with

Table 1. *Calcium-binding proteins with known bonding affinity*

Name	Source	Kd (M)	Moles Ca^{++} bound per mole protein	Mol. Wt.
Calsequestrin	Sarcoplasmic Reticulum	4×10^{-4}	43	44,000
Prothrombin	Plasma	5×10^{-5}	3	70,000
Brain CBP-II	Brain; Adrenal medulla	2.5×10^{-5}	1	11,500
DNAse A	Pancreas	1.4×10^{-5}	2	31,000
Insol. Calcium-binding Glyco-protein	Mitochondria	4×10^{-6}	5	67,000
Vitamin-D-Inducible calcium-binding protein	Avian duodenal mucosa	3.8×10^{-6}	1	24,000
Elastin	Tendon	1.4×10^{-6}	2	74,000
Troponin C	Myofibrils	1.9×10^{-7}	1	24,000
Soluble calcium-binding Glycoprotein	Mitochondria	1×10^{-7}	3	42,000
Parvalbumins	Fish White muscle	1×10^{-7}	2	11,000
Calcium-binding Glycoprotein	Cartilage	1×10^{-7}	2	200,000

unknown affinity or low affinity in cases where the function of the protein is presumed to involve calcium binding. Calcium binding requirements for proteins performing ion storage functions, for example calsequestrin, are quite different than those of proteins which are calcium receptors, for example troponin. The affinity of a binding protein must also be in the range of ligand concentrations that the protein is exposed to *in situ*. The wide range of binding capacities and affinities are thus reflective of the wide range of functional diversity of these proteins. In some instances calcium binding proteins have been isolated from the same tissue by different laboratories, and we shall consider the methods of isolation and the properties of the isolated proteins in order to establish their identity or non-identity.

IV. Muscle Calcium-Binding Proteins

A. Introduction

The coupling of excitation and contraction in muscle involves the translocation of calcium bound to membranes of the sarcoplasmic reticulum to binding sites on muscle fibers (*65, 66*). Skeletal muscle myofibrils have two classes of binding sites, with affinity constants of $2.1 \times 10^6 M^{-1}$ and $3 \times 10^4 M^{-1}$ respectively; the high affinity binding site has been shown to be troponin (*67*). No other myofibrilar proteins bind calcium with this affinity, and one may conclude that troponin contributes calcium sensitivity to muscle fibers (*67, 68*). The calcium binding activity of sarcoplasmic reticulum has been attributed to calsequestrin (*69*), and to other proteins which may in fact be identical to calsequestrin or which may represent aggregates of calsequestrin. Transport of extracellular calcium to the storage sites on sarcoplasmic reticulum is accomplished by a Ca^{++}—Mg^{++} ATPase of the sarcoplasmic reticulum membrane.

B. Calsequestrin

Before the isolation of calsequestrin, the focus of attention had been directed toward the Ca^{++}-activated ATPase of sarcoplasmic reticulum as being the key to the calcium economy of muscle. This ATPase is instrumental in Ca^{++} transport (*70, 71*), but when the ATPase was found to be incapable of physiologically significant calcium storage (*72*) several laboratories were led to look at other protein components of sarcoplasmic reticulum as potential calcium storage sites. *Masoro* and his-co-workers first attempted to isolate the binding protein, employing a method which involved extracting a membrane fraction prepared from sarcoplasmic reticulum with buffer containing 10 *m*M sodium dodecyl sulfate (SDS) and, following centrifugation to remove material not solubilized, to remove the detergent by dialysis (*73*). The proteins remaining soluble following dialysis (95% of the membrane protein) were fractionated on columns of Sepharose 4B in Tris buffer containing 8 *m*M NaCl but no detergent. The authors mention that attempts to solubilize the proteins of sarcoplasmic reticulum without resorting to extraction with detergents met with no success. These attempts included sonication and delipidation with chloroform-methanol. Gel filtration of the SDS-solubilized proteins produced an elution profile characterized by a small peak at the void volume of the column and a large peak which was cleanly separated from the first peak. The second peak, which the authors call fraction 2, accounted for more than 90% of the protein applied to the column. Acrylamide gel electrophoresis of fraction 2 following SDS-mercaptoethanol-induced dissociation produced one protein band, and the conclusion was made that the protein of fraction 2 is comprised of subunits of 6500—10,000 daltons. This protein was found to bind calcium at two sites (*66*). The high affinity site has an association constant of $2 \times 10^5 M^{-1}$ and binds one mole of calcium per subunit, while the low affinity site binds two moles of calcium per subunit with an affinity constant of $1 \times 10^4 M^{-1}$. The high affinity site showed a sharp *p*H optimum of about 7, while calcium binding to the low affinity site increased with *p*H in a linear manner. It is uncertain whether subunit aggregation is a *sine qua non* of binding activity. It is of considerable interest to distinguish between calcium storage sites on the interior of the membrane and the transport ATPase. Fraction 2 had no Ca^{++}—Mg^{++} ATPase activity, but some activity was found following incubation of fraction 2 with phospholipids extracted from sarcotubular membranes.

MacLennan used a similar extraction procedure, extracting rabbit sarcoplasmic reticulum with buffers containing deoxycholate and removing detergent by dialysis (*75*). The protein which remained soluble following dialysis was fractionated by ion-exchange chromatography on DEAE-cellulose. The proteins eluted at a KCl concentration of 0.48 M were then subjected to gel filtration on Sephadex G-200. Two peaks were obtained; one at the void volume and a second, assymetrical peak. The material in the leading portion of the second peak was applied to a column of hydroxylapatite and eluted with a linear potassium phosphate gradient. The last peak, which was eluted at 0.33 *m*M phosphate, contained a homogeneous calcium-binding protein, designated calsequestrin by *MacLennan* and *Wong* (*75*). Calsequestrin and the transport ATPase were clearly separable by electrophoresis in SDS-gels (*75*).

Calsequestrin has a molecular weight of 44,000 daltons and is an acidic protein which contains non phosphorus, lipid or sialic acid (*75*). It binds 43 moles of calcium per mole of protein with a dissociation constant of $4 \times 10^{-4}M$. Calsequestrin precipitates at calcium concentrations greater than $10^{-4}M$; presumably this represents aggregation, as the protein will redissolve upon dilution or dialysis to lower the calcium concentration. Binding of calcium was measured over the *p*H range from 6 to 9 and exhibited a sharp decline at values below 7 and a slight increase between 7 and 9. Boiling of calsequestrin solutions for 4 min. resulted in only a 20% loss of binding activity. The order of binding affinity for calsequestrin was $Ca^{++} > Cd^{++} > Sr^{++} > Mg^{++}$. Mn^{++} did not compete with calcium for binding sites on calsequestrin. In contrast to the calcium-binding proteins from intestine and nervous tissue, calsequestrin appears to be a membrane protein solubilized only by detergent treatment.

Gergely and his co-workers have also reported the isolation of a calcium-precipitable protein from sarcoplasmic reticulum and have proposed it as a putative calcium-storage site in muscle (*76, 77*). This protein was solubilized by extraction with Triton X-100 and was precipitated by calcium in electrophoretically homogeneous form. The molecular weight of this protein was reported to be 54,000 daltons (*77*) and it was found to bind 41 moles of calcium per mole of protein, with a dissociation constant, measured in 0.1 M KCl, of 1.3×10^{-3}. Binding of calcium to this protein was inhibited by magnesium. What is the relation between the protein isolated by Gergely and calsequestrin? The temptation to conclude that they are identical is strong, as on the basis of their binding sites and levels in sarcoplasmic reticulum each protein can account for the physiological binding of calcium to sarcoplasmic reticulum, which is about 80—180 n moles per mg of protein (*72, 77*). The two proteins were found to have identical electrophoretic behavior by both *MacLennan* and *Gergely*. The binding data from both groups are not directly comparable as the calsequestrin binding constant was determined at very low ionic strength (5 *m*M Tris-HCl), where Donnan effects may be important, and those for Gergely's protein were done in 0.1 M KCl, an ionic strength which should obviate Donnan binding.

The identity of two other reported binding proteins from sarcoplasmic reticulum is in doubt. Tropocalcin, isolated by *Benson* and *Han* may be identical to calsequestrin (*78*) and cardioglobulin-C (*79*) may be identical to the calcium transport ATPase (*80*), although the cardioglobulin has also been demonstrated in blood plasma (*79*).

C. Troponin

1. Skeletal Muscle Troponin. Troponin, the calcium binding site on muscle fibers, may be dissociated to yield several subunits, only one of which exhibits high affinity calcium binding. The fractionation of troponin from skeletal muscle, which has a molecular weight variously estimated between 35,000 and 80,000 daltons, into two protein components was first demonstrated by *Hartshorne* and *Mueller* (*82*). These workers used low *p*H and high salt to achieve dissociation and named the two proteins troponin A and troponin B. Troponin A, with a molecular weight of 18,500, was found to carry the high-affinity calcium-binding site (*83*). Troponin B, the ATPase inhibitory factor, was heterogeneous, being composed of one component with a molecular weight of 39,000 daltons and another with a molecular weight of 26,000 daltons. Hartshorne's finding that the ATPase-inhibitory and calcium-binding activities of troponin could be resolved was confirmed by several laboratories, but a universal nomenclature for troponin subunits has not yet evolved.

Perry and his co-workers chromatographed purified troponin on columns of SE-Sephadex in 6 m urea, obtaining two peaks (*84*). The first peak, which was not bound to SE-Sephadex, had different electrophoretic properties than the second peak, which was eluted by 0.3 m KCl. The first peak had greater electrophoretic mobility than did the second, and the addition of EGTA to the electrophoretic system increased the mobility of the first peak but did not affect the second. The second peak was found to have the ATPase inhibitory activity of troponin and has been designated as inhibitory factor, and the faster component, as calcium-sensitizing factor. Further purification of the calcium-sensitizing factor has been achieved by chromatography on QAE-Sephadex in buffers containing 6 M urea (*85*). Using a similar approach, *Ebashi* (*20*) and *Wilkinson* (*86*) isolated three subunits of troponin.

Greaser and *Gergely* have also fractionated troponin under dissociating conditions, but have isolated four components (*87*). Previously this group had reported that the isolation procedures employed by others yielded heterogeneous troponin fractions (*88*), and attempted to completely resolve troponin into its subunits by chromatography on DEAE-Sephadex in 50 *m*M Tris-HCl containing 6 M urea. This procedure gener ated five UV-absorbing components (Fig. 1). Each fraction was assayed by SDS-acrylamide electrophoresis, for its activity in inhibiting actomyosin ATPase and for its calcium-binding activity. These data permitted

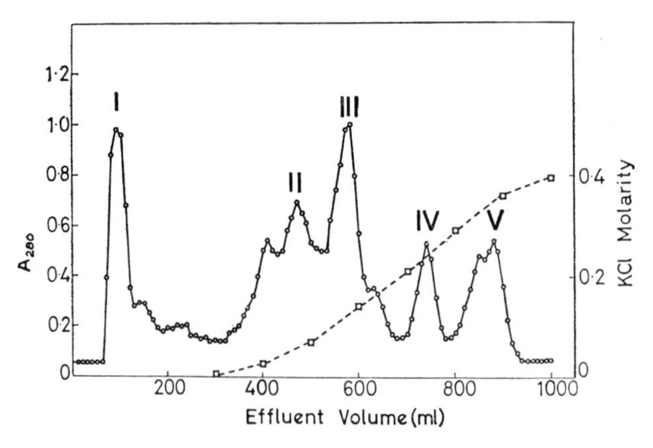

Fig. 1. DEAE-Sephadex chromatography of troponin in 6 M urea. Sixty milliliters of a troponin solution (18.6 mg/ml) dialyzed against a solution containing 6 M urea 50 *m*M Tris (*p*H 8.0) and 1 *m*M DTT were applied to a column (30 × 4 cm) of DEAE-Sephadex A-50. The protein was eluted with a linear gradient of KCl. (0) Absorbance at 280 nm, (□) KCl concentration of effluent. Peak I 14,000 dalton component; Peak II, TN-I; Peak III, TN-T; Peak IV, TN-C; Peak V, Nucleotide. (Reproduced from Reference 87 with permission of Dr. *Greaser, M.*)

the identification of three fractions with calcium binding (TN-C), ATPase inhibition (TN-I) and tropomyosin binding (TN-T), (Table 2). The 14,000 dalton subunit was contaminated with three other proteins, TN-I and TN-C were nearly homogeneous, and TN-T also had other electrophoretic components but could be purified by rechromatography (*89*).

Table 2. *Identification of troponin sub-units*[a]

Peak No.	Mol. Wt.	Calcium Bound (moles/10^5 g protein)	Percent Inhibition of Actomyosin ATPase	Designation
1	14,000	0	45	14,000 dalton component
2	24,000	0	85	TN-I
3	37,000	0	20	TN-T
4	24,000	4.58	0	TN-C
5	No protein	—	—	Nucleotide

[a] Ref. (89).

Unfractionated troponin can bind about two moles of calcium per mole of protein, but only one mole of calcium was found to bind to purified TN-C (*89*). When TN-I, which has no calcium binding activity, was added to TN-C in equimolar amounts two moles of calcium were bound (Table 3). If, in the course of purification of TN-C, calcium is

Table 3. *Calcium binding of mixtures of troponin sub-units*[a]

Sample	Moles Ca^{++}/mole of TN-C
Troponin	~ 2
TN-C	1.06
TN-T + TN-C	1.15
TN-I + TN-C	2.19
TN-T + TN-I + TN-C	2.20

[a] Ref. (89).

added before urea is removed by dialysis the TN-C obtained is also able to bind two moles of calcium per mole; these experiments suggest that the configuration is an important factor regulating the calcium-binding activity of TN-C.

Drabakowski and his co-workers have reported the results of similar experiments, which resolved troponin into four protein components, one of which confers calcium sensitivity upon troponin (*90*). They have proposed adopting the previously suggested nomenclature for the TN-I and TN-C components but propose to call the 35,000 dalton component TN-B (B for binding) rather than TN-T as suggested by *Greaser et al.*, as this material was found to bind not only tropomyosin, but F-actin as well (*91*). The 14,000 dalton protein, peak 1 in most fractionation schemes, is considered by these workers to be an artifact generated by proteolysis during the course of troponin purification and fractionation. This contention is supported by the extreme lability of troponin (*92*) and by trypsin treatment of troponin, which causes the rapid disappearance of TN-I and the concomitant increase in lower molecular weight components (*93*).

The amino acid composition of the calcium binding sub-unit as reported by *Hartshorne* (*83*), *Greaser* (*89*), *Perry* (*85*), and *Kay* (*94*) seems to be in mutual agreement, and we may talk of binding experiments from these and other laboratories with reasonable confidence that they describe properties of the same protein; we will refer to this protein as TN-C. TN-C is an acidic protein with a slightly larger content of glutamic acid than aspartic acid. It has a phenylalanine to tyrosine ratio of about five and contains no tryptophan; as a result TN-C has a flat absorption spectrum from 240 to 270 *n*M (*85*). Evidence for conformational changes in TN-C during calcium binding are provided by several types of experiments. The addition of $CaCl_2$ to a final concentration of 10^{-3}M results in an increase of the molecular weight of TN-C from 22,000 to 40,000 daltons, presumably as the result of aggregation (*94*). Conformational shifts at lower calcium concentrations were demonstrated with circular dichroism studies (*94*). In the absence of calcium TN-C has an α-helical content of 21% while in the presence of 5×10^{-4}M calcium this figure increases to 41% (Fig. 2). The addition of calcium to TN-C also causes an increase in sedimentation coefficient from 1.36S to 1,92S at 5×10^{-5}M $CaCl_2$ (*94*) with no increase in molecular weight, indicating an increased symmetry is conferred upon TN-C by calcium binding. Calcium also causes an increase in fluorescence of TN-C (*95*); fluorescence polarization studies indicate that calcium binding imposes restraints on tyrosine motion within the molecular structure of TN-C (*95*) indicating a greater rigidity of the more ordered helix. Troponin has two classes of calcium binding sites, a high affinity site which binds one mole calcium per mole of protein with a dissociation constant of 0.19×10^{-6}M and a lower affinity site, which binds one mole of calcium per mole of protein with a dissociation constant of 5×10^{-6}M (*83*). A comparison of troponin subunits and their nomenclature of different laboratories is provided by Table 4 (*96*).

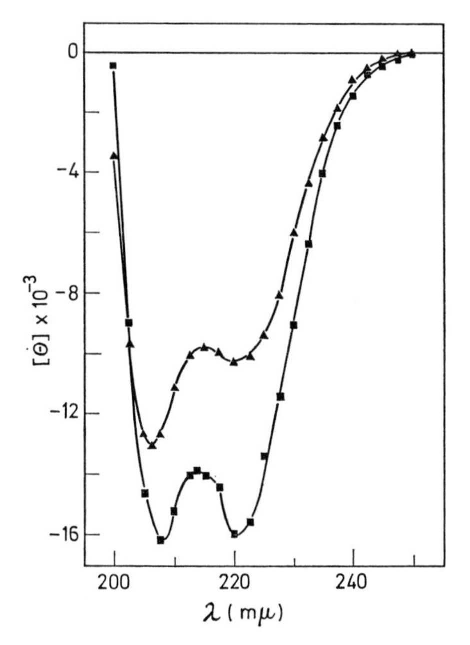

Fig. 2. Far-ultraviolet circular dichroism of troponin-C in 50 *m*M Tris (*p*H 7.6) — 1 *m*M EGTA (▲) and in 50 *m*M Tris (*p*H 7.6) — 5×10^{-4}M $CaCl_2$ (■). (Reproduced from Ref. 94 with permission of *Dr. Kay, C. M.*)

2. *Cardiac Troponin.* The difficulties in isolating undegraded troponin are greater when one attempts to prepare and fractionate this protein from cardiac muscle. This is the result of the high content in heart of myoglobin and cytochrome, which help to oxidize the labile sulfhydryl groups of troponin and the high levels of cathepsin activity (*97*). *Reddy* and *Honig* have used the Hartshorne method to isolate cardiac troponin (*96*) and have found that cardiac and skeletal muscle troponins have similar calcium binding properties but the binding of calcium by the heart protein is more easily perturbed by drugs (*98*). *Greaser et al.* (*89*) have fractionated cardiac troponin into its subunits for comparison with

Table 4. *Troponin nomenclature*[a])

Preparation	Troponin fraction I	TN-I	TN-T	TN-C	Criteria for classification
Hartshorne & Pyun (29)					Ca^{++} binding,
Purified troponin A				X	amino acid comp.
Troponin B		X	X		SDS gel mol wt
Schaub & Perry (50)					
Inhibitory factor	X	X	X		SDS gel mol wt
Schaub et al. (30)					SDS gel mol wt,
EGTA-sensitizing factor		*		X	amino acid comp.
Wilkinson et al. (16)					
14,000 dalton	X				SDS gel mol wt
23,000 dalton		X			SDS gel mol wt, amino acid comp.
37,000 dalton			X		SDS gel mol wt, amino acid comp.
Drabikowski et al. (33)					
I	X				Chromatography, SDS gel mol wt
II		X			Chromatography, SDS gel mol wt
III			X		Chromatography, SDS gel mol wt,
IV				X	Chromatography, SDS gel mol wt
Ebashi et al. (15)					
I			X		Amino acid comp., SDS gel mol wt
II		X		*	Amino acid comp., SDS gel mol wt
III				X	Amino acid comp., SDS gel mol wt

X, major constituent(s); *, suspected contaminant,
[a]) Ref. (96).

the corresponding skeletal muscle proteins. Three subunits were found in cardiac troponin, with molecular weights of 40,000, 30,000 and 20,000 daltons. Only the 20,000 dalton component had an electrophoretic mobility similar to the corresponding skeletal muscle troponin sub-unit; this may indicate that only the calcium-binding subunit is common to both troponins. Confirmation of this must await the assay of the 20,000 dalton cardiac subunit for calcium binding.

D. Fish Parvalbumins

When muscles from amphibia and fish are homogenized at low ionic strength, a group of soluble proteins called myogens is extracted. Fish myogens are characterized by the presence of a group of acidic proteins called the parvalbumins, which have molecular weights of about 11,000 daltons and high levels of phenylalanine, but one or no residues of tyrosine, tryptophan, histidine, arginine, proline, cysteine or methionine (*99*). These proteins exhibit high-affinity calcium binding (*100*), and, largely through the elegant research of *Pechère* and his co-workers, more is known about their structure than any other class of calcium binding proteins. The investigation of parvalbumins culminated with the recent determination of the amino acid sequence and three dimensional structure of crystalline carp parvalbumin by *Kretsinger* and his co-workers (*101*).

Detailed chemical investigation of the parvalbumins began with the demonstration that the low molecular weight myogens of carp white muscle were composed of three components, designated as 2, 3 and 5, which could be separated electrophoresis (*102*). These proteins were not present in red muscle of cod (*102*), demonstrating a protein difference between the two types of muscle. The three cod parvalbumins have been crystallized (*103*) and their amino acid composition has been determined. The three proteins have only small differences in amino acid composition, molecular weight and other physical properties.

Following this study, *Pechère* adopted a new method (*104*) of preparation of the parvalbumins which facilitated the comparison of proteins from various species. The proteins are extracted from white muscle with buffered 0.3M sucrose — 3 *m*M EDTA and salted-out with ammonium sulfate at 70% saturation. Gel filtration on Sephadex G-75 separated the parvalbumins from high molecular weight contaminants, which are excluded from the column and elute in the void volume. In the last step of the procedure, the parvalbumins are fractionated by ion-exchange chromatography on DEAE-cellulose developed with a chloride griadent, Using this method, five parvalbumins were found in carp muscle (*104*); the number of parvalbumins varies from species to species. Hake maucle contains one mafor and two minor parvalbumins, frog has two major and two minor components, and turbot has three parvalbumins. Using this method Bhushana Rao has found two major parvalbumins in cod (*105*) and two in the pike (*106*). Nomenclature of these proteins presents an operational difficulty; *Pechère* (*104*) suggests designating each parvalbumin by its isoelectric point; i.e., Component 4.14.

As a consequence of the low level or absence of tyrosine and the absence of tryptophan, the parvalbumins have a highly unusual UV spectrum, with virtually no absorbance at 280 *n*M (*107*) and several peaks between 250 and 270 *n*M which may reflect buried phenylalanine residues. No nonprotein constituents are present in the parvalbumins, with the exception of calcium, which binds to the extent of 2 moles per mole of protein (*2*); apparently calcium is bound to all members of the group, although this point needs clarification. Binding constants were obtained for the major hake parvalbumin and for two frog parvalbumins (*100*). The hake protein has two high affinity calcium binding sites with a dissociation constant of 0.1×10^{-6}M and three low affinity sites which a dissociation constant of 2×10^{-5}M (*100*). The two frog proteins have similar binding properties, which are close to the binding constant reported for hake parvalbumin (*104*).

Amino acid sequence charts have been constructed for the major parvalbumin of hake (*108*) and for one of the carp proteins (*101*). Electron density maps and the elucidation of the three dimensional structure of the carp protein have been accomplished by *Kretsinger* (*101*). The nature of the calcium binding site was determined from these data; calcium was found to be coordinated by the carboxyl groups of three aspartic acid and one glutamic acid residues. The residues involved in the tetrahedral coordination lie in the sequence Gly-Asp-Ser-Asp-Gly-Asp-Gly-Lys-Ile-Gly-Val-Asp-Glu. No grooves or ridges in the topology of this protein were found and the authors accept this finding as indicating a lack of enzymatic activity of the parvalbumins. This is in concert with the conept that these proteins represent evolutionary precursors of troponin and play a TN-C-like role in lower vertebrates (*107*). TN-C, however, has a molecular weight twice as great as the parvalbumins and it has been suggested that troponin arose via duplication of the genes coding for parvalbumin (*109*).

V. Vitamin D-Inducible Calcium-Binding Proteins

A. Avian Proteins

Experimental Vitamin D deficiency decreases the intestinal absorption of calcium. Treatment of rachitic chicks with Vitamin D restores normal levels of calcium absorption. This effect of Vitamin D is blocked by prior administration of actinomycin D (*110, 111*), indicating that macromolecular synthesis is a requisite to the expression of Vitamin D activity. This indication led to the finding of calcium-binding activity in the supernatant fraction of homogenates of Vitamin D-treated, but not rachitic chicks (*112*), (Fig. 3). *Wasserman* and *Taylor* and their co-workers

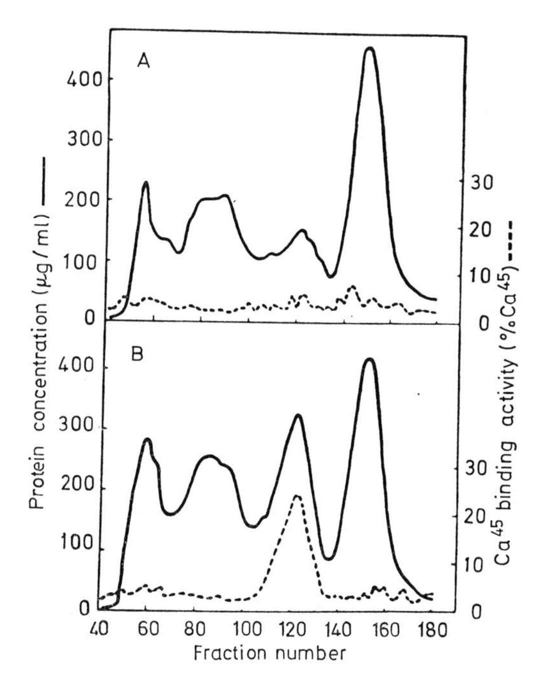

Fig. 3. Distribution of proteins and calcium-binding activity in fractions of rachitic (A) and Vitamin D_3 (B) supernatant after gel filtration on Sephadex G-100. Elution was performed with Tris buffer (*p*H 7.4). (Reproduced from Ref. 113 with permission of *Dr. Wasserman, R. H.*)

have, in a comprehensive study, isolated and purified this protein from chick intestinal mucosa (*113, 114*). This portein is Vitamin D-inducible and appears to participate in calcium transport across the intestinal epithelium.

To isolate the protein, day old chicks were placed on a Vitamin D-deficient diet (*112*) and, at the age of 4 weeks, the rachitic birds were given a single dose of 500 IU of Vitamin D_3, orally or by intramuscular injection. Three days later the animals were sacrificed and the duodenal mucosa was homogenized in five volumes of a buffer containing 0.0137 M Tris, 0.12 M NaCl and 4.74 *m*M KCl, *p*H 7.4. The homogenate was centrifuged 20 min. at 37,000 × g and the supernatant was used for the isolation of a calcium-binding protein. Calcium-binding activity was salted-out by 75% saturated ammonium sulfate and the proteins in this fraction were applied to a column of Sephadex G-100. The calcium-binding protein eluted in the first of three UV-absorbing peaks; analytical electrophoresis revealed the presence of several protein bands. The proteins in this fraction were subjected to preparative discontinuous electrophoresis on 15% acrylamide gel. This step yielded a peak of calcium-binding protein which migrated as a single band during analytical electrophoresis in 10% acrylamide gel at *p*H 7.9 and at *p*H 8.3. Upon electrophoresis in 3.5% acrylamide at *p*H 8.9 a second faster moving protein component was found. This second band, apparently a minor contaminant, did not appear in preparations made from tissue taken from rachitic chicks. Both bands are said to possess calcium-binding activity and the suggestion has been made that the minor band represents a degradation product of calcium-binding protein, formed during isolation (*114*).

The molecular weight of intestinal calcium-binding protein was determined by two methods. Gel filtration on a column of Sephadex G-100 previously calibrated with molecular weight marker proteins gave a value of 28,000 daltons. Sedimentation equilibrium centrifugation gave values ranging from 24,000 to 28,000 daltons. No data are presently available to indicate if this protein is comprised of two or more sub-units. Tests for lipid, phosphorus, carbohydrate and sialic acid were negative (*114*). The sensitivity of the assay used for phosphorus would have detected 0.2% or less of phosphate; one mole of phosphate per mole of CBP corresponds to 0.2%, indicating that one residue of protein bound phosphorus might not have been detected. Attempts were made to label CBP by injecting ^{32}P-labeled phosphate; this was not successful. The assay of the phosphate content of this protein with a more sensitive assay procedure and a study of the effects of phosphatase treatment on its calcium binding activity would be informative. The finding that CBP appears to bind lysolecithin and that lysolecithin-bound calcium-binding protein has greatly reduced electrophoretic mobility and a reduced ability to bind calcium (*115*) is interesting but difficult to interpret. The altered electrophoretic mobility might reflect the detergent properties of cationic lysolecithin and thus might be a common attribute of acidic proteins. Alternatively, it might be indicative of possible binding of calcium-binding protein CBP to cell membrane lipids *in situ*.

The amino acid analysis of chick intestinal calcium-binding protein revealed a high content of aspartic and glutamic acids and lysine. These features are common to all low molecular calcium-binding proteins.

The chick intestinal calcium-binding protein exhibits high affinity calcium binding of one mole of calcium per mole of protein with a dissociation constant for calcium equal to 3.8×10^{-6} M (*114*). Binding of calcium exhibits a biphasic dependence upon *p*H, with binding maxima at *p*H 6.6 and *p*H 9.6 and a shallow trough at *p*H 7.6 (*116*). Calcium binding shows a relatively small dimunition as the ionic strength of the binding medium was increased by the addition of either NaCl or KCl up to 0.15 M (*116*). The addition of urea produced a reversible inhibition of binding at high urea concentration (> 1 M) (*116*), suggesting a conformational requirement for calcium binding. Circular dichroism studies (*116*), however, suggest that the binding protein, which has about 35–40% α-helical content, does not undergo appreciable changes in conformation as a result of calcium binding. No data were presented in this study to indicate that were the calcium binding protein used in the dichroism study was free of bound calcium.

Calcium binding was found to be antagonized by other divalent cations; the order of affinity being:

$Ca^{2+} > Cd^{2+} > Sr^{2+} > Mn^{2+} > Fe^{2+} > Zn^{2+} > Ba^{2+} > Mg^{2+}$. Co^{2+} and Ni^{2+}

did not appear to compete with calcium for binding to the protein. Sulfhydryl reagents do not inhibit calcium binding, nor do certain compounds containing quarternary ammonium groups such as choline, acetylcholine or propionyl choline. However, acyl groups of more than 13 carbon atoms which are adjacent to a quarternary ammonium group block calcium binding (*116*).

The chick intestinal calcium-binding protein was found in all portions of the small intestine and in kidney (*113*), both in Vitamin D-treated rachitic chicks and in chicks fed a normal diet (*117*). The protein was not found in the avian shell gland (*120*), muscle (*10*), colon (*113*), liver (*113*), pancreas (*114*), bone (*114*), or blood plasma (*114*). A specific antiserum was prepared by immunizing rabbits with the chick protein, and this anti-serum was used for immunofluorescence histochemistry (*114*). Specific immunofluorescence was found in the surface coat-microvillar region of all intestinal epithelial cells. Studies in which chicks were placed on a rachitogenic diet (*117*) demonstrated a good correlation between levels of calcium binding protein in duodenal tissue and the absorption of calcium by ligated duodenal loops, indicating a probable role for the binding protein in the absorption of dietary calcium. There also appears to be a correlation between levels of calcium-binding protein biological calcium demand, as levels of the duodenal protein are higher in younger birds (*117*) and higher in hens during egg production (*117, 119*). Dietary calcium restriction was also found to cause a compensatory increase in duodenal binding protein (*119*). Whether or not the duodenal protein functions as a transmucosal ion carrier remains to be determined. The avian calcium-binding protein is also present in kidney, where it may serve to regulate calcium reabsorption, and thus serum calcium levels (*113, 121*).

B. Mammalian Proteins

Calcium-binding proteins have been found in the intestinal mucosa of several mammalian species. While none of these has been studied as extensively as the avian protein, certain species-related differences have been recognized. *Kallflelz*, in Wasserman's laboratory, reported the existence of two calcium-binding proteins in rat duodenal mucosa (*122*). These two proteins are readily separated by gel filtration on Sephadex G-100. The larger of the two proteins is excluded from the column and is found in homogenates made from tissue taken from animals which were rachitic or from rachitic animals which had been treated with Vitamin D_3. Vitamin D_2-treated animals had none of this higher molecular weight binding protein, but did have lower molecular weight calcium-binding activity which was absent from rachitic or vitamin D_3-treated rats. This study, which suggested the presence of a high molecular weight precursor to the Vitamin D_2-inducible calcium binding protein was followed by two further reports of the rat protein. *Ooizumi* (*123*) has isolated two duodenal calcium-binding proteins and reports that the molecular weight of the larger is greater than 100,00 daltons, and that of the smaller protein is 24,000 daltons. These proteins were isolated by procedures similar to those of *Wasserman* (*113*); the use of heat-treatment (60 °C for 5 min.) to coagulate inactive proteins, followed by gel filtration and ion-exchange chromatography of the active principle. Ooizumi's data seem to be in general agreement with those of *Wasserman* but distinctly different from the report of *Drescher* and *DeLuca* who used an isolation procedure (*124*), employing neither heat treatment nor ammonium sulfate salting-out. Isolation of a homogeneous protein was accomplished by two procedures, one using Sephadex G-100 followed by CM-Sephadex chromatography, and a second technique of repeated gel filtration on Biogel columns. The molecular weight of this calcium-binding protein was 13,000 daltons as determined by gel filtration and gel electrophoresis on polyacrylamide containing 9 M urea. Equilibrium centrifugation gave a value of 8,200 daltons. The difference between these values is most probably due to the ellipsoidal nature of the protein. The low yields of this protein (about 20 μg per animal) and its lability have prevented detailed investigation of its chemical properties and calcium-binding activity. It has been possible to label the protein by the injection of ^{3}H- or ^{14}C-amino acid mixtures however, and the evidence obtained by this technique indicates that Vitamin D causes the conversion of a precursor protein to calcium-binding protein (*125*). The precursor has a molecular weight only slightly greater than the calcium binding protein (*125*). Several factors may explain the higher molecular weights reported for the binding proteins by *Ooizumi*. *DeLuca* mentions that this procedure avoids the use of heat or ammonium sulfate, potential aggregating agents. It is also true that DeLuca's procedure is essentially calcium free, whereas *Ooizumi* determines molecular weights in the presence of 5×10^{-6} M $CaCl_2$ (*123*), which can cause aggregation of other calcium-binding proteins (see Section VII B).

Wasserman and his co-workers have also purified calcium-binding proteins from canine (*126*) and bovine (*127*) duodenal mucosa. The canine protein has a greater electrophoretic mobility than the chick protein and also differs from the chick protein in its resistance to trypsin-induced loss of binding activity and lack of immunological cross-reactivity to the chick protein antiserum. The extract of bovine duodenum contains two calcium binding proteins. The major calcium-binding component is an acidic protein of molecular weight 11,000. These properties are similar to those of the porcine and bovine brain calcium-binding proteins (see Sections VII C and D) and a detailed comparison of the duodenal and brain proteins will doubtless prove to be interesting.

An intestinal Vitamin D-inducible calcium binding protein was found in monkeys (*128*); this protein, unlike those from the dog and cow, has a lower electrophoretic mobility than the chick protein, but no chemical data other than this observation have been reported. Reports on the presence of two calcium-binding proteins from human intestinal mucosa (*129*, *130*) are very preliminary and aside from the indication that one of the proteins has a molecular weight of about 20,000 they provide little solid information. The same or a similar protein was also isolated from human kidney (*131*), and it appears probable that the calcium-binding protein recently found in lactating bovine and rat mammary gland (*132*) s identical to the intestinal protein.

VI. Prothrombin

A. Introduction

The clotting of blood takes place when fibrinogen is converted to fibrin, an insoluble protein. Thrombin is required for this proteolytic event to occur and thrombin is, in turn, derived from a precursor, prothrombin. The conversion of prothrombin to thrombin requires calcium (*133*), phospholipid (*134*), and Factor V (*135*). In addition, Vitamin K is needed for the clotting process and it is believed to act at some stage of prothrombin synthesis, probably at a post-translational site (*136*). The administration of the Vitamin K antagonist dicumarol results in the synthesis of a biologically inactive prothrombin (*137*). A key feature of prothrombin's activity is its ability to bind calcium and this property appears absent in the dicumarol-induced protein. Interest in this area has centered about attempts to determine the change in prothrombin structure induced by dicumarol and to relate such change to a deficiency in binding activity. The isolation of blood clotting factors is made difficult by the proteolytic properties of thrombin; heroic measures are often used to circumvent this difficulty.

B. Chemical Properties and Calcium Binding

Most isolation procedures for prothrombin are modifications of the method introduced by *Moore* (*138*). Barium chloride is slowly added to citrated bovine or canine blood plasma and a barium citrate precipitate with adsorbed prothrombin is collected by centrifugation. The precipitation is repeated two times and proteins are eluted from the final barium citrate precipitate by dialysis against 0.2 m EDTA. Prothrombin is next salted out at between 50 and 67% saturated ammonium sulfate. The prothrombin-containing solution is adjusted to *p*H 5.35 and centrifuged to remove protein impurities. The supernatant is adjusted to *p*H 4.60 to effect the isoelectric precipitation of prothrombin. Additional impurities are removed by absorption on kaolin and on bentonite and the prothrombin-containing solution is stored at -85 °C (*138*). This procedure was modified by *Ingwall* and *Scherage* (*139*) who further purified prothrombin prepared by Moore's method, using chromatography on DEAE-Sephadex A-50 eluted with a linear salt gradient; electrophoretically pure prothrombin can be prepared by this method. Several variations on this procedure have been introduced, including the use of different chromatographic elution (*140, 141*) conditions. Proteolytic losses plague workers in this area and the use of buffers containing 1 *m*M diisopropylfluorophosphate to inhibit proteolytic enzymes, as suggested by *Hanahan* (*142*), is often practiced (*141*), despite the potential dangers of large-scale preparations containing this most toxic substance.

Purified prothrombin has a molecular weight of 72,000 (*143*). Glutamic and aspartic acids are present in the highest amounts; the phenylalanine to tyrosine ratio is almost 1 (*143*). Prothrombin is a glycoprotein containing sialic acid, mannose, galactose, and N-acetylglucosamine (*135, 137, 138*) and 1 mole of phosphate per mole of protein (*144*). The protein migrates as a single band during electrophoresis in acrylamide gels containing 8 m urea (*144*) or SDS (*143, 144*) but it resolves into three fractions during isoelectric focusing in acrylamide gels containing 7. 5M urea in a *p*H 3 to 7 gradient (*143*), indicating possible microheterogeneity. An alternate explanation would invoke the presence of proteolytic breakdown products of prothrombin. Stenflo reports that prothrombin binds 10—12 moles of calcium per mole of protein (*145*). Three moles of calcium are bound to one class of binding sites with an affinity constant of $2 \times 10^4\ M^{-1}$. The Scatchard plot suggested that these sites exhibit positive cooperativity; a Hill plot of the same data confirmed this effect (*145*). The remaining calcium was bound with much lower affinity; this may represent non-specific binding. The cooperativity of calcium binding has also been found by *Nelsestun* and *Suttie* who report that four moles of calcium are bound per mole of protein (*146*), with a wide *p*H optimum of binding over the range from *p*H from 8 to 9.

C. Dicumarol-Induced Prothrombin

Biologically inactive prothrombin, which appears following dicumarol administration, does not adsorb to barium citrate or barium sulfate (*140, 144*), which remove normal prothrombin. The supernatant from this adsorption step is fractionated by chromatography on DEAE-Sephadex followed by gel filtration on Biogel P-200 and a second DEAE-Sephadex step (*144*). An alternative procedure employs chromatography on DEAE-Cellulose, DEAE-Sephadex, hydroxylapatite, and recycling on Sephadex G-100 (*140*). Both methods yield a product which is electrophoretically homogeneous. This protein cross-reacts with antibodies to normal prothrombin (*143*), has an identical amino acid composition and identical carbohydrate and phosphorous contents (*144, 145*). This protein does not have the clotting activity of prothrombin (*140, 144*), however, and this may be related to its diminished calcium-binding activity. Binding studies with dicumarol-induced prothrombin indicate that it binds less than one mole of calcium per mole of protein at a calcium concentration of 1 *m*M (*146*). The structural basis for the difference in activity has been under active investigation in the laboratories of both *Suttie* and *Stenflo*.

Even though the normal and dicumarol-induced prothrombins have similar reaction with prothrombin antibody, identical amino acid and carbohydrate composition and similar peptide maps following tryptic digestion (*143*), the two proteins have slight differences in mobility when they are subjected to electrophoresis in the presence of calcium (*140, 143*). That a conformational difference between the two proteins exists is indicated by differences in quantitative precipitin curves run with and without calcium; normal prothrombin had a far greater immuneprecipitation with calcium than with EDTA, whereas dicumarol-induced prothrombin showed no such difference (*143*). Stenflo postulates that the altered conformation is due to an abnormal pairing of disulfide bonds in dicumarol-induced prothrombin (*143*). This contention is supported by the finding that there are slight differences in electrophoretic mobility of the two proteins, or cyanogen bromide fragments from the two proteins, in 8 M urea, but no such differences are seen after reduction and alkylation of their disulfide bonds. It has not yet been possible to relate the structural alteration and loss of calcium-binding activity in molecular terminology.

VII. Calcium-Binding Proteins in Nervous Tissue

A. Introduction

The release of vesicular stores of acetylcholine and norepinephrine which occurs when appropriate neurons fire is a calcium-dependent process (*147, 148*). The precise nature of the role of calcium in the stimulus-secretion coupling is poorly understood, and may be similar mechanistically to the calcium-dependent release of hormones, such as insulin and the peptide hormones of the pituitary gland (*140*). *Katz* and *Miledi* have largely been responsible for the promulgation and acceptance of the calcium hypothesis of stimulus-secretion coupling. According to this concept, extra-cellular calcium, which enters the cell during depolarization, combines with specific receptor molecules on the inner surface of the axonal membrane and on the surface of the synaptic vesicles (*150*). The resulting neutralization of negative charges on the vesicle and plasma membranes allows the vesicles to collide with and then fuse with the plasma membrane, bringing about a release of vesicle contents by exocytosis. Calcium attached to specific membrane receptors is called 'active calcium' by these authors. *Berl*, in a recent extension of this hypothesis (*151*) speculates that the calcium-sensitive sites on the plasma membrane and vesicle correspond to neurin and stenin, subunits of neurostenin, a contractile protein from brain which is similar to actomyosin. Studies with injection of ^{45}Ca into squid giant axons have demonstrated that intracellular calcium of neurons exists almost entirely in bound form (*152*); this finding strengthens the case for macromolecular calcium receptors in neurons. Three calcium-binding proteins which appear to be specific to nervous tissue have been isolated to date — the S-100 protein and two structurally similar acidic proteins.

B. S-100 Protein

If mammalian brain is homogenized in aqueous buffers and the resulting extract is brought to saturation with ammonium sulfate, a brain specific acidic protein, called S-100 protein, remains in solution (*153*). Several laboratories have reported that S-100 protein is heterogeneous (*154,—*

156), and *Levine* has suggested that the heterogeneity is the result of different states of aggregation of the three subunits of S-100 (*157*). The subunits of S-100 all have a molecular weight of 7,000 — one subunit can be separated from the other two by ion-exchange chromatography on DEAE-Sephadex in 8 M urea. The calcium-binding activity of S-100 has been investigated by several methods. Equilibrium dialysis indicates that the protein binds eight to ten moles of calcium per mole of protein with a dissociation constant of 1 *m*M, while enhancement of S-100 fluorescence by calcium was found to follow an S-shaped curve, indicating that there are two classes of binding sites with a dissociation constant of 0.65 *m*M. If S-100 protein is subjected to acrylamide gel electrophories in a series of gels containing increasing amounts of calcium, a progressive increase in electrophoretic mobility and an increase in the number of protein bands from one to five was observed over a range of calcium concentrations from 0 to 1.6 *m*M. Substitution of magnesium for calcium results in no changes in electrophoretic mobility, indicating the specificity of the calcium effect. The multiple bands produced by the inclusion of calcium were shown to be immunologically identical, so it appears as if calcium can produce appreciable aggregation of S-100.

The finding that S-100 appears largely confined to glia, coupled with its relatively low affinity calcium-binding make it appear unlikely that this protein is the neuronal calcium receptor of *Katz* and *Miledi*.

C. Brain CBP I

Aqueous extracts of pig brain were found to contain calcium-binding activity which was salted-out by ammonium sulfate between 60% and 100% of saturation (*158*). When the proteins of this fraction were chromatographed, using step-wise elution from ECTEOLA-cellulose, two peaks of calcium-binding activity were found. One calcium binding protein was not retained by ECTEOLA cellulose in 5 *m*M Tris buffer, *p*H 7.4. We will designate this protein as brain calcium-binding protein I (Brain CBP-I). When the ECTEOLA cellulose fraction containing brain CBP-I is incubated with $^{45}CaCl_2$ and subjected to gel filtration on columns of Sephadex G-75, a single peak of protein-bound radioactivity was observed. The fractions comprising the protein bound radioactivity were pooled and fractionated by gradient elution on DEAE-cellulose. Five peaks were generated and only one of these had appreciable calcium-binding activity. This peak was found to contain electrophoretically homogeneous CBP-I.

Brain CBP-I has a molecular weight of 13,000 and contains no constituents other than amino acids. It does not display electrophoretic heterogeneity in the presence of calcium and may easily be distinguished from S-100 by its lack of solubility in saturated ammonium sulfate and lower electrophoretic mobility. The small amounts of this protein which càn be isolated from brain and its lability have thus far prevented the accurate determination of calcium binding parameters.

D. Brain CBP II

The second calcium-binding protein present in the soluble extract of pig brain is eluted from ECTEOLA cellulose by 50 *m*M Tris HCl *p*H 7.4 containing 0.32 M NaCl (*159*). This protein will be designated brain CBP-II for the sake of clarity of this discussion. The ECTEOLA-cellulose fraction containing this protein was incubated with $^{45}CaCl_2$ and subjected to gel filtration on Sephadex G-75. A single peak of protein-bound radioactivity was found, and no free calcium peak was detected (Fig. 4).

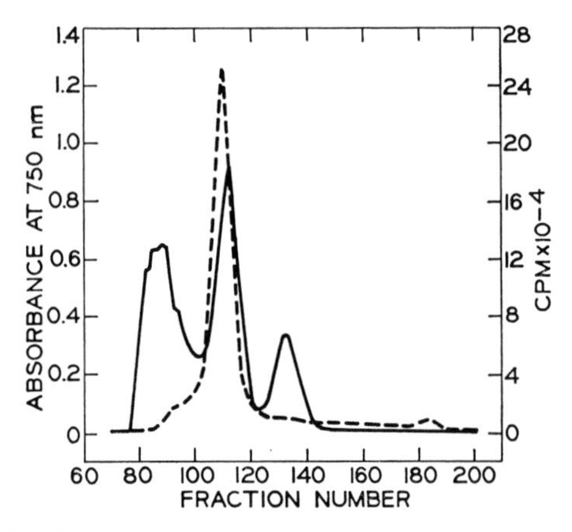

Fig. 4. Gel filtration (Sephadex G-75) of the soluble proteins from pig brain which are eluted from ECTEOLA cellulose at a NaCl concentration between 0.1 M and 1 M. The proteins in this fraction were concentrated in a Diaflo apparatus and incubated with 25 μCi $^{45}CaCl_2$ before gel filtration. Absorbancy at 750 *n*M following Lowry reaction (—); radioactivity (---). From Ref. (*159*)

This fraction when analyzed by electrophoresis in 15% acrylamide gels was found to contain three major protein bands and at least five minor components. Only one band of protein-bound radioactivity was found when duplicate gels were sliced and radioactivity was determined. The calcium binding protein was purified to homogeneity by chromatography on hydroxylapatite.

Treatment of brain CBP-II with alkaline phosphataes resulted in the loss of calcium binding activity and a marked shift to electrophoretic mobility. This indication of protein-bound phosphate in brain CBP-II was confirmed by chemical analysis which indicated the presence of ten moles of phosphate per mole of protein. No carbohydrate or lipid moieties were detected.

Calcium binding was assayed by ultrafiltration and was found to exhibit a *p*H optimum at about 7.4. One class of binding sites is present which bind one mole of calcium per mole of protein with a dissociation constant equal to 2.5×10^{-5}M. Calcium binding was competitively inhibited by other divalent cations; the binding affinity being Ca^{++}, Mn^{++}, Sr^{++}, Ba^{++}, Mg^{++} (*160*). Lanthanum salts irreversibly precipitate brain CBP-II. Those cations which are the most effective inhibitors of calcium binding to this protein are also the most active for *in vitro* calcium-dependent release of neurotransmitters.

One calcium-binding protein has been detected in aqueous extracts of adrenal medulla. This protein has been isolated and purified to homogeneity (*161 a*). The adrenal protein and brain CBP-II have identical molecular weight, amino acid composition, peptide composition following tryptic digestion, and calcium-binding properties. Brain CBP-I and CBP-II are not found in liver, kidney parotid gland, lung, adrenal cortex or duodenal mucosa, and the possibility that CBP-II is present only in adrenergic neurons has been suggested (*161 a*).

A calcium-binding protein has recently been isolated from the optic lobes of squid and cuttlefish by *Alema* and his co-workers (*161 b*).
This protein is acidic, has a molecular weight estimated at between 13,000 and 15,000 daltons, and binds calcium with a dissociation constant of 2.5×10^{-5}M. With the aid of immunological methods these workers have shown that the calcium-binding protein is eight or nine times more concentrated in the axoplasm of squid giant axon than in optic lobes of the same organism. It is also present at high levels in the sheath of this axon, but this might represent axoplasmic contamination. The similarities between chemical and physical properties of this protein and brain CBP-II suggest that they may in fact be the same material with species differences accounting for the very few inconsistencies in amino acid composition.

VIII. Mitochondrial Calcium-Binding Proteins

A. Calcium Accumulation by Mitochondria

Mitochondria actively transport calcium and the accumulation of calcium is reported to stimulate mitochondrial functions, including steroid hydroxylation (*162*). Calcium transport into mitochondria is inhibited by ruthenium red (*163*), a polysaccharide stain. This finding has prompted several laboratories to attempt the isolation of calcium-binding glycoproteins from mitochondria in an effort to identify the molecular constituents of the "calcium carrier". Although several calcium binding glycoproteins have been isolated from mitochondria their functional significance in calcium transport is yet to be established.

B. Insoluble Calcium-Binding Factor

The extraction of rat liver mitochondria with distilled water solubilized calcium-binding activity which was salted out by ammonium sulfate at 10% of saturation (*164*). The salting-out step appears irreversible, as efforts to redissolve the calcium-binding activity in aqueous media were not successful. Electrophoresis in SDS-gels yielded one major protein component and 4 minor bands. The major component had a molecular weight of 67,000. Calcium-binding studies were performed by equilibrium dialysis of suspensions of the insoluble factor prepared by sonication for brief time intervals. Scatchard plots revealed two classes of binding sites, a class of low affinity sites with a dissociation constant of 5×10^{-4}M and a capacity of more than 900 *n* moles per mg of protein, and a class of high affinity binding sites with a dissociation constant of 4×10^{-6}M and a capacity of from 48 to 70 *n* moles of calcium per mg of protein. Calcium binding was found to be antagonized by lanthanum and ruthenium red but not by uncoupling agents or respiratory inhibitors. Chemical analysis demonstrated the presence of hexosamines, sialic acid, lyso-phosphatidyl choline, free fatty acids and phosphoprotein phosphorus. The calcium-binding activity was very labile; all activity was lost during two hours of dialysis in the presence of low concentrations of calcium. The possibility that dialysis removes a low molecular weight calcium binding prosthetic group was suggested but no evidence supporting this contention is available.

C. Soluble Calcium-Binding Glycoprotein

Ox liver mitochondria were swollen in hypotonic Tris buffer and then were shrunk in hypertonic sucrose. This technique liberated a mitochondrial glycoprotein which was purified by preparative electrophoresis and found to have a molecular weight of 42,000 (*165*). Calcium binding was determined by flow dialysis and two classes of binding sites were found. The low affinity sites, which bind about 600 *n* moles of calcium per mg of protein have a dissociation constant of 10^{-5}M. The high affinity sites, which bind about 72 *n* moles of calcium per mg of protein have a dissociation constant of 10^{-7}M. Calcium binding was found to be inhibited by lanthanum and ruthenium red. Non-protein constituents included sialic acid, galactosamine, glucosamine, glucose, xylose and phospholipids.

It has been suggested that the chemical properties and calcium-binding activities of the soluble glycoprotein are similar to those of the insoluble mitochondrial glycoprotein (*164*), and that the insoluble factor is a denatured or aggregated form of the soluble calcium-binding protein. To validate this hypothesis it will be necessary to reconcile the extreme lability of calcium-binding activity of the insoluble protein with the findings that the soluble protein retains considerable activity after one hour exposure to a temperature of 100 °C, and also with the observation that the higher molecular weight glycoprotein fails to disaggregate during electrophoresis in SDS gels.

IX. Calcium-Binding Proteins from Mineralized Tissue

A. Elastin

Elastin and the mitochondrial glycoprotein isolated by *Lehninger* (Section VIII B) are the only known insoluble calcium-binding proteins. The solubility properties of elastin derive from its highly cross-linked state. In the cross-linking of elastin chains lysine residues are converted to desmosine, isodesmosine and lysinonorleucine, which are found in no other proteins (*166*). Lysine residues are therefore lost, and upon hydrolysis and assay elastin is found to have no detectable lysine (*167*). If animals are made copper deficient by dietary restriction and disulfide feeding (*168*) the amount of elastin, but not collagen, in aorta is decreased, while the amount of soluble proline in extracts of aorta is increased (*169*). This was shown to be due to the failure of elastin to cross-link and the concomitant production of a soluble elastin (*170*) containing 46 lysines per 1000 residues (*2*). Calcium binding has been studied only on insoluble elastin, which binds two moles of calcium per mole of protein with an apparent affinity constant of 7×10^{-5}M (*171*). *Urry* (*172*) has proposed that the high glycine content (65% of all residues) of elastin favor the formation of β-turns similar to structures in the calcium-binding antibiotics valinomycin and gramicidin. He proposes that calcium is coordinated to oxygen atoms of neutral amino acids and that calcium binding gives rise to positively charged loci on elastin. These positive charges can then bind phosphate which becomes the fuocs for elastin calcification. It would be useful to have comparative binding data for soluble and insoluble elastin as a first step in testing this theory.

B. Glycoprotein from Cartilage

The calcification of pre-osseous cartilage to produce bone requires calcification sites and it has been proposed that the cartilage glycosamino glycan-protein complexes (PP-L; protein polysaccharides) may be the initiation sites which support calcification (*173*). The heterogeneity of PP-L is well established; bovine nasal cartilage PP-L, for example, has been fractionated to give five subfractions, using a scheme of salting-out by a variety of mono-and divalent cations (*174*). The PP-L fractions differ from one another with respect to their uronic acid, sialic acid, protein and amino acid composition (*174*) as well as in their ability to bind calcium (*173*). A detailed study of calcium binding to one glycoprotein isolated from PP-L has recently been reported by *Vittur* (*175*). This homogeneous material, which has a molecular weight of 200,000 (*174*), binds calcium at two classes of binding sites. The high affinity sites bind two moles of calcium per mole with a dissociation constant of 10^{-7}M. The low affinity sites bind 600 moles of calcium per mole with a dissociation constant of 10^{-4}M. Calcium binding exhibited a biphasic variation with *p*H, with optima at about 6.5 and *p*H 8.2 (*174*). High affinity binding was not inhibited by either lantanum (10 μm) or butacaine (100 μm), both of which inhibit binding by the mitochondrial glycoprotein isolated by *Gazzotti et al.* (Section VIII C).

C. Sialoprotein from Bone

The investigation of metal binding proteins in bone came not because of a postulated role in mineralization or calcium transport, but rather as possible binding sites for actinide elements, which are deposited in bone. The actinide elements bind at regions which are rich in carbohydrate and the possibility of binding to glycoproteins was raised (*176*). Bone sialoprotein (BSP) is the designation of a glycoprotein which was isolated from pulverized bone powder by extraction with either EDTA (*178*) or phosphate buffer (*178*). BSP is separated from other bone glycoproteins by chromatography on Amberlite CE 50, precipitation with cetylpyridinium chloride and precipitation of contaminating glycoproteins with $MgCl_2$ (*179*). BSP thus isolated is an electrophoretically homogeneous protein with a molecular weight estimated by sedimentation velocity at between 21,800 and 25,000; the two values were obtained from EDTA and phosphate extractions (*180*). BSP is a highly acidic protein containing 40 per cent carbohydrate (*181*) and from 2.2 to 3.8 moles of serine-bound phosphate (*178*). The carbohydrate in BSP is accounted for by sialic acid, galactose, mannose, fucose, glucosamine, and galactosamine.

The binding of Ca^{++} and Y^{+++} to BSP has been studied by *Williams* and *Peacocke* (*182*) and by *Chipperfield* (*183*). The data obtained by these workers indicate that calcium is bound to carboxyl groups and with much less avidity than is yttrium. Sialic acid appears not to participate in metal binding, as removal of this moiety does not affect binding parameters (*182*). As the binding of calcium by BSP and polyglutamic acid is the same order of magnitude (*181*), and despite the suggestion that BSP may function to transport calcium ions from cells in bone (*181*), it would seem that BSP has a dubious claim at best to classification as a calcium-binding protein in the strict sense of this term.

X. Calcium Metalloenzymes

A. α-Amylase

Calcium stabilizes α-amylase and is routinely added to buffers used in the purification of this enzyme (*184*). Crystalline amylase from various organisms contains firmly bound calcium; *B. subtili* amylase contains three moles of calcium per mole of protein (*184*). Amylase from human saliva or hog pancreas contains one or two moles of bound calcium per mole of enzyme (*184*). In contrast to its non-catalytic role in DNase, calcium is required for the enzymatic function of amylase (*185*). When amylase is stripped of calcium by electrodialysis (*185*), enzyme activity is lost at a slower rate than bound calcium. *Hsiu et al.* (*185*) interpret this finding as an indication that calcium is not bound at the active site, but rather confers stability upon the enzyme by forming buried intramolecular bridges essential to the maintenance of an optimal configuration. Such bridges are regarded as analogous to disulfide bridges and it is of interest to note that *B. subtilis* amylase has no disulfide bonds and contains four moles of bound calcium, whereas amylase from human saliva binds only one mole of calcium, but does contain disulfide bridges, suggesting that calcium may form intramolecular bonds which can be substituted for disulfide bridges (*185*). In addition to the different amylases in various species (*184*) it is known that there are differences in the amino acid composition of pancreatic and parotide amylases (*187*) and that there are at least five parotide amylase isoenzymes (*188*), It will no doubt be informative to learn how slight structural changes in these proteins affect their calcium binding activity.

B. Pancreatic Deoxyribonuclease A

Pancreatic DNase activity is stabilized against proteolytic attack by calcium (*189*); the addition of calcium also provokes a tryptophan-dependent shift in the UV spectrum of the enzyme (*190*). Early work had indicated that the enzyme might be heterogeneous (*191*); *Moore* has verified this, first by fractionating DNase into two (*190*) and later into four fractions (*192*). The fractionation procedure employed chromatography on columns of DEAE-cellulose and phosphocellulose; the four fractions obtained were designated A, B, C and D. Fractions A, B and C are present in the molar ratio 4:1:1 and fraction D represents a minor component (*192*). The stabilizing effect that calcium had indicates that the enzyme might be a calcium-binding protein and an attempt was made to demonstrate ^{45}Ca binding (*189*). When DNase was incubated with $^{45}CaCl_2$ and chromatographed on a short column of Sephadex G-25 at low ionic strength, no protein bound radioactivity was found (*189*). When DNase A is incubated with $^{45}CaCl_2$ and subjected to gel filtration on Sephadex G-25 in buffers containing $^{45}CaCl_2$, however, appreciable binding is found and binding constants can be determined (*193*). Using this method, two classes of binding sites were found. The high affinity site binds two moles of calcium per mole of protein with a dissociation constant of 1.4×10^{-5}M and the low affinity site binds three moles of calcium per mole of protein with a dissociation constant of 2×10^{-4}M at *p*H 7.5 (*193*). At *p*H 5.5 one calcium is bound with high affinity, and whereas at *p*H 7.4 Mg^{++} and Mn^{++} compete for the binding of calcium to one of the two strong sites, this competition is not seen at *p*H 5.5, indicating non-equivalence of the environment of the two sites. It is of interest to compare the failure of DNase to retain bound calcium during gel filtration with complete retention of bound calcium by brain CBP-II during gel filtration (see Section VIID), as these proteins are reported to have similar binding constants.

C. Thermolysin

The thermal stability of thermolysin, a peptidase from the thermophillic bacterium *Bacillus thermoproteolyticus*, is apparently conferred by calcium bound to the enzyme (*194*), which has a molecular weight of 35,000 (*195*), and binds one mole of zinc (*195*) and three moles of calcium per mole of protein (*196*). Thermolysin contains no non-protein constituents (*197*) and consists of a single peptide chain containing neither cysteine nor cystine (*197*). Thermolysin has a high content of aspartic acid and glycine, and unlike most calcium-binding proteins it contains almost three times as much tyrosine as phenylalanine, and three tryptophan residues (*197*). The presence of two methionine residues made possible a simple cyanogen bromide cleavage scheme, with the subsequent complete sequence determination of the enzyme (*197*). Thermolysin was crystallized and x-ray diffraction studies have led to a 2.3 Å resolution electron density map and a determination of the three dimensional structure (*198*) of the protein. Two of the three bound calcium ions were located close to one another and are surrounded by three aspartic acid and two glutamic acid residues; the electron density distribution indicates that both of these calcium ions may be involved in octahedral coordinate bonding (*199*). The remaining calcium ion appears bound to two aspartic acid residues. The binding of calcium of thermolysin thus resembles that of calcium binding in carp parvalbumin (Sect. IV D).

XI. Other Calcium-Binding Proteins

A. Calcium-Binding Protein from Wheat Flour

Water extracts of spring wheat flour contain calcium binding activity. The addition of calcium and a phospholipid mixture to such extracts precipitates a crystalline ternary complex of calcium, phospholipid and a calcium-binding protein (*200*). The protein of this complex has one major component, as determined by gel electrophoresis. This protein binds calcium in the absence of phospholipid and has recently been isolated for chemical investigations (*201*). The molecular weight, as determined by SDS-gel electrophoresis, was found to be 16,500. It is an acidic protein with chemical features common to other low molecular weight calcium-binding proteins, including a high lysine content, the absence of tryptophan and a tyrosine to phenylalanine ratio of 0.5. Neither phosphorus nor carbohydrate moieties were detected. After exhaustive dialysis the protein retained 0.6 mole of calcium per mole of protein, indicating the presence of a single binding site. No suggestion of a possible function of this protein has been made nor is there a detailed comparison between it and low molecular weight mammalian calcium-binding proteins.

B. Vitellogenin

The administration of estrogen to the clawed toad *Xenopus laevis* causes the induction of a calcium-binding glycoprotein which is found in the serum of this animal (*202*). This protein has been called xenoprotein (*202*), SLPP (*203*), or lipophosphoprotein (*204*), and more recently the name vitellogenin has been adopted (*205*). The novel isolation scheme used for purifying vitellogenin employs precipitating the protein from the sera of estrogen-treated toads with dimethylformamide (*202*); virtually pure vitellogenin is obtained. There is no precipitate when DMF is added to sera from toads which have received no estrogen. The molecular weight of this protein has been estimated to be 6×10^5 daltons by gel filtration (*206*). It has a high content of glutamic acid and serine (*207*), 12 per cent lipid (*206*), and contains hexose, hexosamine, sialic acid, calcium and phosphorous (*206*). If $^{45}CaCl_2$ is injected into estrogen-treated toads and the serum proteins are fractionated by electrophoresis, the vitellogenin band contains appreciable radioactivity (*202, 206*). The radioactive calcium also remains bound to calcium after DMF precipitation of vitellogenin (*206*) and after dialysis (*206*). It is therefore curious that gel filtration on Sepharose 4B is reported to remove bound calcium from vitellogenin (*206, 208*). While no estimates of binding affinity for calcium are available, one may readily calculate from the calcium content of purified protein that one mole of vitellogenin binds *126* moles of calcium. The physiological role of vitellogenin is thought to be that of precursor to the yolk proteins lipovitellin and phosvitin (*207*); the calcium binding status of these proteins have not been established.

C. Hemocyanin

Calcium causes aggregation of hemocyanin with an increase of sedimentation coefficient from 19S at 10^{-5}M calcium to 100S at 10^{-2}M (*209*). Gel filtration study of calcium binding indicates that each mole of hemocyanin (mol. wt 50,000 daltons) binds 20 moles of calcium (*209*). The association constant for this binding is 75 (*209*). No data are available to indicate the cation specificity of binding, but it seems safe to say that one can not consider this protein to be a calcium binding protein in any strict sense.

XII. Conclusions

The proteins which have been described fulfill a wide variety of physiological functions. Apparently there exist correspondingly large differences in their chemical structures and binding affinities. The mode of calcium binding is known without ambiguity for only parvalbumin and thermolysin. In these proteins oxygen coordination of calcium via amino acid carboxyl groups appears responsible for ligand bonding. Calcium has a coordination number of eight for oxygen and it might be assumed that calcium would be bound to all high affinity sites via oxygen coordination. The finding that removal of phosphate from brain CBP II destroys the calcium binding properties of this protein may indicate that binding by phosphate may be important in some instances, however, The amino acid composition of representative calcium binding proteins is given in Tables 5—10. Again these data make generalization difficult, but one is impressed by the fact that many of these proteins have not only high levels of, lysine, glycine, aspartic and glutamic acids but low levels of tyrosine, tryptophan and the sulfur-containing amino acids.

Table 5. *Amino acid composition of calcium-binding proteins from muscle*

	Moles Amino Acid Per 100 Moles of Protein			
Amino Acid	Calsequestrin[a]	Troponin C[b]	Hake Parvalbumin[b]	Carp Parvalbumin[d]
Aspartic Acid	18.9	14.2	11.5	14.6
Threonine	2.8	3.4	4.6	3.9
Serine	3.6	4.1	4.8	5.7
Proline	5.1	1.2	0	0.8
Glutamic Acid	18.4	21.3	9.2	8.9
Glycine	3.8	7.7	11.2	8.3
Alanine	6.6	7.8	17.5	19.0
Valine	6.9	4.3	3.7	3.8
Cysteine	0.8	0.6	0.9	0
Methionine	1.5	5.6	0.9	0
Isoleucine	5.1	5.5	6.6	5.4
Leucine	8.9	5.5	7.4	8.5
Tyrosine	1.5	1.1	0	0.9
Phenylalanine	6.1	5.7	8.8	9.3
Lysine	6.6	7.0	11.4	10.2
Histidine	1.5	0.7	0.9	0
Arginine	1.8	4.2	0.9	0.9
Tryptophan	—	—	0	—

[a]) Ref. 75; [b]) Ref. 89; [c]) Ref. 107; [d]) Ref. 103.

Table 6. *Amino acid composition of avian Vitamin D-inducible calcium-binding protein*[a])

Amino Acid	Amino Acid Residues per 100 Moles of Protein
Aspartic Acid	14.0
Threonine	4.1
Serine	4.2
Proline	1.5
Glutamic Acid	15.6
Glycine	7.1
Alanine	7.4
Valine	2.8
Cysteine	1.5
Methionine	2.0
Isoleucine	4.6
Leucine	11.9
Tyrosine	4.1
Phenylalanine	5.4
Lysine	9.7
Histidine	1.6
Arginine	2.5
Tryptophan	—

[a]) Ref. 114.

Table 7. *Amino acid composition of prothrombin and dicumarol-induced prothrombin*[a])

	Moles Amino Acid Per 100 Moles of Protein	
Amino Acid	Prothrombin	Dicumarol-Induced Prothrombin
Aspartic Acid	10.0	10.1
Threonine	5.0	5.1
Serine	6.8	6.7
Proline	6.1	6.2
Glutamic Acid	12.4	12.5
Glycine	8.0	8.0
Alanine	5.7	6.1
Valine	5.9	6.2
Cysteine	3.6	3.4
Methionine	1.4	1.3
Isoleucine	3.3	3.5
Leucine	7.7	7.6
Tyrosine	3.1	2.9
Phenylalanine	3.5	3.3
Lysine	5.6	5.9
Histidine	1.8	1.8
Arginine	7.2	6.8
Tryptophan	2.7	2.5

[a]) Ref. 143.

Table 8. *Amino acid composition of calcium-binding proteins from nervous tissue*

	Moles Amino Acid per 100 Moles of Protein			
Amino Acid	S-100[a]	CBP-I[b]	CBP-I[c]	Cuttlefish CBP[d]
Aspartic Acid	11.6	11.5	14.6	15.5
Threonine	3.7	8.7	6.1	6.9
Serine	5.8	7.8	4.9	5.2
Proline	—	0	1.0	2.6
Glutamic Acid	19.1	11.5	21.3	20.7
Glycine	5.1	8.1	8.1	6.9
Alanine	6.6	7.2	8.1	6.9
Valine	7.7	5.9	6.3	4.3
Cysteine	1.9	1.5	0	—
Methionine	2.6	1.0	1.6	4.3
Isoleucine	3.2	4.9	5.2	5.2
Leucine	9.5	8.1	7.1	6.0
Tyrosine	1.4	1.0	1.0	1.0
Phenylalanine	6.5	5.4	5.1	4.3
Lysine	9.4	11.6	7.1	6.0
Histidine	4.3	2.1	1.6	1.0
Arginine	1.0	2.3	1.0	3.4
Tryptophan	0.6	0.8	0	

a) Ref. 153; b) Ref. 158; c) Ref. 159; d) Ref. 161b.

Table 9. *Amino acid composition of elastin and bone sialoprotein*

	Moles Amino Acid Per 100 Moles of Protein	
Amino Acid	Elastin[a]	Bone Sialoprotein[b]
Aspartic Acid	0.7	17.4
Threonine	1.0	11.1
Serine	1.0	8.0
Proline	12.8	4.9
Glutamic Acid	1.8	22.9
Glycine	32.2	12.2
Alanine	21.7	4.0
Valine	13.7	3.1
Cysteine	—	1.3
Methionine	—	—
Isoleucine	2.7	2.5
Leucine	6.6	2.9
Tyrosine	0.6	1.2
Phenylalanine	3.4	2.3
Lysine	0.4	2.9
Histidine	0.1	1.2
Arginine	0.7	1.2
Tryptophan	—	1.1
Hydroxyproline	0.7	0

[a]) *Franzblau, C.* and *Lent, R. W.*: Brookhaven Symp. Biol. *21*, 358 (1969).
[b]) Ref. 181.

Table 10. *Amino acid composition of DNase A and thermolysin*

	Moles Amino Acid Per 100 Moles of Protein	
Amino Acid	DNase A[a]	Thermolysin[b]
Aspartic Acid	12.3	14.0
Threonine	5.6	7.9
Serine	11.2	8.6
Proline	3.5	2.5
Glutamic Acid	7.5	7.0
Glycine	3.6	11.4
Alanine	8.4	8.9
Valine	10.0	7.0
Cysteine	1.5	0
Methionine	1.5	0.6
Isoleucine	4.4	5.7
Leucine	8.7	5.1
Tyrosine	5.9	8.9
Phenylalanine	4.3	3.2
Lysine	3.4	3.5
Histidine	2.2	2.5
Arginine	4.5	3.5
Tryptophan	1.5	1.0

a) Ref. 189; b) Ref. 197.

The lability of this class of proteins makes it important to consider the details of both isolation and calcium-binding assay, and perhaps the general lability problem is a reflection of a structural similarity associated with metal binding. One may hope that additional proteins of this class will be studied by x-ray diffraction analysis and that it will be possible then to make more generalized statements relating their structures and binding abilities.

XIII. References

1. *Manery, J. F.:* In Mineral Metabolism III, p. 405 (Ed. *Colmar, C. L.*, and *Bronner, F.*). Academic Press 1969.
2. *Rasmussen, H., Tenenhouse, A.:* Proc. Nat. Acad. Sci. *59*, 1364 (1968).
3. *Cameron, L. E., Le'John, H. B.:* J. Biol. Chem. *247*, 4729 (1972).
4. *Lehninger, A. L., Carofoli, E., Rossi, C. S.:* Adv. Enzymol. *29*, 259 (1968).
5. *Perris, A. D.:* In Cellular Mechanisms for Calcium Transfer and Homeostasis, p. 101. (Ed. *Nichols, G., Jr.*, and *Wasserman, R. H.*) New York: Academic Press 1971.
6. *Casteneda, M., Tyler, A.:* Biochem. Biophys. Res. Commun. *33*, 783 (1968).
7. *Whitney, R. B., Sutherland, R. M.:* J. Cell Physiol. *80*, 329 (1972).
8. *Grodsky, G. M.:* Diabetes *21* (Suppl. 2), 584 (1972).
9. *Vale, W., Burgus, R., Guillemin, R.:* Experientia *13*, 853 (1967).
10. *Samli, M. H., Geschwind, I. I.:* Endocrinology *82*, 225 (1968).
11. *Douglas, W. W., Poisner, A. M.:* J. Physiol. (London) *172*, 19 (1964).
12. *Dicker, S. E.:* J. Physiol. (London) *185*, 429 (1966).
13. *Schofield, J. G.:* Nature *215*, 1382 (1967).
14. *Zor, V., Lowe, I. P., Bloom, G., Field, J. B.:* Biochem. Biophys. Res. Commun. *33*, 649 (1968).
15. *Douglas, W., Poisner, A. M.:* Nature *196*, 379 (1962).
16. *Douglas, W. W., Rubin, R. P.:* J. Physiol. (London) *159*, 40 (1961).
17. *Katz, B., Miledi, R.:* Proc. Roy. Soc. B.: *161*, 496 (1965).
18. *Bulbring, E., Tomita, T.:* In Calcium and Cellular Function, p. 249. (Ed. *Cuthbert, A. W.*) London: St. Martin's Press 1970.
19. *Takagi, K., Takayanagi, I., Liao, C. S.:* Europ. J. Pharmacol. *19*, 330 (1972).
20. *Ebashi, S., Wakabayashi, T., Ebashi, F.:* J. Biochem. (Tokyo) *69*, 441 (1971).
21. *Sandow, A.:* Ann. Rev. Physiol. *32*, 87 (1970).
22. *Konijn, T. M., Van de Meene, J. G. C., Bonner, J. T., Barkley, D. S.:* Proc. Nat. Acad. Sci. *58*, 1152 (1967).
23. *Linskens, H., Kroh, M.:* Curr. Top. Develop. Biol.: *5*, 89 (1970).
24. *Galston, A. W.:* Science *163*, 1288 (1969).
25. *Bowness, J. M.:* Clin. Orthop. Rel. Res. *59*, 233 (1968).
26. *Schraer, H., Schraer, R.:* In Cellular Mechanisms for Calcium Transfer and Homeostasis, p. 351. (Ed. *Nichols, G., Jr.* and *Wasserman, R. H.*) New York: Academic Press 1971.
27. *Gevers, W., Krebs, H. A.:* Biochem. J. *98*, 720 (1966).
28. *Kimmich, G. A., Rasmussen, H.:* J. Biol. Chem. *244*, 190 (1969).
29. *Esnouf, M. P., Macfarlane, R. G.:* Adv. Enzymol. *30*, 255 (1968).
30. *Weisenberg, R.:* Science *177*, 1104 (1972).
31. *Rasmussen, H., Nagata, N.:* In Calcium and Cellular Function, p. 198. (Ed. *Cuthbert, A. W.*) London: St. Martin's Press 1970.
32. *Bradham, L. S.:* Biochem. Biophys. Acta *276*, 431 (1972).
33. *Dunham, E. T., Glynn, I. M.:* J. Physiol. (London) *156*, 274 (1961).
34. *Melgar, F., Goldthwait, D. A.:* J. Biol. Chem. *243*, 4409 (1968).
35. *Jungalwala, F. B., Freinkel, N., Dawson, R. M. C.:* Biochemistry *123*, 19 (1971).
36. *Gorini, L., Fromageot, C.:* Compt. Rend. *229*, 559 (1949).
37. *Hsiu, J., Fischer, E. H., Stein, E. A.:* Biochemistry *3*, 61 (1964).
38. *Hodgkin, A. L., Keynes, R. D.:* J. Physiol. (London) *138*, 253 (1957).
39. *Lehninger, A. L.:* Biochem. J. *119*, 129 (1970).
40. *Cittadini, A., Scarpa, A., Chance, B.:* Biochem. Biophys. Acta *291*, 246 (1973).

41. *Poisner, A. M., Hava, M.:* Mol. Pharmacol. *6,* 407 (1970).
42. *Piccinini, F., Galatulas, I., Galli, C., Pomarelli, P., Cova, D.:* Europ. J. Pharmacol. *10,* 328 (1970).
43. *Gielen, W.:* Z. Naturforsch. *21,* 1007 (1966).
44. *Carr, C. W., Chang, K. Y.:* In Cellular Mechanisms for Calcium Transfer and Homeostasis, p. 41. (Ed. *Nichols, G., Jr.* and *Wasserman, R. H.*) New York: Academic Press 1971.
45. *Chau-Wong, M., Seeman, P.:* Biochem. Biophys. Acta *241,* 473 (1971).
46. *Harnach, F., Coolidge, T. B.:* Anal. Biochem. *6,* 447 (1963).
47. *O'Reilly, R. A., Ohms, J. I., Motley, C. H.:* J. Biol. Chem. *244,* 1303 (1969).
48. *Calissano, P., Moore, B. W., Friesen, A.:* Biochemistry *11,* 4318 (1969).
49. *Smith, Q. T., Lindenbaum, A.:* Calc. Tissue Res. *7,* 290 (1971).
50. *Bray, G. A.:* Anal. Biochem. *1,* 279 (1960).
51. *Hartshorne, D. J., Pyun, H. Y.:* Biochem. Biophys. Acta *229,* 698 (1971).
52. *McPhie, P.:* In Methods in Enzymology XXII, p. 23. (Ed. *Jakoby, W. B.*). New York: Academic Press 1971.
53. *Briggs, F. N., Fleishman, J.:* J. Gen. Physiol. *49,* 131 (1965).
54. *Scatchard, G.:* Ann. N. Y. Acad. Sci. *51,* 660 (1949).
55. *Hughes, T. R., Klotz, I. M.:* In Methods of Biochemical Analysis, III, p. 265. (Ed. *Glick, D.*). New York: Interscience Publishers 1956.
56. *Colowick, S. P., Womack, F. C.:* J. Biol. Chem. *244,* 774 (1969).
57. *Hummel, J. P., Dreyer, W. J.:* Biochim. Biophys. Acta *63,* 530 (1962).
58. *Price, P. A.:* J. Bil. Chem. *247,* 2895 (1972).
59. *Flexner, L. B.:* J. Biol. Chem. *121,* 615 (1937).
60. *Blatt, W. F., Robinson, S. M., Bixler, H. J.:* Anal. Biochem. *26,* 151 (1973).
61. *Paulus, H.:* Anal. Biochem. *32,* 91 (1969).
62. *Brooks, J. B., Siegel, F. L.:* Unpublished work.
63. *Heyde, E.:* Anal. Biochem. *51,* 61 (1972).
64. *Steinhardt, J., Beychok, S.:* In The Proteins II, p. 139. (Ed. *Neurath, H.*) New York: Academic Press 1964.
65. *Sandow, A.:* Pharmacol. Rev. *17,* 265 (1965).
66. *Ebashi, S., Endo, M.:* Prog. Biophys. Mol. Biol. *18,* 213 (1968).
67. *Fuchs, F., Briggs, F. N.:* J. Gen. Physiol. *51,* 655 (1968).
68. *Ebashi, S., Kodama, A., Ebashi, F.:* J. Biochem. *64,* 465 (1968).
69. *MacLennan, D. H., Wong, P. T. S.:* Proc. Nat. Acad. Sci. *68,* 1231 (1971).
70. *Ebashi, S., Lipmann, F.:* J. Cell Biol. *14,* 389 (1962).
71. *Hasselbach, W., Makinose, M.:* Biochem. Z. *333,* 518 (1963).
72. *MacLennan, D. H.:* J. Biol. Chem. *245,* 4508 (1970).
73. *Yu, B. P., Masoro, E. J.:* Biochemistry *9,* 2909 (1970).
74. *Bertrand, H. A., Masoro, E. J., Ohnishi, T., Yu, B. P.:* Biochemistry *10,* 3679 (1971).
75. *MacLennan, D. H., Wong, P. T. S.:* Proc. Nat. Acad. Sci. *68,* 1231 (1971).
76. *Ikemoto, N., Streter, F. A., Gergely, J.:* Arch. Biochem. Biophys. *147,* 571 (1971).
77. — *Bhantnagar, G. M., Nagy, B., Gergely, J.:* J. Biol. Chem. *247,* 7835 (1972).
78. *Benson, E.:* Personal communication.
79. *Hajdu, S., Posner, C. J., Leonard, E. J.:* Circ. Res. *22,* 358 (1971).
80. — Personal communication.
81. *Schaub, M. C., Perry, S. V.:* Biochem. J. *115,* 993 (1969).
82. *Hartshorne, D. J., Mueller, H.:* Biochem. Biophys. Res. Commun. *31,* 647 (1968).
83. — *Pyun, H. Y.:* Biochem. Biophys. Acta *229,* 698 (1971).

84. *Schaub, M. C., Perry, S. V.*: Biochemistry *115*, 993 (1969).
85. — — *Hacker, W.*: Biochem. J. *126*, 237 (1972).
86. *Wilkinson, J. M., Perry, S. V., Cole, H. A., Trayer, I. P.*: Biochem. J. *127*, 215 (1972).
87. *Greaser, M. L., Gergely, J.*: J. Biol. Chem. *246*, 4226 (1971).
88. *Drabikowski, W., Dabrowska, R., Barylko, B., Graeser, M., Gergely, J.*: Proc. Third Int. Biophys. Congr., Cambridge, p. 193 (1969).
89. *Greaser, M. L., Yamaguchi, M., Brekke, C., Potter, J., Gergely, J.*: Cold Spring Harbor Symp. Quant. Biol. *37*, 235 (1971).
90. *Drabikowski, W., Dabrowska, R., Barylko, B.*: FEBS Ltrs. *12*, 148 (1971).
91. — *Nowak, E., Barylko, B., Dabrowska, R.*: Cold Spring Harbor Symp. Quant. Biol. *37*, 245 (1972).
92. *Ebashi, S., Ebashi, F.*: J. Biochem. *55*, 604 (1964).
93. *Drabikowski, W., Rafatowska, V., Dabrowska, R., Szpacenko, A., Barylko, B.*: FEBS Ltrs. *19*, 259 (1971).
94. *Murray, A. C., Kay, C. M.*: Biochemistry *11*, 2622 (1972).
95. *Van Eerd, J. P., Kawasaki, Y.*: Biochem. Biophys. Res. Commun. *47*, 859 (1972).
96. *Greaser, M. L., Gergely, J.*: J. Biol. Chem. *248*, 2125 (1973).
97. *Reddy, Y. S., Honig, C. R.*: Biochem. Biophys. Acta *275*, 453 (1972).
98. *Honig, C. R., Reddy, Y. S.*: Fed. Proc. *30*, 497 (1971).
99. *Pechère, J. F.*: Comp. Biochem. Physiol. *24*, 289 (1968).
100. *Benzonana, G., Capony, J. P., Pechère, J. F.*: Biochem. Biophys. Acta *278*, 110 (1972).
101. *Nockolds, C. E., Kretsinger, R. H., Coffee, C. J., Bradshaw, R. A.*: Proc. Nat. Acad. Sci. *69*, 581 (1972).
102. *Hamoir, G., Konosu, S.*: Biochem. J. *96*, 85 (1965).
103. *Konosu, S., Hamoir, G., Pechère, J. F.*: Biochem. J. *96*, 98 (1965).
104. *Pechère, J. F., Demaille, J., Capony, J. P.*: Biochem. Biophys. Acta *236*, 391 (1971).
105. *Bhushana Rao, K. S. P., Focant, B., Gerday, Ch., Hamoir, G.*: Comp. Biochem. Physiol. *30*, 33 (1969).
106. — *Gerday, Ch.*: Comp. Biochem. Physiol. *44B*, 1113 (1973).
107. *Pechère, J. F., Capony, J. P., Ryden, L.*: Eur. J. Biochem. *23*, 421 (1971).
108. — — — *Demaille, J.*: Biochem. Biophys. Res. Commun. *43*, 1106 (1971).
109. *Kretsinger, R. H.*: Nature, New Biol. *240*, 85 (1972).
110. *Eisenstein, R., Passavoy, M.*: Proc. Soc. Exptl. Biol. Med. *117*, 77 (1964).
111. *Taylor, A. N., Wasserman, R. H.*: Nature *205*, 248 (1965).
112. *Wasserman, R. H., Taylor, A. N.*: Science *152*, 791 (1966).
113. *Taylor, A. N., Wasserman, R. H.*: Arch. Biochem. Biophys. *119*, 536 (1967).
114. *Wasserman, R. H., Corradino, R. A., Taylor, A. N.*: J. Biol. Chem. *243*, 3978 (1968).
115. — Biochem. Biophys. Acta *203*, 176 (1970).
116. *Ingersoll, R. J., Wasserman, R. H.*: J. Biol. Chem. *246*, 2808 (1971).
117. *Wasserman, R. H., Taylor, A. N.*: J. Biol. Chem. *243*, 3987 (1968).
118. *Taylor, A. N., Wasserman, R. H.*: J. Histochem. Cytochem. *18*, 107 (1970).
119. *Bar, A., Hurwitz, S.*: Comp. Biochem. Physiol. *41B*, 735 (1972).
120. *Corradino, R. A., Wasserman, R. H., Pubols, M. H., Chang, S. I.*: Arch. Biochem. Biophys. *125*, 378 (1970).
121. *Taylor, A. N., Wasserman, R. H.*: Amer. J. Physiol. *223*, 110 (1972).
122. *Kallfelz, F. A., Taylor, A. N., Wasserman, R. H.*: Proc. Soc. Exptl. Biol. Med. *125*, 54 (1967).

123. *Ooizumi, K., Moriuchi, S., Hosoya, N.:* J. Vitaminol. *16*, 228 (1970).
124. *Drescher, D., DeLuca, H. F.:* Biochemistry *10*, 2302 (1971).
125. — — Biochemistry *10*, 2308 (1971).
126. *Taylor, A. N., Wasserman, R. H.:* Fed. Proc. *27*, 675 (1968).
127. *Fullmer, C. S., Wasserman, R. H.:* Fed. Proc. *31*, 693 (1972).
128. *Wasserman, R. H., Taylor, A. N.:* Proc. Soc. Exptl. Biol. Med. *136*, 25 (1971).
129. *Menczel, J., Eilon, G., Steiner, A., Karaman, C., Mor, E., Ron, A.:* Israel J. Med. Sci. *7*, 396 (1971).
130. *Alpers, D. H., Lee, S. W., Avioli, L. V.:* Gastroent. *62*, 559 (1972).
131. *Piazolo, P., Schleyer, M., Franz, H. E.:* Z. Physiol. Chem. *352*, 1480 (1971).
132. *Baumane, V., Valiniece, M., Pastukhov, M. V.:* Latv. P. S. R. Zinat. Akad. Vestis *1*, 133 (1972).
133. *Barton, P. G., Hanahan, D. G.:* Biochem. Biophys. Acta *187*, 319 (1967).
134. *Esnouf, M. P., MacFarlane, R. G.:* Adv. Enzymol. *30*, 255 (1968).
135. *Jobin, F., Esnouf, M. P.:* Biochem. J. *102*, 666 (1967).
136. *Shah, D. V., Suttie, J. W.:* Proc. Nat. Acad. Sci. *68*, 1653 (1971).
137. *Stenflo, J.:* Acta Chem. Scand. *24*, 3762 (1970).
138. *Moore, H. C., Lux, S. E., Malhotra, O. P., Bakerman, S., Carter, J. R.:* Biochem. Biophys. Acta *111*, 174 (1965).
139. *Ingwall, J. S., Scheraga, H. A.:* Biochemistry *8*, 1860 (1969).
140. *Stenflo, J., Ganrot, P. O.:* J. Biol. Chem. *247*, 8160 (1972).
141. *Nelsestuen, G., Suttie, J. W.:* J. Biol. Chem. *247*, 6096 (1972).
142. *Jackson, C. M., Johnson, T. F., Hanahan, D. J.:* Biochemistry *7*, 4492 (1968).
143. *Stenflo, J.:* J. Biol. Chem. *247*, 8167 (1972).
144. *Nelsestuen, G. L., Suttie, J. W.:* J. Biol. Chem. *247*, 8167 (1972).
145. *Stenflo, J., Ganrot, P. O.:* Biochem. Biophys. Res. Commun. *50*, 98 (1973).
146. *Nelsestuen, G. L., Suttie, J. W.:* Biochemistry *11*, 4961 (1972).
147. *Katz, B., Miledi, R.:* Proc. Roy. Soc. B *161*, 496 (1965).
148. *Douglas, W. W.:* Brit. J. Pharmacol. *34*, 451 (1968).
149. *Matthews, E. K.:* In Calcium and Cellular Function, p. 163. (Ed. *Cuthbert, A. W.*). London: St. Martin's Press 1970.
150. *Katz, B., Miledi, R.:* J. Physiol. (London) *195*, 481 (1968).
151. *Berl, S., Puszkin, S., Nicklas, W. J.:* Science *179*, 441 (1972).
152. *Hodgkin, A. L., Keynes, R. D.:* J. Physiol. (London) *138*, 253 (1957).
153. *Moore, B. W.:* Biochem. Biophys. Res. Commun. *19*, 739 (1965).
154. *Calisanno, P., Moore, B. W., Friesen, A.:* Biochemistry *8*, 4318 (1964).
155. *McEwen, B. S., Hyden, H.:* J. Neurochem. *13*, 823 (1966).
156. *Uyemura, K., Vincendon, G., Gombos, G., Mandel, P.:* J. Neurochem. *18*, 429 (1971).
157. *Dannies, P. S., Levine, L.:* J. Biol. Chem. *246*, 6276 (1971).
158. *Wolff, D. J., Siegel, F. L.:* Arch. Biochem. Biophys. *150*, 578 (1972).
159. — — J. Biol. Chem. *247*, 4180 (1972).
160. — *Huebner, J., Siegel, F. L.:* J. Neurochem. *19*, 2855 (1972).
161a. *Brooks, J., Siegel, F. L.:* J. Biol. Chem. In Press.
161b. *Alemà, S., Calissano, P., Rusca, G., Giuditta, A.:* J. Neurochem. *20*, 681 (1973).
162. *Peron, F., Galdwell, B. V.:* Biochim. Biophys. Acta *143*, 532 (1967).
163. *Moore, C. L.:* Biochem. Biophys. Res. Commun. *42*, 298 (1971).
164. *Comez-Puyou, A., de Gomez-Puyou, M. T., Becker, G., Lehninger, A. L.:* Biochem. Biophys. Res. Commun. *47*, 814 (1972).
165. *Gazzotti, G., Vasington, F. D., Carafoli, E.:* Biochem. Biophys. Res. Commun. *47*, 808 (1972).
166. *Partridge, S. M.:* Fed. Proc. *25*, 1023 (1966).

167. *Petruska, J. A., Sandberg, L. B.:* Biochem. Biophys. Res. Commun. *33*, 222 (1968).
168. *Shields, G. S., Coulson, W. F., Kimball, D. A., Carnes, W. H., Cartwright, G. E., Wintrobe, M. M.:* Am. J. Pathol. *41*, 603 (1962).
169. *Weissman, N., Shields, G. S., Carnes, W. H.:* J. Biol. Chem. *238*, 3115 (1963).
170. *Smith, D. W., Brown, D. M., Carnes, W. H.:* J. Biol. Chem. *247*, 2427 (1972).
171. *Tosatti, M. P. M., Gotte, L., Moret, V.:* Calc. Tiss. Res. *6*, 329 (1971).
172. *Urry, D. W.:* Proc. Nat. Acad. Sci. *68*, 810 (1971).
173. *Smith, Q. T., Lindenbaum, A.:* Calc. Tiss. Res. *7*, 290 (1971).
174. *Pal, S., Doganges, P. T., Schubert, M.:* J. Biol. Chem. *241*, 4261 (1966).
175. *Vittur, F., Pugliarello, M. C., de Bernard, B.:* Biochem. Biophys. Res. Commun. *48*, 143 (1972).
176. *Vaughn, J. M.:* In The Biochemistry and Physiology of Bone, p. 729. (Ed. *Bourne, G. H.*). New York: Academic Press 1956.
177. *Herring, G. M.:* Proc. 1st Europ. Bone and Tooth Symp., p. 263. Oxford: Pergamon Press 1963.
178. *de B. Andrews, A. T., Herring, G. M.:* Biochem. Biophys. Acta *101*, 239 (1965).
179. *Herring, G. M.:* Biochem. J. *107*, 41 (1968).
180. *Williams, P. A., Peacocke, A. R.:* Biochem. Biophys. Acta *101*, 327 (1965).
181. *Herring, G. M., de B. Andrews, A. T., Chipperfield, A. R.:* In Cellular Mechanisms for Calcium Transfer and Homeostasis, p. 63. (Ed. *Nichols, G.* and *Wasserman, R. H.*) New York: Academic Press 1971.
182. *Williams, P. A., Peacocke, A. R.:* Biochem. J. *105*, 1177 (1967).
183. *Chipperfield, A. R.:* Biochem. J. *118*, 36P (1970).
184. *Valee, B. L., Stein, E. A., Summerwell, W. N., Fischer, G. H.:* J. Biol. Chem. *234*, 2901 (1959).
185. *Hsiu, J., Fischer, E. U., Stein, E. A.:* Biochemistry *3*, 61 (1964).
186. *Stein, E. A., Junge, J. M., Fischer, E. H.:* J. Biol. Chem. *235*, 371 (1960).
187. *Sanders, T. G., Rutter, W. J.:* Biochemistry *11*, 130 (1972).
188. *Keller, P. J., Kauffman, D. L., Allan, B. J., Williams, B. C.:* Biochemistry *10*, 4867 (1971).
189. *Price, P. A., Liu, T.-Y., Stein, W. H., Moore, S.:* J. Biol. Chem. *244*, 917 (1969).
190. *Poulos, T. L., Price, P. A.:* J. Biol. Chem. *246*, 4041 (1971).
191. *Laskowski, M., Sr.:* In Procedures in Nuclei Acid Research, p. 85. (Ed. *Cantoni, G. L.* and *Davies, D. R.*) New York: Harper and Row 1966.
192. *Salnikow, J., Moore, S., Stein, W. H.:* J. Biol. Chem. *245*, 5685 (1970).
193. *Price, P. A.:* J. Biol. Chem. *247*, 2895 (1972).
194. *Feder, J., Garrett, L. R., Wildi, B. S.:* Biochemistry *10*, 4552 (1971).
195. *Petra, P. H.:* In Methods in Enzymology *19*, 460 (1970).
196. *Latt, S. A., Holmquist, B., Vallee, B. L.:* Biochem. Biophys. Res. Commun. *37*, 333 (1969).
197. *Titani, K., Hermodson, A., Ericsson, L. H., Walsh, K. A., Neurath, H.:* Nature New Biology *238*, 35 (1972).
198. *Matthews, B. W., Jansonius, J. N., Colman, P. M., Schoenborn, B. P., DuPourque, D.:* Nature New Biology *238*, 37 (1972).
199. *— Colman, P. M., Jansonius, J. N., Titani, K., Walsh, K. A., Neurath, H.:* Nature New Biology *238*, 41 (1972).
200. *Fullington, J. G.:* Cer. Sci. *15*, 309 (1970).
201. — Unpublished work.
202. *Munday, K. A., Ansari, A. Q., Oldroyd, D., Akhtar, M.:* Biochem. Biophys. Acta *166*, 748 (1968).
203. *Wallace, R. A., Jared, D. W.:* Science *160*, 91 (1968).

204. *Follett, B. K., Redshaw, M. R.:* J. Endocrinol. *40,* 439 (1968).
205. *Pan, M. L., Bell, W. J., Telfer, W. H.:* Science *165,* 393 (1969).
206. *Ansari, A. Q., Dolphin, P. J., Lazier, C. B., Munday, K. A., Akhtar, M.:* Biochem. J. *122,* 107 (1971).
207. *Follett, B. K., Nicholls, T. J., Redshaw, M. R.:* J. Cell. Physiol. *72,* 91 (1968).
208. *Wallace, R. A.:* Biochem. Biophys. Acta *215,* 176 (1970).
209. *Klarman, A., Shaklai, N., Daniel, E.:* Biochem. Biophys. Acta *257,* 150 (1972).

Received April 24, 1973

Structure and Bonding: Index Volume 1-17

ST. JOHN FISHER COLLEGE LIBRARY
0 1220 0010126 3